SUSTAINABLE HORTICULTURE

Volume 1:

Diversity, Production, and Crop Improvement

Innovations in Horticultural Science

SUSTAINABLE HORTICULTURE
Volume 1:
Diversity, Production, and Crop Improvement

Edited by

Debashis Mandal, PhD

Amritesh C. Shukla, DPhil, DSc

Mohammed Wasim Siddiqui, PhD

AAP APPLE ACADEMIC PRESS

Apple Academic Press Inc. 3333 Mistwell Crescent Oakville, ON L6L 0A2 Canada	Apple Academic Press Inc. 9 Spinnaker Way Waretown, NJ 08758 USA

International Standard Book Number-13: 978-1-77188-646-8 (Hardcover)
International Standard Book Number-13: 978-1-315-14793-2 (eBook)

Library and Archives Canada Cataloguing in Publication

Sustainable horticulture / edited by Debashis Mandal, PhD, Amritesh C. Shukla, DPhil, DSc, Mohammed Wasim Siddiqui, PhD.

(Innovations in horticultural science)
Includes bibliographical references and indexes.
Contents: Volume 1: Diversity, production, and crop improvement.
Issued in print and electronic formats.
ISBN 978-1-77188-646-8 (v. 1 : hardcover).--ISBN 978-1-315-14793-2 (v. 1 : PDF)

1. Sustainable horticulture. I. Mandal, Debashis, editor II. . Shukla, Amritesh C., editor III. Siddiqui, Mohammed Wasim, editor IV. Series: Innovations in horticultural science

G155.C6H67 2018   338.4'79151   C2018-903208-1   C2018-903209-X

CIP data on file with US Library of Congress

Apple Academic Press also publishes its books in a variety of electronic formats. Some content that appears in print may not be available in electronic format. For information about Apple Academic Press products, visit our website at **www.appleacademicpress.com** and the CRC Press website at **www.crcpress.com**

CONTENTS

INNOVATIONS IN HORTICULTURAL SCIENCE

About the Series

Editor-in-Chief:

Dr. Mohammed Wasim Siddiqui Assistant Professor-cum-Scientist
Bihar Agricultural University | www.bausabour.ac.in
Department of Food Science and Post-Harvest Technology
Sabour | Bhagalpur | Bihar | P. O. Box 813210 | INDIA
Contacts: (91) 9835502897
Email: wasim_serene@yahoo.com | wasim@appleacademicpress.com

The horticulture sector is considered as the most dynamic and sustainable segment of agriculture all over the world. It covers pre- and postharvest management of a wide spectrum of crops, including fruits and nuts, vegetables (including potatoes), flowering and aromatic plants, tuber crops, mushrooms, spices, plantation crops, edible bamboos etc. Shifting food pattern in wake of increasing income and health awareness of the populace has transformed horticulture into a vibrant commercial venture for the farming community all over the world.

It is a well-established fact that horticulture is one of the best options for improving the productivity of land, ensuring nutritional security for mankind and for sustaining the livelihood of the farming community worldwide. The world's populace is projected to be 9 billion by the year 2030, and the largest increase will be confined to the developing countries, where chronic food shortages and malnutrition already persist. This projected increase of population will certainly reduce the per capita availability of natural resources and may hinder the equilibrium and sustainability of agricultural systems due to overexploitation of natural resources, which will ultimately lead to more poverty, starvation, malnutrition, and higher food prices. The judicious utilization of natural resources is thus needed and must be addressed immediately.

Climate change is emerging as a major threat to the agriculture throughout the world as well. Surface temperatures of the earth have risen significantly over the past century, and the impact is most significant on agriculture. The rise in temperature enhances the rate of respiration, reduces cropping periods, advances ripening, and hastens crop maturity, which adversely affects crop productivity. Several climatic extremes such as droughts, floods, tropical cyclones, heavy precipitation events, hot extremes, and heat waves cause a negative impact on agriculture and are mainly caused and triggered by climate change.

In order to optimize the use of resources, hi-tech interventions like precision farming, which comprises temporal and spatial management of resources in horticulture, is essentially required. Infusion of technology for an efficient utilization of resources is intended for deriving higher crop productivity per unit of inputs. This would be possible only through deployment of modern hi-tech applications and precision farming methods. For improvement in crop production and returns to farmers, these technologies have to be widely spread and adopted. Considering the abovementioned challenges of horticulturist and their expected role in ensuring food and nutritional security to mankind, a compilation of hi-tech cultivation techniques and postharvest management of horticultural crops is needed.

This book series, Innovations in Horticultural Science, is designed to address the need for advance knowledge for horticulture researchers and students. Moreover, the major advancements and developments in this subject area to be covered in this series would be beneficial to mankind.

BOOKS IN THE SERIES

- **Spices: Agrotechniques for Quality Produce**
  Amit Baran Sharangi, PhD, S. Datta, PhD, and Prahlad Deb, PhD

- **Sustainable Horticulture, Volume 1: Diversity, Production, and Crop Improvement**
  Editors: Debashis Mandal, PhD, Amritesh C. Shukla, PhD, and Mohammed Wasim Siddiqui, PhD

- **Sustainable Horticulture, Volume 2: Food, Health, and Nutrition**
  Editors: Debashis Mandal, PhD, Amritesh C. Shukla, PhD, and Mohammed Wasim Siddiqui, PhD

- **Underexploited Spice Crops: Present Status, Agrotechnology, and Future Research Directions**
  Amit Baran Sharangi, PhD, Pemba H. Bhutia, Akkabathula Chandini Raj, and Majjiga Sreenivas

ABOUT THE EDITORS

Debashis Mandal, PhD

Assistant Professor, Department of Horticulture, Aromatic and Medicinal Plants, Mizoram University, Aizawl, India

Debashis Mandal, PhD, is Assistant Professor in the Department of Horticulture, Aromatic and Medicinal Plants at Mizoram University in Aizawl, India. Dr. Mandal started his academic career as Assistant Professor at Sikkim University, India. He had post-doctoral research experience as a project scientist at the Precision Farming Development Centre, Indian Institute of Technology, Kharagpur, India. He has done research on sustainable hill farming for the past seven years and has published 25 research papers, in reputed journals, three book chapters and two books. He is a working member of the global working group on lychee and other Sapindaceae crops with the International Society for Horticultural Science (ISHS), Belgium, and as Lead Editor (Hort.) of the *International Journal of Bio-resource and Stress Management*. In addition, he is a consultant horticulturist to the Department of Horticulture and Agriculture (Research and Extension), Government of Mizoram, India, where he works on various research projects. He has visited many countries for his work, including Thailand, China, Nepal, Bhutan, Vietnam, South Africa, and South Korea.

Amritesh C. Shukla, DPhil, DSc

Professor, Department of Botany, University of Lucknow, India

Amritesh C. Shukla, DPhil, DSc, is currently Professor in the Department of Botany, University of Lucknow, India. He was formerly Professor in the Department of Horticulture, Aromatic and Medicinal Plants, Mizoram University, Aizawl, India. He began his academic career as Associate Professor at the same university. He has more than 20 years of experience on natural products and drug development and standardized various scientific methods, viz., MSGIT, MDKT, MS-97, NCCLS (2002–2003). He has developed six commercial herbal formulations and holds four patents. He has published

75 research papers and 15 book chapters. He has also authored six books and has handled many externally funded research projects. He is Fellow of five national and international scientific societies and also Visiting Professor at the University of British Columbia, Canada, and the University of Mauritius. In addition, Dr. Shukla is working as an Editor, Associate Editor, and editorial board member of many internationally reputed journals, including the *American Journal of Food Technology, Journal of Pharmacology & Toxicology*, and the *Research Journal of Medicinal Plants*. He has also been a visiting scientist at universities in Australia, Germany, China, and the USA.

Mohammed Wasim Siddiqui, PhD
Assistant Professor and Scientist in the Department of Food Science and Post-Harvest Technology, Bihar Agricultural University, Sabour, India

Mohammed Wasim Siddiqui, PhD, is Assistant Professor and Scientist in the Department of Food Science and Post-Harvest Technology, Bihar Agricultural University, Sabour, India, and author/co-author of 31 peer-reviewed research articles, 26 book chapters, two manuals, and 18 conference papers. He has 11 edited books and an authored book to his credit, published by Elsevier, CRC Press, Springer, and Apple Academic Press. Dr. Siddiqui established the international peer-reviewed *Journal of Postharvest Technology*. He is Editor-in-Chief of two book series (Postharvest Biology and Technology, and Innovations in Horticultural Science), published by Apple Academic Press. Dr. Siddiqui is also Senior Acquisitions Editor at Apple Academic Press, New Jersey, USA, for Horticultural Science. He has been serving as an editorial board member and active reviewer of several international journals.

Dr. Siddiqui has received several grants and respected awards for his research work from a number of organizations. He is an active member of the organizing committees of several national and international seminars, conferences, and summits. He is one of the key members to establish the World Food Preservation Center (WFPC), LLC, USA, and is currently an active associate and supporter.

Dr. Siddiqui acquired a BSc (Agriculture) degree from Jawaharlal Nehru Krishi Vishwa Vidyalaya, Jabalpur, India. He received his MSc

(Horticulture) and PhD (Horticulture) degrees from Bidhan Chandra Krishi Viswavidyalaya, Mohanpur, Nadia, India, with specialization in Postharvest Technology. He is a member of Core Research Group at the Bihar Agricultural University (BAU), which provides appropriate direction and assistance in prioritizing research.

LIST OF CONTRIBUTORS

Amrapali A. Akhare
Biotechnology Center, Department of Agricultural Botany, Dr. Panjabrao Deshmukh Krishi Vidyapeeth, Akola–444104, Maharashtra, India

Thejangulie Angami
ICAR-RC for NEH Region, Arunachal Pradesh Centre, Basar, Arunachal Pradesh, India

C. Aswath
Division of Ornamental Crops, ICAR – IIHR, Bangalore, India

S. G. Bharad
Department of Horticulture, Dr. Panjabrao Deshmukh Krishi Vidyapeeth, Akola–444104, Maharashtra, India

T. Bharathimeena
ICAR–Central Island Agricultural Research Institute, Port Blair–744101, Andaman and Nicobar Islands, India

R. K. Bhattacharyya
Professor Horticulture, Assam Agricultural University, Jorhat, Assam, India,

Karma Diki Bhutia
Sikkim University, 6th Mile, Tadong, Sikkim, India

Sangay Gyampo Bhutia
Sikkim University, 6th Mile, Tadong, Gangtok East Sikkim, India

Pooja Bohra
ICAR–Central Island Agricultural Research Institute, Port Blair–744101, Andaman and Nicobar Islands, India

K. L. Chadha
Adjunct Professor, IARI and President, The Horticultural Society of India, New Delhi, India

Sanjay Chadha
Department of Vegetable Science and Floriculture, CSK Himachal Pradesh Krishi Vishvavidyalaya, Palampur, Himachal Pradesh, India

Narong Chomchalow
Chairman, Thailand Network for the Conservation and Development of Landraces of Cultivated Plants, Bangkok, Thailand

Y-Y. Chou
AVRDC – The World Vegetable Center, P.O. Box 42, Shanhua, Tainan 74199, Taiwan

Megha H. Dahale
Department of Horticulture, Dr. Panjabrao Deshmukh Krishi Vidyapeeth, Akola–444104, Maharashtra, India

V. Damodaran
ICAR–Central Island Agricultural Research Institute, Port Blair–744101, Andaman and Nicobar Islands, India

R. G. Dani
Vice Chancellor, Dr. Panjabrao Deshmukh Krishi Vidyapeeth, Akola–444104, Maharashtra, India

B. Das
ICAR-RC for NEH Region, Tripura Centre, Lembucherra, West Tripura – 799210, India

Khumbar Debbarma
Department of Agricultural Entomology, Faculty of Agriculture, Bidhan Chandra Krishi Viswavidyalaya, Mohanpur–741252, Nadia, West Bengal, India

Akoijam Ranjita Devi
Department of Spices and Plantation Crops, Faculty of Horticulture, Bidhan Chandra Krishi Viswavidyalaya, Mohanpur–741252, Nadia, West Bengal, India

Angom Sarjubala Devi
Department of Environmental Science, Mizoram University, India

D. Durga Devi
Department of Fruit Crops, Horticultural College and Research Institute, TNAU, Coimbatore, Tamil Nadu – 641 003, India

Hidangmayun Lembisana Devi
ICAR Research Complex for NEH Region, Tripura Centre, Lembuchera, West Tripura, India

Yengkokpam Ranjana Devi
Biochemistry Laboratory, College of Agricultural Engineering and Post Harvest Technology, Central Agricultural University, Ranipool, Gangtok, Sikkim–737135, India

A. W. Ebert
AVRDC – The World Vegetable Center, P.O. Box 42, Shanhua, Tainan 74199, Taiwan

R. M. Gade
Vasantrao Naik College of Agricultural Biotechnology, Yavatmal, Maharashtra, India,

S .J. Gahukar
Biotechnology Center, Department of Agricultural Botany, Dr. Panjabrao Deshmukh Krishi Vidyapeeth, Akola–444104, Maharashtra, India

Vasant P. Gandhi
Centre for Management in Agriculture, Indian Institute of Management, Ahmedabad, India

R. K. Gautam
ICAR–Central Island Agricultural Research Institute, Port Blair–744101, Andaman and Nicobar Islands, India

Bikash Ghosh
Department of Fruits and Orchard Management, Faculty of Horticulture, Bidhan Chandra Krishi Viswavidyalaya, Mohanpur–741252, Nadia, West Bengal, India

B. Gopichand
Department of Forestry, Mizoram University, Aizawl, India

Mary Chinneithiem Haokip
Department of Spices and Plantation Crops, Faculty of Horticulture, Bidhan Chandra Krishi Viswavidyalaya, Mohanpur–741252, Nadia, West Bengal, India

T. K. Hazarika
Department of Horticulture, Aromatic and Medicinal Plants, School of Earth Sciences and Natural Resources Management, Mizoram University, Tanhril, Aizawl, Mizoram

Vanlalremruati Hnamte
Department of Spices and Plantation Crops, Faculty of Horticulture, B.C.K.V., Mohanpur–741252, West Bengal, India

K. S. Thingreingam Irenaeus
Department of Horticulture, College of Agriculture, Tripura

Priyanka Irungbam
Department of Agronomy, Faculty of Agriculture, Bidhan Chandra Krishi Viswavidyalaya, Mohanpur–741252, Nadia, West Bengal, India

I. Jaisankar
ICAR–Central Island Agricultural Research Institute, Port Blair–744101, Andaman and Nicobar Islands, India

Sangeeta Kanwar
Department of Vegetable Science and Floriculture, CSK Himachal Pradesh Krishi Vishvavidyalaya, Palampur, Himachal Pradesh, India

L. Kenyon
AVRDC – The World Vegetable Center, P.O. Box 42, Shanhua, Tainan 74199, Taiwan

Khatemenla
Department of Horticulture, Assam Agricultural University, Jorhat-13, Assam, India

R. M. Vijaya Kumar
Department of Fruit Crops, Horticultural College and Research Institute, TNAU, Coimbatore, Tamil Nadu – 641 003, India

S. Kumar
AVRDC – The World Vegetable Center, P.O. Box 42, Shanhua, Tainan 74199, Taiwan

S. Senthil Kumar
School of Agriculture, Lovely Professional University, Phagwara, Punjab, India.

Subhasis Kundu
Department of Fruits and Orchard Management, Faculty of Horticulture, Bidhan Chandra Krishi Viswavidyalaya, Mohanpur–741252, Nadia, West Bengal, India

R. Lalfakzuala
Department of Forestry, Mizoram University, Aizawl, India

Lalngaihwami
Department of Plant Pathology, Assam Agriculture University, Jorhat-13, Assam, India

F. Lalnunmawia
Department of Botany, Mizoram University, Aizawl, India

F. Lalriliana
Department of Forestry, Mizoram University, Aizawl, India

S-W. Lin
AVRDC – The World Vegetable Center, P.O. Box 42, Shanhua, Tainan 74199, Taiwan

Anurup Majumder
Bidhan Chandra Krishi Viswavidyalaya, Krishi Viswavidyalaya P.O., Nadia, West Bengal–741 252, India

Indrani Majumder
Department of Fruits and Orchard Management, Bidhan Chandra Krishi Viswavidyalaya, Mohanpur, West Bengal (741252), India

T. Mandal
Department of Floriculture and Landscaping, Faculty of Horticulture, BCKV, Mohanpur–741252, West Bengal, India.

S. Manivannan
Sikkim University, 6th Mile, Tadong, Sikkim, India

Puthem Mineshwor
Biochemistry Laboratory, College of Agricultural Engineering and Post Harvest Technology, Central Agricultural University, Ranipool, Gangtok, Sikkim–737135, India

S. K. Mitra
Chair, Section Tropical and Subtropical Fruits, International Society for Horticultural Science, Belgium

Krajairi Mog
Bidhan Chandra Krishi Viswavidyalaya, Krishi Viswavidyalaya P.O., Nadia, West Bengal–741 252, India

Deepa N. Muske
Biotechnology Center, Department of Agricultural Botany, Dr. Panjabrao Deshmukh Krishi Vidyapeeth, Akola–444104, Maharashtra, India

P. K. Nagre
Department of Horticulture, Dr. Panjabrao Deshmukh Krishi Vidyapeeth, Akola–444104, Maharashtra, India

Namita
Division of Floriculture and Landscaping IARI, Pusa, New Delhi–12, India

Lop Phavaphutanon
Department of Horticulture, Faculty of Agriculture at Kamphaeng Saen, Kasetsart University, Thailand

Anjana Pradhan
Sikkim University, 6th Mile, Tadong, Sikkim, India

S. Pradhan
Department of Floriculture and Landscaping, B.C.K.V., Mohanpur–741252, West Bengal, India

Mahadev Pramanick
Department of Agronomy, Faculty of Agriculture, Bidhan Chandra Krishi Viswavidyalaya, Mohanpur–741252, Nadia, West Bengal, India

V. M. Prasad
Department of Horticulture, Sam Higginbottom Institute of Agriculture, Technology and Sciences, Allahabad–211007, U.P., India

V. K. Jayaraghavendra Rao
ICAR- National Academy of Agricultural Research Management (NAARM), Hyderabad, India

E. Reang
College of Agriculture, Lembucherra, West Tripura–799210, India

Y. T. N. Reddy
Division of Fruit Crops, Indian Institute of Horticultural Research, Hessarghata Lake Post, Bangalore – 560089, India

K. Riazunnisa
Department of Biotechnology and Bioinformatics, Yogi Vemana University, Kadapa, YSR District, Andhra Pradesh, India

Rokozeno
Department of Entomology, Assam Agricultural University, Jorhat-13, Assam, India

Debjit Roy
Department of Fruits and Orchard Management, Faculty of Horticulture, Bidhan Chandra Krishi Viswavidyalaya, Mohanpur–741252, Nadia, West Bengal, India

S. Dam Roy
ICAR–Central Island Agricultural Research Institute, Port Blair–744101, Andaman and Nicobar Islands, India

Lalbiakdiki Royte
Department of Environmental Science, Mizoram University, India

Prativa Sahu
ICAR-Indian Institute of Water Management, Bhubaneswar, Odisha, India

L. Saikia
Department of Horticulture, Assam Agricultural University, Jorhat-13, Assam, India

K. Sakthivel
ICAR–Central Island Agricultural Research Institute, Port Blair–744101, Andaman and Nicobar Islands, India

Trudy A. Sangma
Department of Horticulture, Assam Agricultural University, Jorhat-13, Assam, India

Sukamal Sarkar
Department of Agronomy, Bidhan Chandra Krishi Viswavidyalaya, Mohanpur, West Bengal (741252), India

Sayan Sau
Department of Fruits and Orchard Management, Bidhan Chandra Krishi Viswavidyalaya, Mohanpur, West Bengal (741252), India

R. Schafleitner
AVRDC – The World Vegetable Center, P.O. Box 42, Shanhua, Tainan 74199, Taiwan

Laxuman Sharma
Sikkim University, 6th Mile, Tadong, Sikkim, India

H.-C. Shieh
AVRDC – The World Vegetable Center, P.O. Box 42, Shanhua, Tainan 74199, Taiwan

G. S. Shinde
Department of Horticulture, Dr. Panjabrao Deshmukh Krishi Vidyapeeth, Akola–444104, Maharashtra, India

S. S. Sindhu
Division of Floriculture and Landscaping IARI, Pusa, New Delhi–12, India

Akashdeep Singh
Department of Vegetable Science and Floriculture, CSK Himachal Pradesh Krishi Vishvavidyalaya, Palampur, Himachal Pradesh, India

Babita Singh
Division of Floriculture and Landscaping IARI, Pusa, New Delhi–12, India

Devi Singh
Department of Horticulture, Sam Higginbottom Institute of Agriculture, Technology and Sciences, Allahabad, U.P., India

Sh. Herojit Singh
Bidhan Chandra Krishi Viswavidyalaya, Krishi Viswavidyalaya P.O., Nadia, West Bengal–741 252, India

Umakant Singh
Department of Horticulture, Sam Higginbottom Institute of Agriculture, Technology and Sciences, Allahabad–211007, U.P., India

V. B. Singh
Department of Horticulture, SASRD, Nagaland University, Medziphema, Nagaland, India

Prem Na Songkhla
Director, Kehakaset Academic Institute of Agriculture, Bangkok, Thailand

R. Srinivasan
AVRDC – The World Vegetable Center, P.O. Box 42, Shanhua, Tainan 74199, Taiwan

Pabitra Subba
Sikkim University, 6th Mile, Tadong, Gangtok East Sikkim, India

G. Sai Sudha
Department of Biotechnology and Bioinformatics, Yogi Vemana University, Kadapa, YSR District, Andhra Pradesh, India

K. Soorianatha Sundaram
Department of Fruit Crops, Horticultural College and Research Institute, TNAU, Coimbatore, Tamil Nadu – 641 003, India

Deeki Lama Tamang
Sikkim University, 6th Mile, Tadong, Gangtok East Sikkim, India

Tawnenga
Department of Botany, Pachhunga University College, Aizawl, India

Thaneshwari
Division of Ornamental Crops, ICAR – IIHR, Bangalore, India

Rocky Thokchom
Uttar Banga Krishi Viswavidyalaya, Pundibari, Cooch Behar, West Bengal–736165, India

Vanlalruati
Department of Floriculture and Landscaping, Faculty of Horticulture, BCKV, Mohanpur–741252, West Bengal, India

Balaji Vikram
Department of Horticulture, Sam Higginbottom Institute of Agriculture, Technology and Sciences, Allahabad, U.P., India

Ajit Arun Waman
ICAR–Central Island Agricultural Research Institute, Port Blair–744101, Andaman and Nicobar Islands, India

J. F. Wang
AVRDC – The World Vegetable Center, P.O. Box 42, Shanhua, Tainan 74199, Taiwan

G. S. Yadav
ICAR-RC for NEH Region, Tripura Centre, Lembucherra, West Tripura – 799210, India

Jenny Zoremtluangi
Department of Floriculture and Landscaping, Faculty of Horticulture , BCKV, Mohanpur–741252, Nadia, West Bengal, India

LIST OF ABBREVIATIONS

AI	artificial insemination
AIC	Agricultural Insurance Company
AIFOM	All India Federation of Organic Farming
ANI	Andaman & Nicobar Islands
ANOVA	analysis of variance
APEDA	Agricultural and Processed food products Export Development Authority
BSI	Botanical Survey of India
CA	Certification Agency
CA	controlled atmospheric
CaCV	capsicum chlorosis virus
CCOF	California Certified Organic Farmers
ChiRSV	chilli ringspot virus
ChiVMV	chili venial mosaic virus
CMS	cytoplasmic male sterility
CMV	cucumber mosaic virus
CPRI	Central Potato Research Institute
CRD	completely randomized design
CTV	citrus tristeza virus
DAP	days to spike emergence
DCIP	2,6-dichloroindophenols dye
DFV	Desai Fruits and Vegetables
DPPH	2,2-diphenyl-1-picrylhydrazyl
DSB	double stranded break
dsRNA	double-stranded RNA
EERI	Earthquake Engineering Research Institute
ELISA	enzyme-linked immunosorbent assay
EMS	ethyl methanesulfonate
EU	European Union
FCRD	Factorial Completely Randomized Design
FMCG	fast-moving consumer goods
FYM	farm yard manure

GCMMF	Gujarat Cooperative Milk Marketing Federation
GCV	genotypic co-efficient of variation
GMOs	genetically modified organisms
GRAS	generally recognized as safe
gRNA	guide RNA
HACCP	Hazard Analysis and Critical Control Points
HFIL	Heritage Foods India Limited
HPMC	Himachal Pradesh Fruit Processing and Marketing Corporation
IBA	indole butyric acid
IBD	International Business Division
ICAR	Indian Council of Agricultural Research
ICPN	International Chili Pepper Nursery
IFOAM	International Federation of Organic Agriculture Movements
IMO	Institute for Market Ecology
IPM	integrated pest management
ISPN	International Sweet Pepper Nursery
ITS	internal transcribed spacer
IUCN	International Union for Conservation of Nature and Natural Resources
IX	instability index
K	potassium
LDI	Lakshadweep Islands
MLO	mildew-resistance locus
MSL	mean sea level
N	nitrogen
NARES	National Agricultural Research and Extension System
NC	neemcake
NER	north eastern region
NGS	next generation sequencing
NHM	National Horticulture Mission
NOP	National Organic Program
NPOF	National Project on Organic Farming
NPOP	National Programme on Organic Production
OBV	oryctes baculovirus
OD	optical density
OFPA	Organic Foods Production Act

P	phosphorus
PARPH	pimaricin, ampicilin, rifamicin, penta-chloronitrobenzene and hymexazol
PCR	polymerase chain reaction
PCV	phenotypic co-efficient variances
PEGs	polyethylene glycols
PepYLCV	pepper yellow leaf curl virus
PeVYV	pepper vein yellows virus
PMMV	pepper mild mottle virus
PRSV	papaya ringspot virus
PSB	phosphate solubilizing biofertilizer
PSB	phosphorus solubilizing bacteria
PVMV	pepper veinal mottle virus
PVY	potato virus Y
RBD	randomized block design
RDF	recommended dose of fertilizer
rDNA	ribosomal DNA
RNAi	RNA interference
SALT	Sloping Agricultural Land Technology
SCAR	sequenced characterized amplified region
SDI	spatial data infrastructure
SL	strigolactonem
SODs	superoxide dismutases
TALENs	transcription activator-like effector nucleases
TEV	tobacco etch virus
TNRV	tomato necrotic ring virus
ToMV	tomato mosaic virus
TPS	true potato seed
TSS	total soluble solids
TSWV	tomato spotted wilt virus
TYLCTHV	tomato yellow leaf curl Thailand virus
UGS	urban green spaces
UKS	Uttam Krishi Sevak
UNFCCC	United Nations Framework Convention on Climate Change
USDA	US Department of Agriculture
VC	vermicompost

WBTDCC	West Bengal Tribal Development Co-operative Corporation Limited
WHC	water holding capacity
ZFNs	zinc finger nucleases

PREFACE

Global food demand is expected to have doubled by 2050, while the production environment and natural resources are continuously shrinking and deteriorating. Horticulture, a major sector of agriculture, needs to take part in enhancing crop production and productivity in parity with agricultural crops to meet the emerging food demand. There are projections that demand for food grains would increase from 192 million tonnes in 2000 to 345 million tonnes in 2030. Hence, in the next 15 years, production of food grains needs to be increased at the rate of 5.5 million tonnes annually. The demand for high-value commodities (such as horticulture, dairy, livestock and fish) is increasing faster than food grains—for most of the high-value food commodities demand is expected to increase by more than 100% from 2000 to 2030. These commodities are all perishable and require, different infrastructure for handling, value-addition, processing and marketing.

Asia, the major crop-producing continent, thought to be the global super power because of its increasing skilled and energetic human resources and faster developing technological and economic growth, has to play a major role to meet the projected global food demand. However, the developing Asiatic countries are more or less facing the common problem of diminution of cultivable land with massive rapid urbanization. Apart from that, the contemporary production limitations—depleting land fertility, unequal cross-subsidy and, more predominantly, vagaries of climate change—put forth a tough task to perform.

India, the second most important Asiatic food grain producer, is facing a more grim production situation. The average size of land holding declined to 1.32 ha in 2000–01, from 2.30 ha in 1970–71, and absolute number of operational holdings increased from about 70 million to 121 million. If this trend continues, the average size of holding in India would be mere 0.68 ha in 2020, and would be further reduced to a low of 0.32 ha in 2030. This is a very complex and serious problem, when the share of agriculture, including horticulture, in gross domestic product is declining, average size of land-holding is contracting and fragmenting, and numbers of operational holdings are increasing. In addition, annually, India is losing nearly 0.8 million tonnes of nitrogen, 1.8 million tonnes of phosphorus and 26.3 million tonnes of potassium—the deteriorating quality and health of soil is something to be checked. Problems are further aggravated by imbalanced application of nutrients (especially nitrogen, phosphorus and potash) and excessive mining of micronutrients, leading to deficiency of macro- and micro-nutrients in the soils. Similarly, the water-table is lowering steeply in most of the irrigated areas, and water quality is also deteriorating, due to leaching of salts and other pollutants.

Amidst this situation, fulfilling the target with a sustainable model of crop production is really an enormous endeavor. There is an earnest need to develop promising technologies and management options to raise productivity to meet the growing food demand in this situation of deteriorating production environment at the lowest cost; and to develop appropriate technologies, to create required infrastructures, and to evolve institutional arrangements for production, postharvest and marketing of high-value and perishable commodities and their value-added products. Improved agro-techniques, quality planting materials, improved varieties, climate resilient production models, involvement of information technology and biotechnology, improved postharvest handling-storage, and marketing are the key issues on which to focus on to bring about the desired metamorphosis in global horticulture. However, to achieve the goal with a sustainable production environment (i.e., sustainability in terms of economy-ecology-society), is the greatest challenge to meet.

The International Symposium on Sustainable Horticulture, organized by the Department of Horticulture, Aromatic and Medicinal Plants, Mizoram University, Aizawl, India, from 14–16th March, 2016, provided a platform for the exchange of ideas and research experience and discussed several facets of sustainable horticulture with the thematic areas such as management of genetic resources and biodiversity conservation; production technology of horticultural crops; crop improvement and biotechnology; plant protection in horticultural crops; postharvest technology and value addition; trade, marketing, entrepreneurship development and extension; and horticulture for food, health and nutrition.

In this regard, this research compendium has been categorically divided to form two volumes; *Sustainable Horticulture, Volume 1: Diversity, Production, and Crop Improvement*, and *Sustainable Horticulture, Volume 2: Food, Health, and Nutrition*. The first volume outlines the contemporary trends in sustainable horticulture research, covering such topics as crop diversity, species variability and conservation strategies, production technology, tree architecture management, plant propagation and nutrition management, organic farming, new dynamics in breeding, and marketing of horticulture crops. The second volume depicts the research trends in sustainable horticulture comprising postharvest management and processed food production from horticulture crops, crop protection and plant health management, and horticulture for human health and nutrition.

We extend our sincere thanks to the contributors, reviewers, and Apple Academic Press for their efforts and contributions. It is hoped that these book volumes will be useful for students, teachers, scientists, extension workers, and researchers in horticulture and allied disciplines.

—Debashis Mandal
Amritesh C. Shukla
Wasim Siddiqui

PART I

GENETIC RESOURCES AND BIODIVERSITY CONSERVATION

CHAPTER 1

CITRUS GERMPLASM AT ORIGINAL HOME WARRANTS DOCUMENTATION AND CONSERVATION

R. K. BHATTACHARYYA

Professor Horticulture, Assam Agricultural University, Jorhat, Assam, India, E-mail: ranjitkb2010@gmail.com

CONTENTS

Citrus is the collective generic term comprising a number of species and varieties of fruits, known the world over for their characteristic flavor and attractive range of colors. The orient or the places of origin for most citrus fruits are believed to be the southern slope of the Himalayan region, the entire North Eastern Region (NER) of India, and adjacent China. The citrus orient has rich diversity. The northeast India is the richest center of diversity for citrus and includes many types of citrons, lemons and mandarins, *Citrus indica,* papeda, pummelos, and their hybrids. Sour orange, rough lemon, and *C. megaloxycarpa* have been found to grow widely in semi-wild conditions in NERs of India. The region can boast of producing certain quality of citrus, viz., "Khasi Mandarin" or "Assam Sumthira," "Assam Lemon," "Gol Nemu" (a rough lemon), "Bira Jara" (a citron), a number of pink- and white-fleshed pummelos, etc. (Bhattacharyya and Baruah, 1998).

1.1 GERMPLASM WEALTH IN THE REGION

The wild, semi-wild, and cultivated forms of citrus germplasm of the region are listed below.

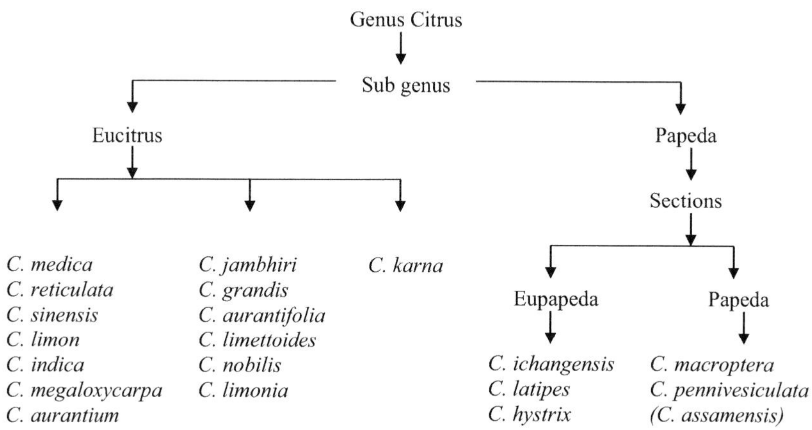

1.2 SPECIFIC CHARACTERISTICS OF THE GERMPLASM

1.2.1 CITRUS MEDICA (CITRON, JARA TENGA)

Citron is a native of northeastern India and is found growing wild in the region. Plants are monoembryonic, and leaf petioles are wingless. Its sweet albedo is eaten raw, and the fruits are used for the preparation of pickles. The thick, whitish, fleshy albedo portion of most of the citron cultivars is invariably sweet in taste. Juice content in the vesicles is scanty and acidic in taste. "Bira Jara" of Assam is the real citron of commercial value. In addition, some citron cultivars are "Ban Jara" and "Mitha Jara" found in Brahmaputra valley, "Jaare Jamir" in Barak valley, "Soh-Madeh" or "Soh-Monong" in Khasi hills, and "Tuma-Hauthor" in Karbi Anglong.

Some citron × lemon hybrids of the regions are "Godha Pati Nimbu," "Kata Jamir," "Pati Nimbu," "Pani Jamir," etc.

1.2.2 C. RETICULATA (MANDARIN)

Khasi mandarin is the most important commercial citrus of the region and exhibits some morphological variations in fruit characteristics in varied locations. Some synonyms of the cultivar in the region are "Soh-Niamtra," "Burnihat Orange," "Shella Orange," "Lushai Orange," "Jatinga Orange," "Karimganj Orange," "Sumthira," and "Kamala." The fruits are loose skinned, smooth surface, usually 10 segments, orange-colored abundant juice, sweetness and acidity well blended, green cotyledons, and polyembryonic.

1.2.3 C. SINENSIS (SWEET ORANGE)

Poor form of sweet oranges available in this region are "Soh-Nianriang" and "Soh-Biatra" of Meghalaya, "Tasi" of Arunachal Pradesh, and "Mitha Chakala" of Assam with acidulously sweet fruits of limited commercial value grown even at higher attitudes of 1500 m. Plants are vigorous and upright, heavy bearer, tight-skinned fruits, white cotyledons, and polyembryonic.

1.2.4 C. LIMON (LEMONS)

"Assam Lemon" is the principal lemon cultivar of commercial importance. Some varied forms available in the region are "Nayachangney" (with scant juice), "Soh-Long," and "Soh-Synteng." The plants are prolific bearer, pink-tinged flower, long elliptic to oblong-ovate fruit, crystal white juice vesicle with acidic juice, and usually seedless fruit.

1.2.5 C. INDICA (INDIAN WILD ORANGE)

The species is grown in Sesatgiri areas of Garo hills in Meghalaya and locally known as "Memang Narang." Plants are bushy and grow near marshy areas in forests. The plants have inedible small fruits with slimy juice of acidic taste, bold seeds, and unpleasant aroma.

1.2.6 C. MEGALOXYCARPA (SOUR PUMMELOS)

The varied forms available in the region are "Bor tenga," "Hukma Tenga," "Holong Tenga," and "Jamir Tenga." Fruits are very acidic with medicinal properties.

1.2.7 C. JAMBHIRI (ROUGH LEMON)

"Gol Nemu" is the popular cultivar grown as a lemon substitute. Fruits are spherical or globose, juice vesicle light yellowish white, juicy, acidic, seeds a few to none, and polyembryonic. Some varied forms of the region are "Soh-Myndong," "Soh-Jhalia," and "Mitha Tulia" (acidless).

1.2.8 C. GRANDIS (PUMMELOS)

Bhattacharyya and Dutta (1956) described nine different forms of pummelos from Assam with white and pink-tinged juice vesicles. Pummelos are known in this region as "Rabab Tenga," "Dawadi," "Burni," "Zemabawk," "Aijwal," and "Khasi Pummelos." Fruits are largest among citrus with subglobose, spherical, broadly obovoid to short pyriform shaped, spongy white or pink-tinged albedo, leaves with broad petiolar wing, and monoembryonic.

1.2.9 C. AURANTIFOLIA (LIMES)

Poor forms of limes available in the region are "Abhayapuri Lime" and "Karimganj Lime" with limited commercial value. Fruits are oval-elliptical to roundish, white-colored flower, and greenish-white juice vesicle.

1.2.10 C. LIMETTOIDES (SWEET LIME)

"Soh Jew" is a type available in the region and is believed to be a sour mutant of the species.

1.2.11 C. NOBILIS (KING ORANGE)

"Jeneru Tenga" of Assam belongs to this group and is a heavy bearer. Fruits size is similar to that of "Khasi Mandarin."

1.2.12 C. LIMONIA (RANGPUR OR SYLHET LIME)

"Sinduri Nemu Tenga" of Assam is a true representative of this species.

1.2.13 C. KARNA (KHATTA NIMBU)

"Soh-Sarkar" is the type available in the region with acidic fruits.

1.2.14 C. AURANTIUM (BITTER-SWEET ORANGE)

"Karunjamir," "Gondh-Huntra," and "Chakala Tenga" (Tita Chakala) are the types available in the region with vigorously growing plants and juicy fruit having slight bitter principle in juice.

1.2.15 C. ICHANGENSIS (ICHANG PAPEDA)

Originated in the Ichang region of southwest China, this species is known as "Ketsa-Shuffu" in the region. It is a cold-tolerant plant with nonedible fruit and scant juice.

1.2.16 C. LATIPES (KHASI PAPEDA)

"Soh-Shyrkhoit" is the Khasi name of the species. Fruits have moderate juice with acidic property.

1.2.17 C. HYSTRIX

Plants are small bushy, with leaves having very broad petiolar wing and fruit with scant juice.

1.2.18 C. MACROPTERA

"Satkara" (in Meghalaya, Mizoram, and Tripura), "Harebo" (in Manipur), and "Tithkara" (in Assam) are the available types. Plants are tall, thorny, with edible fruits that are highly juicy and acidic.

1.2.19 C. PENNIVESICULATA (SYN. C. ASSAMENSIS DUTTA AND BHATTACHARYYA)

This variety resembles "Satkara." The plants are vigorous, with greenish juice vesicle. It is known as "Ada Jamir" as described by Bhattacharyya and Dutta (1956) and is believed to be a variety of the species. Fruits and leaves of "Ada Jamir" are distinct with ginger-like aroma. Fruits are edible, acidic, and juicy.

1.3 PROBABLE CITRUS HYBRIDS

Some natural citrus hybrids of the region are as follows:

1.3.1 SOH KHYLLAH

A probable natural hybrid of *C. reticulata* and *C. grandis* found in Khasi hills.

1.3.2 HASH KHULI

A probable natural hybrid of *C. grandis* and "Ada Jamir." Fruits are similar to pummelos in size and aroma resembles that of "Ada Jamir."

1.3.3 SARBATI

A probable natural hybrid of *C. medica* and *C. aurantium*. Plant characteristics, flavor, color, and fruit rind texture are similar to those of citron, while leaf shape, size, and taste of albedo resemble that of sour orange.

1.3.4 DEWA TENGA

A probable hybrid of *C. aurantium* and *C. grandis.* The leaf shape, size, and aroma resemble the former and the fruit and flower size resemble the latter.

1.3.5 NICHOLOS ROY

A probable hybrid of *C. grandis* and *C. limon.* Fruits are attractive and juicy with low acidity.

1.4 CONCERN TOWARD EXISTENCE OF CITRUS GERMPLASM

In the region, a few citrus variants, namely "Sumthira" or Khasi Mandarin; "Assam Lemon"; "Gol Nemu," a rough lemon; and "Rabab Tenga," the pink and white-fleshed pummelos, are found to a certain extent in cultivated forms in homestead gardens or *baris*. Other citrus species are found in wild or semi-wild conditions or in *baris* in neglected or dilapidated conditions. Systematic collection, morpho-biochemical characterization, thorough assessment of the indigenous citrus genetic resources to exploit potentialities as commercial fruits with nutraceutical and antioxidant properties, and conservation of the germplasm are of utmost importance.

In citrus, germplasm are evaluated for their use as rootstocks and as a source of resistance in breeding program for their improvement as well as for exploiting them for developing various byproducts needed in food industry.

A good number of citrus germplasm are in the state of extinct because of the problem caused by dreaded disorders, leading to their decline. In order to conserve the invaluable citrus germplasm of the region, measures need to be taken to combat the decline by controlling biotic factors like citrus diseases and pests and abiotic factors like malnutrition and other physiological disorders like fruit drop, fruit splitting, and granulation.

1.5 NBPGR-NOTIFIED ENDANGERED CITRUS SPECIES OF NER

The NBPGR (National Bureau of Plant Genetic Resources) had already enrolled the following indigenous citrus species of NER (North East Region, India) as endangered in view of which intensive care is of utmost importance to safeguard their existence.

- *C. assamensis*
- *C. indica*
- *C. latipes*
- *C. megaloxycarpa*
- *C. macroptera*

1.6 LIST OF THREATENED CITRUS SPECIES OF NER

The following citrus species found in NER which were earlier documented by Bhattacharyya and Dutta (1956), are now considered to be threatened as they are at present in a state of extinction.

1.7 NECESSITY FOR CONSERVATION OF EXISTING CITRUS SPECIES

- In order to save useful and invaluable germplasm, before they become extinct, there is a strong growing need to establish citrus-specific conservation centers in NER.
- Some citrus germplasm are being maintained in various regions as a working collection. In the NE Region, Citrus Research Station, AAU, Tinsukia, NBPGR Substation, Umium, and ICAR Research Complex for the NEH Region have field gene banks.
- However, attempts should be made to preserve the native habitat of citrus species of the region as in situ conservation. This is of prime importance to conserve and maintain available citrus species in situ in *baris* by farmers.

Threatened Citrus Species	Scientific Name
1. Pati lebu	*C. limon*
2. Jara tenga	*C. medica*
3. Kata Jamuri	*C. limon*
4. Elachi lebu	*C. limon*
5. Soh-synteng	*C. limon*
6. Pani Jamir	*C. limon*
7. Kata Jamir	*C. jambhiri*
8. Soh myndong	*C. jambhiri*
9. Sinduri Nemu Tenga	*C. jambhiri*
10. Gol Nemu	*C. jambhiri*
11. Mitha Chakala	*C. sinensis*
12. Soh-sarkar	*C. karna*
13. Abhayapuri lime	*C. aurantifolia*
14. Karimganj lime	*C. aurantifolia*
15. Mitha kaghzi	*C. limetta*
16. Jeneru tenga	*C. nobilis*
17. Indian Wild Orange	*C. indica*
18. Soh-niangriang	*C. sinensis*
19. Mitha Chakala	*C. sinensis*
20. Tita Chakala or Chakala tenga	*C. aurantium*
21. Karun Jamir	*C. aurantium*
22. Gondh-Huntra	*C. aurantium*
23. Bor tenga	*C. megaloxycarpa*
24. Hukma tenga	*C. megaloxycarpa*
25. Holong tenga	*C. megaloxycarpa*
26. Jamir tenga	*C. megaloxycarpa*
27. Ichang papeda or Ketsa-shupfu	*C. ichangensis*
28. Khasi papeda	*C. latipes*
29. Satkara	*C. macroptera*
30. Tithkara	*C. macroptera*
31. Ada Jamir	*C. assamensis*
Probable Natural Hybrids:	
32. Soh Khyllah	*C. reticulata* × *C. grandis*
33. Hash Khuli	*C. grandis* × *C. assamsensis*
34. Sarbati	*C. medica* × *C. aurantium*
35. Dewa tenga	*C. aurantium* × *C. grandis*
36. Nicholos Roy	*C. grandis* × *C. limon*

1.8 VARIOUS WAYS OF CITRUS GERMPLASM CONSERVATION

Germplasm conservation may be done by adopting the following methods (Ghosh, 1999):

a. **In situ conservation**: attempts should be made to preserve the native habitat of citrus species of the region.
b. **Field gene bank**: Citrus germplasm are maintained in various regions as a working collection. In the NE Region, Citrus Research Station, AAU, Tinsukia, NBPGR Substation, Umium, and ICAR Research Complex for the NEH Region have field gene banks.
c. **Seed conservation**: Many citrus species are known to have high degree of polyembryony, and for this reason, seed conservation can be used without the risk of genetic instability.
d. **In vitro conservation**: Maintenance of germplasm in field gene bank requires large area and funds and is also affected by pests and diseases. Hence, in vitro conservation of germplasm may be a useful method. This technique is also useful for revitalization of citrus germplasm as meristem culture techniques eliminates many viruses.

1.9 MEASURES TO CHECK LOSS OR EROSION OF EXISTING CITRUS GERMPLASM

As a working collection, some citrus germplasm are being maintained in various regions. Perhaps, attempts should be made in order to preserve the native habitat of citrus species of the region as in situ conservation. This is of prime importance to conserve and maintain available citrus species in situ in *baris* by farmers. Cultivars like "Bira Jara," "Ban Jara," "Mitha Jara," "Jeneru Tenga," "Pati Nemu," "Godha Pati Nemu," "Pani Jamir," "Kata Jamir," "Mitha Chakala," "Chakalatenga," "Bor tenga," "Hukum tenga," "Holong tenga," "Jamir tenga," "Abhayapuri Lime," "Karimganj Lime," "Sinduri Nemu Tenga," "Tithkara," "Ada Jamir," etc. are at present available only in certain specific restricted pockets in insignificant numbers. There is a growing need to have rigorous awareness program for the common people for in situ conservation of the invaluable threatened citrus species as well as for adopting measures to multiply them using suitable propagation techniques. Farmers at large may be imparted effective training program in order to combat citrus decline by controlling biotic factors like citrus diseases and pests and abiotic factors like malnutrition and other physiological disorders.

Farmers may be requested to inform about the availability of the threatened citrus species to the Citrus Research Station, AAU, Tinsukia, NBPGR Substation, Umium, and ICAR Research Complex for the NEH Region or to the Department of Horticulture, Assam Agricultural University, Jorhat, so that measures may be taken for multiplication of the species as well as for conservation in field gene banks.

1.10 DOCUMENTATION OF EXISTING CITRUS GERMPLASM

Documentation of citrus germplasm of the region is poor. Except the pioneer classic documentation on the varied citrus in the region (Bhattacharyya and Dutta, 1956) made by S.C. Bhattacharyya, the then Horticultural Development Officer, Government of Assam, and S. Dutta, the then Officer-in-charge, Government Citrus Fruit Research Station, Burnihat, Assam, in 1956, no systematic approach has so far been made to document the existing citrus germplasm of the region. A few information on the genetic diversity and germplasm situation of NER had, however, been gathered (Hazarika, 2012; NBPGR, 2012; Bhattacharyya, 2013). Detail documentation of the citrus species of the region is the need of the hour.

KEYWORDS

- **citrus**
- **diversity**
- **endangered**
- **threatened**
- **germplasms**
- **North East Region**

REFERENCES

Bhattacharyya, R. K., (2013). Citrus germplasm situation in north-east India. *Science & Technology Journal, Mizoram University,* 1(1): 1-6.

Bhattacharyya, R. K., & Baruah, B. P., (1998). Citrus. In: *A Textbook on Pomology*, vol. III, Ed., Chatopadhyay, T. K., Kalyani Publishers, New Delhi. pp. 1–71.

Bhattacharryya, S. C., & Dutta, S., (1956). Classification of citrus fruits of Assam. *Scientific Monograph*, No. *20*, ICAR, New Delhi.

Ghosh, S. P., (1999). Citrus germplasm in India and its conservation strategy. *Intensive Agriculture*, January-February XXXVI, 8–11, 22.

Hazarika, T. K., (2012). Citrus genetic diversity of north-east India, their distribution, eco geography and eco biology. Genetic Resources and Crop Evolution, 59(6):1267-1280.

CHAPTER 2

MEDICINAL ORCHIDS OF MIZORAM AND HABITAT APPROACH FOR CONSERVATION

VANLALRUATI,[1] T. MANDAL,[1] S. PRADHAN,[1] and VANLALREMRUATI HNAMTE[2]

[1]*Department of Floriculture and Landscaping, B.C.K.V., Mohanpur–741252, West Bengal, India*

[2]*Department of Spices and Plantation Crops, Faculty of Horticulture, B.C.K.V., Mohanpur–741252, West Bengal, India*
E-mail: maruathmar@gmail.com

CONTENTS

ABSTRACT

Orchids are highly diverse, habitat specific, and occupy about 8% of global floriculture trade. Mizoram has 253 species of orchids (Botanical Survey of

India (BSI)), out of which 10 species have medicinal importance. Mostly, the tubers of these orchids are used in medicine. Many of these orchids face the extreme danger of extinction due to overexploitation and habitat destruction. Orchids require a special kind of environment and habitat, and unique microclimatic conditions to survive and perpetuate. Studies on quantitative information are vital for conservation plan in a particular area and to understand the ecology of the species. The present study, thus, aims to identify the habitat types preferred for medicinally important orchids. An extensive sampling was conducted for medicinal orchids in the state of Mizoram where 10 medicinal orchid species belonging to eight genera were recorded in different habitats, viz., tropical evergreen and semi-evergreen forests, tropical moist deciduous forests, and montane subtropical pine forests. *Paphiopedilum hirsutissimum, Paphiopedilum villosum*, and *Renanthera imschootiana* are highly endangered in the state. Among the three habitats, montane subtropical pine forests was found to be the most suitable habitat for medicinal orchids of Mizoram.

2.1 INTRODUCTION

The abundance of many orchid species is believed to have fallen to critical levels in recent years (Kull et al., 2006). Orchids are subjected to high levels of threats, through both natural and anthropogenic causes (Kull et al., 2006). Of the 259 species of endemic orchid in India, about 72 species are vulnerable and 26 species are endangered (Nayar, 1996). Studies at various times have shown that many known brands of herbal medicines use substitutes for some medicinal orchids due to their unavailability and one such example is that of *Eulophia dabia.* It is so rare today that it has been substituted by *S. nepalense* (Jalal and Rawat, 2009). Orchids (*Orchid* spp.) occupy a wide range of habitats and are characterized by distinct floral morphology, pollination mechanism, and association with unique fungal partners (Dressler, 1990). In India, orchids are represented by 1,129 species and 184 genera (Karthikeyan, 2000) and show maximum diversity in the eastern Himalaya, including the North Eastern Region, Western Ghats, eastern Himalaya, and eastern part of Western Himalaya (Kumaon Himalaya).

Mizoram encompasses an area of 21,081 sq. km., comprising eight districts. Three forests types are known to occur in the state, viz., tropical evergreen and semi-evergreen forests, tropical moist deciduous forests, and montane subtropical pine forests located between 21°58" N to 24°35" N latitude and 91°15" E to 93°29" E with varying altitudes from 722 m to 2000 m asl (Thapa and Sahoo, 2003). The climate is influenced by the monsoon pattern of rainfall. Generally, the average annual rainfall ranges between 200 and 250 cm. Fischer (1938) reported about 151 species of orchids from Lushai Hills, presently in Mizoram. However, BSI, North East Region listed 253 species of orchids in Mizoram, out of which 8 species have medicinal importance (Table 2.1). Mostly, the tubers of these orchids are used in medicine. Many of these orchids face the extreme danger of extinction due to overexploitation and habitat destruction. Orchids require a special kind of environment, habitat, and unique microclimatic conditions to survive and perpetuate. Studies on their habitat is important to conserve reliant species;

TABLE 2.1 Medicinal Uses of Orchid Species by the Mizo Tribes of Mizoram

Species	Types	Medicinal uses
Acampa papillosa	Roots	Rheumatism
Aerides multiflorum	Plant	Juice externally applied to remove worms in the body; root used for rheumatism, neuralgia, uterine disease, etc.
Aerides odoratum	Plant	Anti-tuberculosis
Anoectochilus formosinus	Leaves	Faucitis, tonsillitis, hypertension and cancer
Cymbidium aloifolium	Plant/pod	Emetic
Dendrobium ariaeflorum	Stem	Extracts as narcotics.
Dendrobium denudans	Stem	Extracts as narcotics.
Dendrobium nobile	Seeds	Fresh and dried stem as stomachic, analgesic
Ephimeranthan macrei		Rheumatism, stomachic and external application for skin diseases
Eria tomentosa		Juice as medicinal bath for ague
Eulophia nuda	Tuber	Vermifuge and scrofula
Habenaria		Hemipligia, paralytic infections, chronic diarrhoea, diabetes
Malaxis Acuminata	Pseudobulb	Bleeding diathesis, burning sensation, fever and phthisis
Pholidota imbricata	Pseudobulb	Finely macerated in mustard oil and applied on joints for rheumatic pains
Ryncostylis retusa	Plants	As emollient for softening/soothing
Vanda corulea	Leaf	Leaf juice for diarrhoea, dysentery and external application for skin diseases

Source: Adapted from Thapa and Sahoo (2003).

protect and restore wild species; and prevent their extinction, fragmentation, or reduction in range. Habitat preference of orchids should also be identified wherein habitat-wise conservation strategies can be applied. The present study, thus, aims to identify the habitat types preferred for medicinally important orchids.

2.2 HABITAT AND ECOLOGY

Orchids are herbaceous plants distributed from high mountains to tropical rain forests. Their habitats are directly influenced by temperature, rainfall, altitude, humidity, and soil condition besides the microclimatic conditions, and the survival of individual species depends on the availability of suitable conditions (Chowdhery, 2001). Depending upon their occurrence, different vegetational types of orchid habitats can be recognized:

2.2.1 TROPICAL EVERGREEN AND SEMI-EVERGREEN FORESTS

These forests occur in regions with heavy rainfall and are highly rich in floral diversity, and they are multistoried (Chowdhery, 2001). The top canopy is composed of lofty trees-like species of *Bauhinia variegate, Bischofia javanica, Bursesa serrata, Dipterocarpus turbinatus, Duabanga sonnesatioides, Dysoxylum procerum, Engelhardtia spikata, and Erythrina suberosa* (Thapa and Sahoo, 2003). Such tall trees with their dense close canopy form a dense and dark humid environment, which is excellent for the luxuriant growth of epiphytic species. As a result, these forests are occupied by epiphytes and epiphytic climbers. The following genera with medicinal potential are found in this region, viz., *Aerides multiflorum, Eria tomentosa, Eulophia nuda, Dendrobium ariaeflorum, Rhynchostylis retusa, etc.*

2.2.2 TROPICAL MOIST DECIDUOUS FORESTS

Apart from tropical evergreen and semi-evergreen forests, tropical deciduous forests are also found in areas where rainfall is less. The important trees of these forests include *Actinodaphne macroptera, Bischofin javanica, Castenopsis* sp., *Juglans regia, Lagerstroemia flos reginae, Lagerstroemia religiosa, Michelia oblonga, Quercus* sp. etc. (Thapa and Sahoo, 2003). In the absence of sufficient rainfall and humidity, the number of epiphytic medicinal orchids are low compared to those in evergreen forests. The following genera with medicinal potential are found in this region, viz., *Aerides, Eria, Dendrobium denudans, Ephimeranthan macrei, etc.*

2.2.3 MONTANE SUBTROPICAL PINE FORESTS

The subtropical zone has cooler and humid climate and is characterized by trees that are bushy in appearance and shorter than those of tropical forests (Chowdhery, 2001). The common tree species are *Bischofia javanica, Canarium* sp., *Castenopsis indica, Kadsura roxburghianum, Michelia oblonga, Phoebe goalpareusis, Pterospermum* sp., *Quercus* sp., and *Terminalia mysiocarpa* (Thapa and Sahoo, 2003). The following genera with medicinal potential are found in this region, viz., *Habenaria, Cymbidium aloifolium, Dendrobium denudans, Cymbidium aloifolium, Pholidota imbricata,* and *Vanda coruelia.* The humus-rich forests floor that remains moist and shaded also provide very favorable substratum for terrestrial medicinal orchids like *Phaius*. The stones and boulders that are covered with a thick layer of moss can be seen densely covered with species of *Coelogyne* spp. *and Dendrobium nobile.*

The occurrence of specific mycorrhizal fungus in the microclimate might also influence the habitat of the orchids (Hegde, 1982). Soil requirements, freedom from competition, mycorrhiza, acidity, soil temperature, and solar exposure are the environmental factors that categorize orchid habitats (Case, 1962). Species with specific habitat requirements have greater tendencies of extinction than the species with a broad habitat range (Samant et al., 1996). Among the three habitat types, montane subtropical pine forests had the maximum diversity to help the growth of medicinal orchids. Species such as *Ephimeranthan* generally prefer a canopy that has less than 30% exposure to sun. *M. acuminate and Eulophia nuda* forms colonies in shady places, moist ground, and in the areas that are wet and mossy. *Habenaria* spp. too prefer moist localities, but this species generally grows in a scattered way.

2.3 PRESENT STATUS AND CONSERVATION

The old age practice of shifting cultivation or "Jhuming" is the single major factor for large-scale depletion of natural forest cover in Mizoram, thereby affecting many orchid species in their natural habitat (Singh et al., 1990). As a result, 55% of the orchids of Mizoram today fall under different categories of threat as envisaged in the "Red Data Book" published by the International Union for Conservation of Nature and Natural Resources (IUCN). Based on herbarium records, literature, and authors' own field observations, the orchids of Mizoram have been classified as common, vulnerable, rare, threatened, and extinct following Lucas and Synge (1980). Of the 253 taxa, only 112 species are fairly widely distributed in the state with larger populations, whereas 97% or 40% of the total species belongs to rare category, while the remaining 15% of the orchids fall under vulnerable and threatened category. In the rare category, the taxa like *Cymbidium cochleare, Cymbidium macrorhizon, Dendrobium chrysotoxum, D. parishii, Galeola falconeri, G. lindleyana. Paphiopedilum hirsutissimum, P. spicerianum, P. villosum, and Vanda parishii* are under greater threat due to habitat destruction or continued overexploitation. *Paphiopedilum charlesworthii* has been recorded only from Mizoram and Burma (Pradhan, 1976), has not been collected again, and appears to have vanished from its Indian haunts. *Arundina graminifolia Hochr, Dendrobium densiflorum Lindl, Paphiopedilum hirsutum Pfitz, Paphiopedilum spicerianum Pfitz, Renanthera imschootiana Rolfe, and Vanda coerulea Griff.* are the list of endangered species in Mizoram.

In situ preservation, through the establishment of sanctuaries and reserve forests, is the most effective means of conservation for such a relatively large population. The orchid sanctuaries at Sairep, near Lunglei, and Ngopa, near Champhai, set up by the State Forest Department, are positive steps in this direction. Similarly, the wildlife sanctuaries at Ngengpui, in Chhimtuipui district and Dampa, in Aizawl district, also provided conservation of rich orchid habitats. Nonetheless, the steps taken so far are not sufficient to offset the threat to the rich orchid diversity met within the state, and it is desirable to establish additional sanctuaries in the orchid-rich habitats in different eco-climatic zones like Blue Mountains, Mamit, North Vanlaiphai, Saithah, Ngurtlang, Zote-tlang, and Dampa tlang for preserving the orchid flora for posterity (Singh et al., 1990).

Another viable means for preservation of selected taxa in their ex-situ conservation in botanical gardens, where they can be rehabilitated and cultivated not only for future research but also for multiplication and commercial exploitation. BSI has effectively brought under cultivation about 135 species of Mizoram orchids at the National Orchidarium and Experimental Garden, Shillong, viz., *Aerides fieldingii, Anoectochilus griffithii, A. sikkimensis, Coelogyne barbata, D. densiflorum, Renanthera imschootiana, Arundina graminifolia, Vanda Coerulia*, etc. (Singh et al., 1990). The in vitro raised seedlings would, in future, be made available both for rehabilitation in the wild and for commercial purposes so as to release pressure on the depleting natural populations of ornamental species.

2.4 CONCLUSION

The most striking feature of the orchids is that they need a specific microhabitat for their growth and proliferation, which vary in different habitats. In order to prevent their extinction, it is necessary to grow and conserve the species according to their favorable habitat, which acts as a gene pool. The fragile habitats should also be conserved as nature reserves to ensure sustainability of the ecosystem and thereby conserving the fast dwindling germplasm.

KEYWORDS

- **conservation**
- **habitat**
- **orchids**

REFERENCES

Case, F. W., (1962). Growing native orchids of the great lakes region. *Am. Orchid Soc. Bull. 31*, 473–445.

Chowdhery, (2001). Orchid diversity in north-east India. *Journal Orchid Soc. India, 15*(1–2), 1–17.

Dressler, R. L., (1990). *The Orchid: Natural History and Classification.* Harvard University Press, U. S. A.

Fischer, E. C., (1938). *The Flora of Lushai Hills*, Botanical Survey of India. *12*(2), 75–161.

Hegde, S. N., (1982). Observations on the habitat-distribution of orchids of Arunachal Pradesh. *J. Bombay Nat. Hist. Soc. 82*, 114–125.

Jalal, J. S., & Rawat, G. S., (2009). Habitat studies for conservation of medicinal orchids of Uttarakhand, western Himalaya. *African J. of Plant Sci., 3*(9), 200–204.

Karthikeyan, S., (2000). A statistical analysis of flowering plants of India, pp. 201–217. In: *Indian Orchids: Guide to Identification and Culture 2.* N. P. Singh, P. K. Pradhan, U. C., (1979). Faridabad, Thomson Press Ltd., pp. 747.

Kull, T., Kindalmann, P., Hutchings, J., & Primac, B., (2006). Conservation biology of orchids, Introduction to special issue. *Biol. Conserv., 129*(1), 1–3.

Lucas, G., & Synge, H., (1980). *The IUCN Plant Red Data Book*, Berne, Switzerland.

Nayar, M. P., (1996). *"Hotspots" of Endemic Plants of India*, Nepal and Bhutan S. B. Press, Bhutan. Tropical Botanic Garden and Research Institute. Thiruvananthapuram, Kerala.

Pradhan, U. C., (1979). *Indian Orchids: Guide to Identification and Culture.* Volume 1 and II. Primulaceae Books, Kalimpong, West Bengal.

Rao, N., (2004). Medicinal orchid wealth of Arunachal Pradesh. In: *Indian Medicinal Plants of Conservation Concern. 1*(2), 1–10 (http://envis.frlht.org).

Samant, S. S., Dhar, U., & Rawal, R. S., (1996). Conservation of rare endangered Plants: *The Context of Nanda Devi Biosphere Reserve*. In: Ramakrishnan, P. S., (ed.) Conservation and management of biological resources in Himalaya. Oxford & IBH Publishing Co. Pvt. Ltd., New Delhi.

Singh, D. K., Wadhwa, B. M., & Singh, K. P., (1990). A conspectus of orchids of Mizoram: Their status and Conservation. *J. Orchid Soc. India, 4*(1–2), 51–64.

Thapa, H. S., & Sahoo, U. K., (2003). Survey on orchid biodiversity of Mizoram in North – East India. *J. Natcon., 15*(1), 235–245.

CHAPTER 3

VARIETAL WEALTH OF TURF GRASSES AT THE INDIAN AGRICULTURAL RESEARCH INSTITUTE

BABITA SINGH, S. S. SINDHU, and NAMITA

Division of Floriculture and Landscaping IARI, Pusa, New Delhi–12, India, E-mail: bflori17feb@gmail.com

CONTENTS

ABSTRACT

Turf grasses are used in athletic fields, soccer fields, golf courses, cricket fields, and even in parks and home lawns. They are durable, useful, and contribute not only to the esthetic beauty but also to the environment and health of people by abating pollution. Lawn is a natural green-color carpet. It is an important feature of landscape, and a garden without a lawn is not considered complete. Lush green-color lawn provides a great satisfaction to owner and becomes a center of the garden for major activities. Lawn provides a place for taking rest after tiring job of the day, for holding parties and social functions, and for passive and active recreation. It harmonies with the surroundings and provides a natural green color. With increased emphasis on recreation and urbanization in the recent past, high-quality turf has gained prime importance. As the use of turf grass is gaining importance for the aforesaid purposes, it has become imperative to incorporate and initiate research on different aspects of this important area. The Division of Floriculture and Landscaping of IARI (Indian Agricultural Research Institute, New Delhi, India) is maintaining different varieties/species of turf grasses with different landscape use, and they are also evaluated for their morphological characteristics. Presently, we have about 15 turf grass germplasm of different characteristics like warm season, shade tolerant, drought tolerant, and variegated leaves and with different patterns of texture.

3.1 INTRODUCTION

The turf grasses belong to the family Poaceae, and there are more than 600 genera and 7500 species of turf grasses. The family Poaceae is the most ubiquitous family of the higher plant groups found on the Earth (Gould, 1968). Turf grasses are the plants that form a continuous ground cover that persist under regular mowing and traffic (Turgeon, 1980). Turf grasses are increasingly gaining popularity and importance in India. The demand for turf grasses is rising as more and more people are opting for turfs and lawns at home, offices, landscapes, parks, sports field, etc. Turf grasses are broadly divided into two categories. The first is cool season turf grasses that require temperature range of 15.56°C to 23.89°C for optimum growth, like bluegrass (*Poa* sp.), fescue grass (*Festuca* sp.), bentgrass (*Agrostis* sp.), and ryegrass (*Lolium* sp.). The second category of grasses are the warm season grasses. They grow at optimum temperatures of 26.60°C to 35.00°C, and their growth decreases when temperatures drop during winter and ceases completely when soil temperature falls below 10°C. Grasses that fall in this category are bermudagrass (*Cynodon* sp.), zoysia grass (*Zoysia* sp.), Augustine grass (*Stenotaphrum* sp.), bahia grass (*Paspalum* sp.), buffalo grass (*Buchloe* sp.), and carpet grass (*Axonopus* sp.). In India, mostly, warm season grasses are popular except in cooler regions where ryegrass and fescue grass are found. *Cynodon dactylon* grass is adapted to tropical and warm climate and has good density, texture, and wear tolerance, thereby making it suitable for high-quality turf (Romani, 2004; Rodriguez-Fuentes, 2009). *Cynodon dactylon* grass also shows drought and heat tolerance and has the capacity to provide high-quality turf with minimal cost and low water inputs (Macolino et al., 2012; Rimi et al., 2012). *Paspalum notatum* grass is useful for lawns as well as roadsides and for slope stabilization as a low-maintenance grass (Christians, 2004). *St. Augustine grass* (*Stenotaphrum secundatum*) is also known as Charleston *grass* in South Carolina and Buffalo Turf in Australia. It is a medium- to high-maintenance *grass* that forms a thick, carpet-like *sod*, crowding out most weeds and other *grasses. Dichondra* is a small genus of flowering plants in the morning glory *family* in Southern California; in the 1950s and 1960s, it was used as a *grass* substitute for *lawns.* Zoysia grass seed varieties are part of the Japonica grass family, which is known for less plant density, interseeding ability, and cold tolerance. *Love grass* is an attractive mid-sized *grass* for meadow gardens or garden accents. *Dactyloctenium aegyptium*, or Egyptian crowfoot grass is a member of the family Poaceae native in Africa. The plant mostly grows in heavy soils at damp sites (https://en.wikipedia.org/wiki/Lawn).

Lawn is the common feature of private garden, public landscape, and parks and gives the view of a green carpet. Lawns are created for esthetic pleasure as well as for recreational purposes. The most significance of lawn in a garden is that it has ability to stabilize the soil against water and wind erosion. A healthy, well-maintained lawn creates an exciting view and esthetically pleasing. Acceptable turf quality is defined as meeting the appearance and expectations for a particular area and includes traits such as green color, fine leaf texture, high tiller density, and overall esthetic appeal (Hanks, 2005).

3.2 TURF GRASS FUNCTIONAL BENEFITS

3.2.1 SOIL EROSION CONTROL AND DUST STABILIZATION – VITAL SOIL RESOURCE PROTECTION

Turf grasses serve as an inexpensive, durable ground cover as well as in protecting our valuable, nonrenewable soil resources. Perennial turf grasses offer one of the most cost-efficient methods to control wind and water erosion of soil, which is very important in eliminating dust and mud problems around homes, factories, schools, and businesses.

3.2.2 REDUCES RUNOFF FROM PRECIPITATION AND CONTRIBUTES TO FLOOD CONTROL

The dense plant canopy of mowed turf grass is very effective in the entrapment of water and airborne particulate materials as well as in absorbing gaseous pollutants. The high degree of water runoff that occurs from impervious surfaces in urban areas carries many pollutants in the runoff. Turf grasses offer one of the best-known systems for catchment of the runoff water plus the pollutants, especially if proper landscape designs are used (Schuyler, 1987).

3.2.3 ENHANCES GROUND WATER RECHARGE

One of the key mechanisms by which turf grasses control soil erosion is through a superior capability to essentially absorb or trap and hold runoff water. A healthy turfed lawn absorbs rainfall six times more effectively than a wheat field and four times better than a hay field; it is exceeded in this important function only by a virgin forest. This attribute is certainly important in enhancing ground water recharge.

3.2.4 ENVIRONMENTAL ISSUES

Turf grasses function in entrapment and biodegradation of organic chemicals, and in conversion of carbon dioxide emissions. The extensive fibrous root system of turf grasses contributes substantially to soil improvement through organic matter additions derived from atmospheric carbon dioxide via photosynthesis. In this process, a diverse large population of soil microflora and microfauna are supported. These same organisms offer one of the most active biological systems for the degradation of trapped organic chemicals and pesticides. Thus, this turf ecosystem is important in the protection of ground water quality.

3.2.5 ENHANCES HEAT DISSIPATION: TEMPERATURE MODERATION

The overall temperature of urban areas may be as much as 10–12 °F warmer than nearby rural areas. Turf grasses, through the cooling process of evapotranspiration, serve an important function in dissipating the high levels of heat generated in urban areas. For example, a football field has the cooling capacity of a 70-ton air conditioner. The cooling effect of irrigated turfs and landscapes can result in energy savings via reducing the energy input and allied costs required for the mechanical cooling of interiors of adjacent homes and buildings. The transpirational cooling effect of green turfs and landscapes can save energy by reduction in the energy input required for interior mechanical cooling of adjacent homes and buildings (Johns and Beard, 1985).

3.2.6 REDUCES NOISE ABATEMENT, GLARE REDUCTION, AND VISUAL POLLUTION PROBLEMS

The rough surface characteristics of turf grass function help in noise abatement as well as in multidirectional light reflection that reduces glare. A grass area of 70 feet distance on a roadside can abate obtrusive vehicle noises by 40%. Thus, turfs lower the hardness of unwanted noise and lessen the visual stress of glare (Cook and Van Haverbake, 1971; Robinette, 1972).

3.2.7 DECREASES NOXIOUS PESTS AND ALLERGY-RELATED POLLEN

Regularly mowed residential lawns reduce problems of nuisance pests such as snakes, rodents, mosquitoes, and chiggers and also allergy-related pollens produced by many weedy species. As small animals seek haven in taller grass at locations more distant from the house, they also are less likely to invade the house.

3.2.8 PROVIDES SAFETY IN VEHICLE OPERATION/ EQUIPMENT LONGEVITY

Roadside turf grasses are important in highway safety as well as erosion control, in that they function as a stabilized zone for emergency stoppage of vehicles. Turf grasses are also utilized for soil and dust stabilization around airfield runways in order to prolong the operating life of engines, while smaller airstrips utilize turf grasses as the runway surface itself.

3.2.9 LOWERS THE FIRE HAZARD AND PROVIDES SECURITY FOR VITAL INSTALLATIONS

The spacing provided by green lawns serves as a firebreak and as a high visibility zone that discourages unwanted intruders.

3.2.10 BUSINESS AND ECONOMIC IMPROVEMENT

Lawns increase the value of a property by 15% to 20%. They improve curb appeal. Real estate agents will tell you that homes with well-maintained landscapes and turf sell quicker and for higher values. Property managers say that well-maintained landscapes increase tenant satisfaction and lower vacancy rates.

3.2.11 THERAPEUTIC BENEFITS

Gardens (including turf) provide mental health benefits. The Horticultural Therapy Association promotes therapeutic benefits of green space as a technique for rehabilitation. No wonder so many people use landscape pictures as screensavers on computers. It has been shown that looking at images of landscapes and plants helps to reduce stress. Grassy areas in golf courses, cemeteries, parks, and homes can create feelings of peacefulness and remind us that the Earth is alive.

3.3 TURF GRASS SPECIES MAINTAINED AT IARI

3.3.1 BERMUDA GRASS

3.3.1.1 Cynodon dactylon (Poaceae)

3.3.1.1.1 Characteristics

Warm-season grass, fine to medium leaf texture, dark green, dense, and low growing via rhizomes and stolons; some varieties tolerate very low maintenance, while others produce lawns of exceptional beauty when given extra care; root system is extensive and very deep.

3.3.1.1.2 Recommended Usage

Best adapted to hot, dry, or tropical climates – recommended for residential and commercial landscapes, golf courses, sports fields, parks, and recreation areas – ideal for homes with children and pets. Withstand heavy traffic pressure and has ability to recover.

3.3.1.1.3 Temperature Tolerance

Excellent heat tolerance up to 40–45°C and performs best during periods of heat – has a winter dormancy period; turns tan to brown at temperatures below 10–12°C. Poor cold hardiness.

3.3.1.1.4 *Drought Resistance*

Superior – highly drought resistant, but also responds to irrigation in dry periods. Can go into summer dormancy when irrigation is withheld; upon return of moisture supply, will become green again.

3.3.1.1.5 *Shade Adaptation*

Poor – requires full sun for most of the day to grow properly – should not be used in a shady site.

3.3.1.1.6 Varieties of Bermuda grass

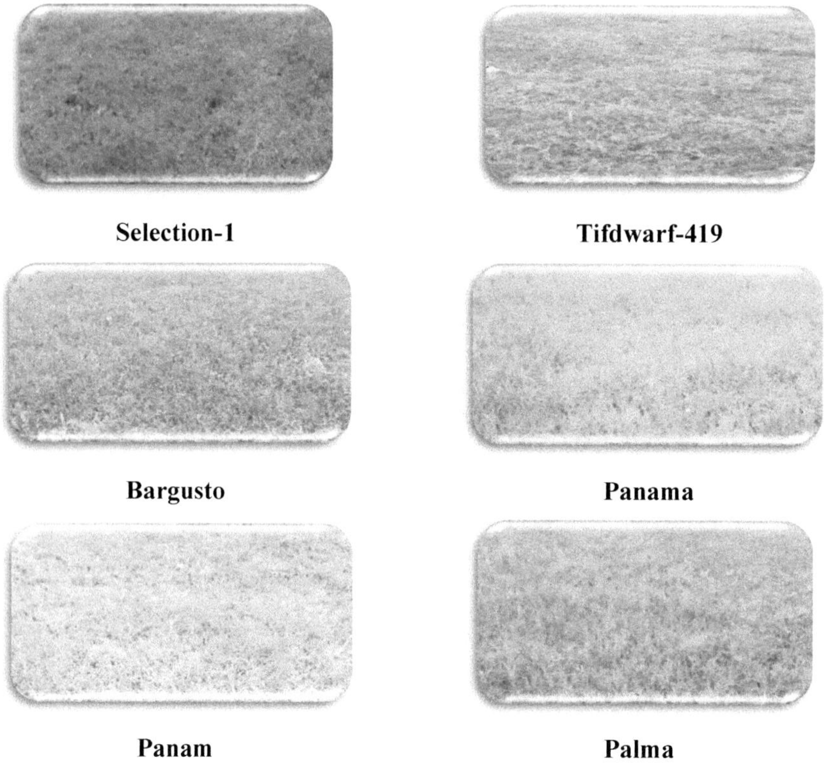

FIGURE 3.1 **(See color insert.)**

3.3.2 ST. AUGUSTINE GRASS

3.3.2.1 Stenotaphrum secundatum (Poaceae)

3.3.2.1.1 Characteristics

Warm-season grass – light to medium green color, coarse leaf texture, creeping growth habit, fast growing; it has low level of maintenance with moderate wear ability.

3.3.2.2 Stenotaphrum secundatum variegata (Striped St. Augustine Grass)

3.3.2.2.1 Characteristics

Variegated St. Augustine grass is strongly and thickly striped in white, and is one of the brightest variegated grasses available.

3.3.2.2.2 Recommended usage

Well adapted to coastal regions with hot, tropical climates – used in residential, commercial, and industrial landscapes.

3.3.2.2.3 Temperature tolerance

Thrives in heat, adjusting well to temperatures up to 40–42°C; goes dormant and turns tan colored during winter when temperatures drop below 10°C.

3.3.2.2.4 Drought resistance

Excellent to fair – wide range in drought avoidance among varieties. It can go into summer dormancy when irrigation is withheld; upon return of moisture, will become green again.

3.3.2.2.5 Shade adaptation

Excellent to poor – varieties show wide range in shade adaptation.

3.3.3 BAHIA GRASS

3.3.3.1 Paspalum notatum (Poaceae)

3.3.3.2 Characteristics

Bahia grass is a warm season turf grass. It can be established from a seed and requires reduced inputs compared to St. Augustine grass. It is a low- to mid-maintenance grass. It is a tough, coarse bladed grass that is capable of surviving in conditions that would destroy most turf grasses. One important characteristic of Bahia grass is its capability to withstand weather extremes.

3.3.3.3 Temperature Tolerance

Thrives in heat, adjusting well to temperatures more than 45°C; goes dormant and turns tan colored during winter when temperatures drop below 10°C.

3.3.3.4 Shade Adaptation

It does not have good tolerance for shade.

3.3.3.5 Wear Resistance

Bahia grass is extremely tolerant to wear and traffic and is commonly used as a roadside grass. It has a coarse leaf texture and provides less cushioning for recreational activities than some other species. Bahia grass grows best in full sun.

St.Augustine St Augustine (Striped) Bahia grass

FIGURE 3.2 (See color insert.)

3.3.4 LOVE GRASS

3.3.4.1 Eragrostis curvula (Poaceae)

3.3.4.1.1 Characteristics

It is a drought tolerant perennial grass and forms a large tuft with long narrow drooping leaves. It can grow up to 30 to 40 inches. Weeping Love grass, a plant introduced from East Africa, is a rapidly growing warm-season bunchgrass that prefers light textured soils.

3.3.4.1.2 Temperature Tolerance

Thrives well in summer season, and adjusts well to temperatures more than 45°C.

3.3.4.1.3 Shade Adaptation

It does not have good tolerance for shade. It requires full sun for growth.

3.3.4.1.4 Recommended Usage

Weeping Love grass can be used as a conservation species on sandy soils to prevent wind erosion. Its quick germination and rapid establishment provide good cover on critical sites. Weeping Love grass can be used in the landscape for mass plantings. The unique drooping appearance, quick establishment, and drought resistance make it a desirable low-maintenance species used for erosion control along hills and highways.

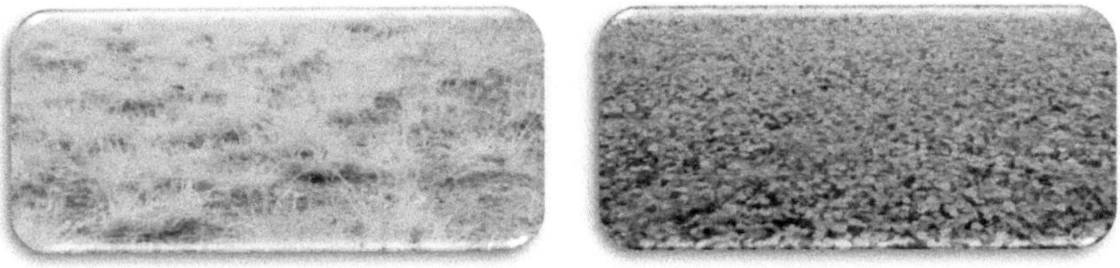

Love grass **Dichondra**

FIGURE 3.3 (See color insert.)

3.3.5 DICHONDRA

3.3.5.1 Dichondra (Convolvulaceae)

3.3.5.1.1 Characteristics

Dichondra is not a turf grass but a fast-growing ground cover. This is a perennial warm season grass with fine texture that has a prostrate or creeping growth habit with circular leaves. It grows very close to the ground (usually not over 2 inches tall).

3.3.5.1.2 **Temperature Tolerance**

It is adapted to warmer climates, but will retain its striking green color during winter temperatures as below 0°C with only slight leaf browning.

3.3.5.1.3 **Shade Adaptation**

It will grow in partial shade, but it grows best in full sun under cool coastal conditions.

3.3.5.1.4 *Recommended Usage*

Dichondra has broad, almost circular leaves and when mowed low gives a thick dense carpet look. It is now used in many ground cover situations where normal grasses may not do as well.

3.3.6 ZOYSIA GRASS

3.3.6.1 Zoysia species (Poaceae)

3.3.6.1.1 *Characteristics*

Zoysia grass is a thick, slow-growing turf. It has a fine-to-medium texture; color varies by cultivar. The zoysia grasses have superior cold tolerance, and give more attractive lawn than St. Augustine grass or centipede grass.

3.3.6.1.2 Zoysia japonica

Korean lawn grass has a coarse texture similar to tall fescue. Its light green leaf is hairy and has a relatively faster growth rate than other zoysia grass species, with excellent cold tolerance. However, this species of zoysia grass does not make as good lawn as other improved cultivars and species make.

3.3.6.1.3 Zoysia matrella

Matrella has a finer leaf texture and is more shade tolerance. Manila grass resembles Bermuda grass and is recommended for a high-quality, high-maintenance lawn, but it has less cold tolerance and seems to be highly susceptible to damage caused by nematodes.

3.3.6.2 Temperature Tolerance

Manila grass (*Zoysia matrella*) is intermediate in cold tolerance and fineness. It is increasingly popular due to good-quality lawn. Japanese lawn grass (*Zoysia japonica*) is coarse in texture but much more cold-hardy than the other species.

3.3.6.3 Shade Adaptation

The zoysia grasses form dense, thick sods of high quality. They have more shade tolerance than Bermuda grass and form a finer, more attractive lawn.

3.3.6.4 Recommended Usage

A highly versatile species, zoysia grasses make ideal lawn grasses in some situations and can be used on golf courses, parks, and athletic fields. They can be grown in all kinds of soils ranging from sands to clays and both acid and alkaline in reaction.

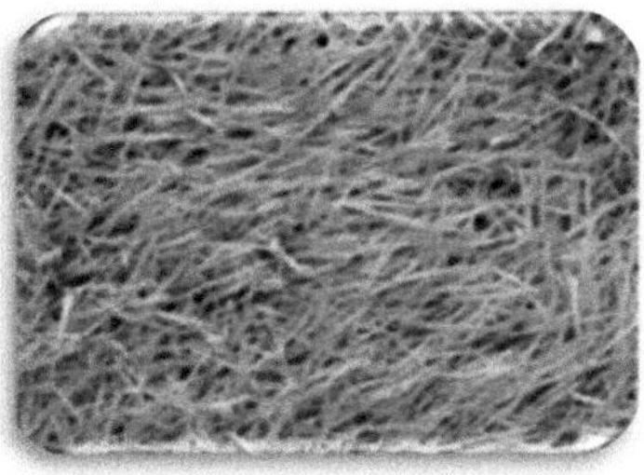

Zoysia matrella

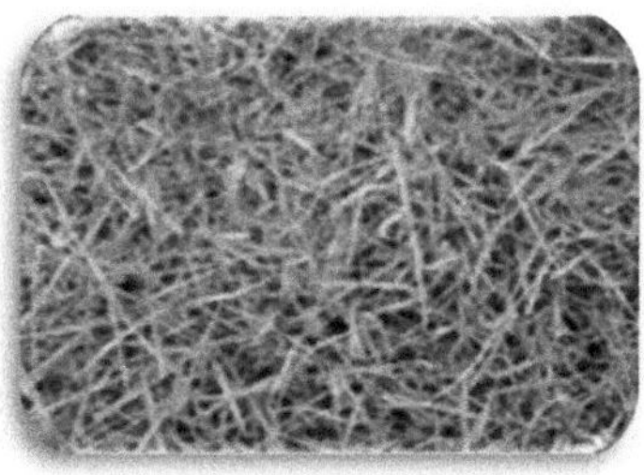

Zoysia japonica

FIGURE 3.4 (See color insert.)

3.3.7 EGYPTIAN CROWFOOT GRASS

Egyptian crowfoot grass (*Dactyloctenium aegyptium,* Poaceae) is a tufted, slightly stoloniferous annual or short-lived drought tolerant perennial grass. It is a multipurpose grass. It is a mainly used as fodder and relished by all classes of ruminants. In semi-arid areas, it makes valuable annual pastures as well as excellent hay. Currently, it can be used as turf grass and is suitable for sports field.

3.4 SUMMARY

For many centuries, people have been willing to devote time and resources to enhance their quality-of-life and recreational opportunities through the use of turf grasses (Beard, 1989). Also, for many centuries turf grasses have played a vital role in protecting our environment, long before it became an issue of major national and international importance to modern societies. Turf grasses are relatively inexpensive, durable ground covers that protect our valuable, nonrenewable soil resource from water and wind erosion. Agricultural operations and similar activities such as construction involve extensive land disruption, in contrast to turfed land areas, which are maintained in a long-term stable state. The future of turf grasses is bright in India (Table 3.1). To support the ever-increasing demand and to meet the requirements of present day, scientific studies on turf culture are essential. The information on turf grass species is meager in India; therefore, to generate information on the available turf grass cultivars and to come up with suitable recommendation with respect to selection of turf grasses is essential. Evaluation of growing behavior of the turf grass is important as it endows us with the knowledge enabling us to make wise choices that will best meet our requirements in a turf at a particular geographical location.

ACKNOWLEDGMENT

The present work was supported by NBPGR (New Delhi) for providing the material.

TABLE 3.1 Performance of Turf Grass Species/Varieties in Delhi Condition

Var./Species	Shoot Length (cm)	Root Length (cm)	Shoot Density (no.100cm^2)	Root Density (no.100cm^2)	Shoot fresh weight (g/100cm^2)	Shoot dry weight (g/100cm^2)	RWC%
Cynodon dactylon var. **Selection-1**	9.50	15.40	507.67	33.50	6.50	1.77	84.93
Cynodon dactylon var. **Bargusto**	9.00	15.00	235.00	20.00	4.50	1.67	86.00
Cynodon dactylon var. **Tifdwarf419**	11.50	16.50	445.00	37.50	6.76	2.00	87.50
Cynodon dactylon var. **Palma**	9.80	13.00	176.50	20.00	4.25	1.39	87.00
Cynodon dactylon var. **Panama**	9.00	14.00	150.00	21.50	3.95	1.08	85.00
Cynodon dactylon var. **Panam**	13.00	14.50	140.50	23.00	4.00	1.00	86.50
Stenotaphrum secundatum	8.50	10.00	45.33	12.00	8.33	3.66	81.00
Stenotaphrum secundatum **variegata**	10.50	9.50	43.50	8.33	9.33	4.00	79.50
Dichondra	7.00	8.00	230.00	12.33	6.83	2.25	83.50
Paspalum notatum	12.50	10.00	35.50	10.00	6.00	1.83	84.00
Zoysia japonica	7.60	8.50	515.67	32.50	6.00	1.86	83.50
Zoysia matrella	9.50	10.00	445.00	30.50	5.67	1.50	81.00
Dactyloctenium aegyptium	13.50	11.50	32.00	13.00	10.50	4.25	80.00
Eragrostis curvula	15.00	9.00	85.50	7.50	8.00	2.35	70.00
Mean	10.42	11.77	220.51	20.11	6.47	2.18	82.81

KEYWORDS

- **morphological characters**
- **turf grass**
- **warm season**

REFERENCES

Beard, J. B., (1989a). The role of gramineae in enhancing malt's quality of life. In: *Symp. Proc. Nat. Comm. Agric. Sci.*, Tokyo, July 1989, *Japanese Sci. Council*, Tokyo, pp. 1–9.

Christians, N., (2004). *Fundamentals of Turf Grass Management*. Published by John Wiley & Sons Hoboken. NJ., 171–188.

Cook, D. I., & Van Haverbeke, D. F., (1971). Trees and shrubs for noise abatement. *Nebraska Agric. Exp. Stn. Res. Bull.*, 246, Lincoln.

Gould, F. W., (1968). *Grass Systematics*. McGraw-Hill, New York.

Hanks, J. D., Waldron, B. L., Johnson, P. G., Jensen, K. B., & Asay, K. H., (2005). Breeding CWG-R crested wheatgrass for reduced-maintenance turf. *Crop Sci., 45*, 524–528.

Johns, D., & Beard, J. B., (1985). A quantitative assessment of the benefits from irrigated turf on environmental cooling and energy savings in urban areas. In: *Texas Turf Grass Research*. Texas Agric. Exp. Stn., PR-4330: College Station, TX, pp. 134–142.

Macolino, S., Ziliotto, U., & Leinauer, B., (2012). Comparison of turf performance and root systems of bermudagrass cultivars and 'Companion' zoysia grass. *Acta. Hort*, vol. *938*, pp. 185–190, ISSN, 0567–7572.

Rimi, F., Macolino, S., & Leinauer, B., (2012). Winter-applied glyphosate effects on spring Greenup of Zoysiagrasses and 'Yukon' Bermudagrass in a transition zone. *Hort. Technology, 22*, 131–136.

Robinette, G. O., (1972). *Plants, People, and Environmental Quality*. U. S. Dep. Interior, National Park Service, and Am. Soc. I-and. Archit. Foundation, Washington, DC.

Romani, M., Piano, E., Carroni, A. M., & Pecetti, L., (2004). Evaluation of native Bermudagrass (*Cynodon dactylon*) germplasm from Italy for the selection of adapted turfgrass cultivars. *Acta Hort., 661*, 381–6.

Schuyler, T., (1987). *Controlling Urban Runoff*: A practical manual for planning and designing urban BMPs. Metropolitan Washington council of governments, Washington, DC.

Turgeon, A. J., (1980). *Turfgrass Management*, Reston Publishing Company, Reston, Virginia, pp. 391.

Wikipedia. https://en.wikipedia.org/wiki/Lawn (accessed Jun 23, 2018).

CHAPTER 4

LOCAL GERMPLASM AND LANDRACES OF VEGETABLES: AVAILABILITY, MARKET PRICE, AND PERCEPTION OF CONSUMER

ANJANA PRADHAN, LAXUMAN SHARMA, S. MANIVANNAN, and KARMA DIKI BHUTIA

Sikkim University, 6th Mile, Tadong, Sikkim, India,
E-mail: anjanamalla@yahoo.com

CONTENTS

ABSTRACT

Vegetable production in the hill farming system has major stakes for nutrition, livelihood, and economic sustainability. Most often, successful cultivation of high-yielding varieties of vegetables in the hills cannot be achieved due to various biotic and abiotic stress factors. In the vegetable production system in hills like Sikkim, people still prefer to grow the local landraces and relish more in consuming these vegetables. Local landraces and germplasm of vegetables stand good for organic farming owing to their wide adaptability to local environment and fair resistance to pest and diseases. The present study envisaged to understand the pattern of availability of these vegetables in the local market and to assess the price fluctuation. Perception of the consumer was also recorded for their choice of vegetable. It is inferred that a policy can be adopted by the local government for expanding the area and market of these vegetables. Some of the vegetables like *Ishkus*, *Lentils*, *Ningro*, *Sisno*, *Palak,* and *Simrayo* are available almost 8 to 9 months a year and fetch as high as other vegetables. Almost 90% of the consumers consume these vegetables owing to their medicinal properties and organic cultivation, and above all, they have consumed it from their childhood.

4.1 INTRODUCTION

Sikkim is blessed by a fertile land that largely supports agriculture. The topography and the climatic condition of Sikkim are favorable for agriculture. Therefore, Sikkim witnesses a high crop yield every year. Vegetable production has been a major prospect of cultivation aspect, which is an income source of farmers for their livelihood and sustainability for raising their livelihood. However, there are several abiotic and biotic stress factors that affect cultivation of high-yielding varieties of vegetables. Farmers are prone to grown local landraces, and they are dependent on consumption of these vegetables. Organic farming performs well with local landraces of vegetable as they showed wide adaptability to local environment and shows fair resistance to pest and diseases.

Markets have the power to influence income generation, food security, and other developmental objectives of a particular area (shodganga.inflibnet.ac.in); in this regard, a study has been conducted in two markets, viz, Lal Bazaar and Tadong of Sikkim, to know the price fluctuation in two markets in different seasons along with seasonal availability of indigenous vegetables and price fluctuation with related commercial vegetables. According to the information on www.sikkimstdc.com, Lal Bazaar portrays the true vibrant color of Sikkim; it is the market where local farmers from around Gangtok congregate to sell their products. Especially on Sundays, Gangtok's Lal Bazaar is a melody of colors where various ethnic groups from different villages gather to haggle, bargain, gossip, or just sell their wares. Tadong bazaar has small shops for selling vegetables. Perception of the consumer was also recorded for their choice of vegetable. It is inferred that a policy can be adopted by the local government for expanding the area and market of these vegetables. Some of the vegetables like Chayote (*Ishkus*), *palak* leaf, stinging nettle, and *simrayo* leaf are available almost 12 months a year and that of Lentil of Chayote and Lentil of pumpkin are found around 8–9 months in the market and fetch as high as other vegetables. Almost 90% of the consumers consume these vegetables owing to their medicinal properties and organic cultivation, and above all, they have consumed it from their childhood.

4.2 METHODOLOGY

The study has been performed based on market study in Lal Bazaar and Tadong area of Sikkim for the period of 8 months based on interviews with sellers of local landraces of vegetables available from April to December 2015. The perception of consumers toward purchasing organic vegetables was based on free listing methods with 20 consumers selected using a simple random sampling technique.

4.3 RESULTS

Market study, perception of consumers, and availability of local landraces of vegetables were studied in Lal Bazaar and Tadong market. Some of the indigenous varieties are enlisted in Table 4.1. Some of the most important crops are discussed, and others are enlisted in the table. The majority of consumers prefer the indigenous landraces or the varieties of different crops that are cultivated by farmers over the last several decades. It has won the farmers' belief regarding tolerance to extreme weather and pests and diseases.

TABLE 4.1 Local Landraces of Sikkim

Sl. No.	Crop	Local Name (in Nepali)	Botanical Name	Family
1	Chayote	Ishkus	*Sechium edule*	Cucurbitaceae
2	Chakote leaf tendril	Ishkus ko munta	*Sechium edule*	Cucurbitaceae
3	Tuber of Chayote root	Ishkush ko Jara	*Sechium edule*	Cucurbitaceae
4	Pumpkin leaf tendril	Farsi leaf tendril	*Cucurbita pepo*	Cucurbitaceae
5	Fern	Chiple Ningro	*Diplazium esculentum*	Drypoteridaceae/ Athyraceae
6		Sawane Ningro	*Diplazium esculentum*	Drypoteridaceae/ Athyraceae
7	Water cress	Simrayo Sag	*Nasturtium officinale*	Cruciferae
8	Fenugreek leaf	Methe Sag	*Trigonella foenum-graecum*	Fabaceae
9	Lamb's quarter/pig weed/goose foot	Bethe sag	*Chenopodium album*	Chenopodiaceae
10	Pig weed	Latte Sag/Lunde Sag	*Amaranthus viridis*	Amaratheceae
11	Spinach leaf	Palak Sag	*Spinacia oleracea*	Chenopodiaceae
12	Stinging nettle	Sisno	*Uritca dioca*	Urticaceae
13	Mustard leaf	Tori Sag	*Brassica campestris var. tori*a	Brassicaceae
14		Nakima	*Tupestra nutans*	Liliaceae
15	String beans	Ghue Bori		
16		Bihi	*Solanum anguivi*	Solanaceae
17	Garden cress	Chamsur	*Lepidium sativum*	Brassicaceae
18	Cherry pepper	Dalle Khursani	*Capsicum frutensis var cerasiforme*	Solanaceae
19	Cassava	Simal tarul	*Manihot esculenta*	Dioscoreaceae
20	Kaywa	Chuche karela	*Cyclanthera pedata*	Cucurbitaceae
21	Leaf mustard shoot	Duku Sag	*Brassica juncea*	Brassicaceae

4.3.1 NINGRO (Diplazium esculentum)

The word *ningro* or *neuro* was derived from the word *niurinu,* which means "to bend" in Nepali language. It needs to be plucked out before the shoot grows into feather-like leaves, which are quite inedible due to their taste.

Before cooking, the *ningros* need to be cleaned properly so that its fuzzy covering is removed. *Ningro* is the young shoots of edible ferns that resemble the spiral end of a fiddle, because of which, it is also known as fiddlehead ferns. One of the most popular food items in Nepal, it mostly grows around mid-April. A popular green leaf vegetable, *ningro* is quite common in the Nepali market during spring. *Ningro* mostly grows in the jungles, shady swamps, riverbanks, and damp fields. It is mostly bought from the market; however, mostly in the Terai region; it is collected from the home garden.

4.3.2 LATTE SAG (*Amaranthus viride*)

It is commonly called as Latte sag in Nepali and is used as a vegetable; it is consumed by all local people of Sikkim. It is preferred either alone or in combination with potato or meat. It grows as a seasonal weed in crop field, waste land, livestock shed areas, organic manure deposit without soil, uncultivated field and home garden of urban and rural areas. It grows at an altitude of 2500–6000 ft and is found in lower hill and upper hill of Sikkim (Tamang and Pradhan, 2014).

4.3.3 BETHU/BETHE SAG

Bethu sag named scientifically as *Chenopodium album* is cooked and consumed as sag in Sikkim. It is also preferred by people to eat with potato or sometimes with meat. It is found at an altitude of 12000 ft and grows as a seasonal weed in uncultivated field, home garden, organic manure compost pit areas, soil without standing water, and fallow land. The tender shoots and young leaves are collected by local farmers to sell in the market (Tamang and Pradhan, 2014).

4.3.4 SISNOO (*Urtica dioica*)

Sisnoo grows in roadside verges, waste areas, river side, and edges of stream, hedgerows, and waste disposal areas. It grows well in an altitude of 4000–9500 ft in mid hill, mid upper hill, and upper hill. It has the stinging hair on the upper surface of stem and leaves, and hence, it has to be boiled and cooked to destroy the stinging effect. Leaves are consumed as soup as well as with meat and pakoras, and it has a nutritive value for reducing high blood pressure (Tamang and Pradhan, 2014).

4.3.5 SIMRAYO (*Nasturtium officinale*)

Simrayo *is* also called as water cress in English. It grows in submerged area, floating in water, or spread over sandy surface in flowing water; near springs; open running waterways; and perennial ditches from the constant flow from kitchen. It also found growing near drinking water source and marshy place of the hills. It is consumed as sag alone or with potato soup. It has medicinal values of prevention of goiter according to the ethnic community of Sikkim. A decoction of plant is taken as blood purifier, vermifuge, and diuretic. It is also used in dry throat, cold in the head, asthma, and tuberculosis. It also has anticancer properties and is considered by local people to control high blood pressure (Tamang and Pradhan, 2014).

4.3.6 CHAMSUR (*Lepidium sativum*)

Chamsur also called garden cress is used as a leafy vegetable in Sikkim. It is consumed alone or mixed with potato. It is highly valued for its medicinal value in the treatment of joint pain.

4.3.7 CHAYOTE (*Sechium edule*)

It is also called *Eskus* in Nepali and sometime considered as poor man's vegetable. Every household grows *Eskus* as its production is in large scale. One plant has the capacity to produce more than a quintal of fruits if it is properly manured. The white- and green-colored ones are grown in this area. Matured fruit is sown in a big pit with an adequate quantity of farm yard manures. Depending upon the altitude, the plant start bearing after May up to December.

The tendrils of this plant can be eaten as vegetable when they are tender during April and are called *Munta*. Further after January, plant dries up, and tuberized portion of root called *Jara* is also being consumed as vegetable. The dried-up plants and deformed fruits are also being utilized as cattle feed supplement. It also noteworthy that the plant part can be constituent of diet of the people throughout the year. The fruit of it can be used later after drying and has also been used by local people in making pickles mixed with other vegetables and spices. Local *Eskus* varieties grown are "Seto," "Haryo," and "Dumsay" (Sharma et al., 2015).

4.3.8 PUMPKIN (Cucurbita pepo)

It is consumed by the local people in their diet as green one or fully ripe one. It is called Pharsi in *Nepali*. They are grown without trellis in maize field, and manuring is also not required. The soft tendril of this plant is called *Munta*; it is also consumed, and its flower is also eaten by local people as pakora. Some of the local varieties cultivated by farmers are *Kalo pharsi, Seto pharsi, Thulo pharsi, Paharey,* and *Hazari.* (Sharma et al., 2015).

4.3.9 CHERRY PEPPER (Dalle Khursani)

Different types of cherry pepper are grown by farmers at the commercial level. The pungency of the plant is very high, and it has got good market value. Large scale production of cherry pepper is also done and fetch higher price in the market (Sharma et al., 2015).

4.3.10 NAKIMA (Tupistra nutans)

Nakima is widely cultivated in village homes in the hilly region, and during the flowering season, the inflorescence is sold in markets. Its inflorescence with bud and open flower is made into curry, which is a relishing food item of Sikkimese traditional cuisine.

4.3.11 LEAF OF MUSTARD SHOOT (Brassica juncea)

These tender, young shoots from mustard plants are cooked to make one of most delicious *tarkaari* that is much loved by Nepali people. When the mustard plant is allowed to mature, they will start forming flowering shoots, which are called "*duku*" in local language. The shoots are picked when they are young and tender as older shoots are too stringy and inedible and are called "*chippeko*" in Nepali. Before cooking, the outer covering and fibers should be peeled from all sides of the shoots. The *dukus* can be eaten cooked alone, but usually the shoots and young mustard are cooked together (Sharma et al., 2015).

Spinacia oleracea

Nasturtium officinale

Sechium edule

Uritca dioca

Brassica juncea

Trigonella foenum-graecum

Brassica campestris var. toria

FIGURE 4.1 (See color insert.) Some of the local landraces of Sikkim.

FIGURE 4.2 (See color insert.) Interview with a seller in Lal Bazaar.

FIGURE 4.3 **(See color insert.)** Interview with a vegetable seller in Tadong.

FIGURE 4.4 **(See color insert.)** Commercial vegetables in Lal Bazaar market.

4.4 SEASONAL AVAILABILITY

Market study of local landraces of vegetables showed that leafy vegetables like *simrayo* and *palak* are available throughout the year. Sisno also called stinging nettle is also available throughout the year. Among the vegetables, Chayote can be consumed throughout the year and its tendril from April to November. Likewise, the tuber of chayote is highly priced and is preferred by consumers from October to February only. *Nakima* (October–November) and *bihi* (March–October) are used as one of the traditional Sikkimese cuisines (Figure 4.5).

The details of the availability of other landraces are presented in the graph (Figure 4.5).

4.5 PERCEPTION OF CONSUMERS

It was revealed that consumers are prone to use organic vegetables for their daily consumption from their health perspective rather than purchasing commercial vegetables from Siliguri. People are fond of going to market on Saturday and Sunday when fresh vegetables from *basti* are brought into these markets. About 20 consumers were interviewed based on the free listing method. The vegetables preferred by them are local chayote, tendril of

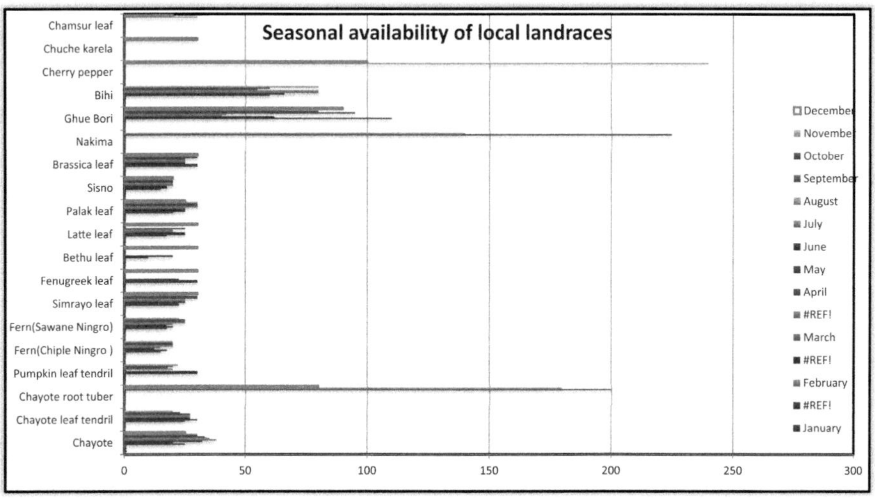

FIGURE 4.5 (See color insert.) Seasonal availability of local vegetables in Sikkim.

chayote, pumpkin also called as *gatta* locally, and tendril of gatta. Among leafy vegetables, they purchased *palak*, *methe*, *latte*, and *tori*. Likewise, *nakima*, *bee*, *ningro*, *sisno*, *ghue bori*, and cherry pepper are their major preference compared with commercial vegetables.

4.6 PRICE COMPARISON WITH RELATED COMMERCIAL VEGETABLES

The price of some of the local vegetables was compared with the commercial vegetable brought from Siliguri to Tadong and Lal Bazaar markets, and it was observed that there was much variability in the price. Comparison was made between cauliflower, chili, and bean of Siliguri vegetables that are sold in our local market with those of local landraces of Sikkim, like chayote, cherry pepper, and *ghue bori* from the market study of price from April to December 2015.

Comparison of commercial cauliflower with local chayote (Table 4.2) showed that the purchase of the local one is more cheaper to consumers in terms of price, health, and organic production (Figure 4.6). Likewise, cherry pepper is available for November onward till March, and its price is higher than that of chili from Siliguri. However, cherry pepper has more taste and pungency, and consumers preferred to buy the local one. Locally available bean called *ghue bori* is more preferred by the local people as it has more taste than snake bean. The price of both is almost the same. However, the price goes little up for *ghue bori* during September to December. But still, the local people are fond of eating it as a vegetable or as a *dal* soup.

4.7 LAL BAZAAR AND TADONG MARKET PRICE VARIATION

According to the study, it was observed that chayote root tuber, cherry pepper, *ghue bori*, and *bihi* fetch higher price than other local landraces. Among them, cherry pepper has the highest market price. The price of Lal Bazaar vegetables is lower than that of Tadong Bazaar vegetables, because Lal Bazaar is a wholesale market and hub where all the local vegetables comes from various *bastis* of Sikkim and Tadong Bazaar is a retail shop where vegetables from Lal Bazaar and other outlet wholesale markets are brought and sold at a higher price (Figure 4.7).

TABLE 4.2 Comparative of Local Vegetables with Commercial Ones

Vegetable(Rs/kg)		April		May		June		July		August		September		October		November		December	
Cauli flower	/Chayote	30	25	50	20	80	35	80	40	80	35	100	35	100	32	60	32	40	22
Chili	/Cherry pepper	60	N/A	60	N/A	60	N/A	120	N/A	120	N/A	120	N/A	90	N/A	80	250	90	100
Snake Beans	/Ghue bori	40	N/A	50	100	60	65	70	45	70	45	70	90	70	90	70	90	80	85

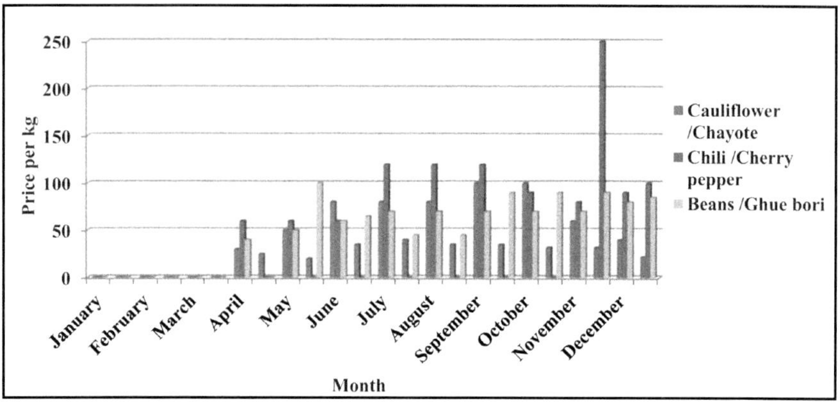

FIGURE 4.6 **(See color insert.)** Comparison of commercial vegetables with local landraces.

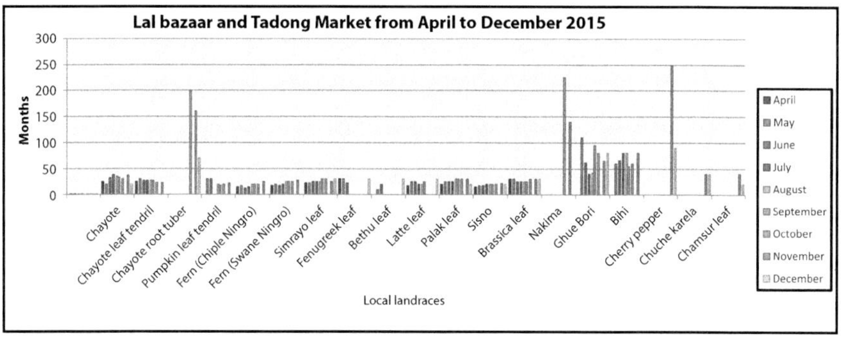

FIGURE 4.7 **(See color insert.)** Lal bazaar and Tadong market price from April to December 2015.

It was also found that cherry pepper, chayote root tuber, *nakima*, *bihi*, and *ghue bori* grow for a certain period only where farmers can fetch higher price (Table 4.3 and Table 4.4). However, the researcher believes that these vegetables can be grown throughout the year and in green houses in offseason for providing more advantages to local farmers.

4.8 CONCLUSION

Local landraces are in demand by consumers compared to the commercial ones. Preference of these local vegetables is due to good taste, health safety,

TABLE 4.3 Lal Bazaar and Tadong Bazaar Market Price from April to December of Local Land Races (Rs/Kg)

Vegetables	April		May		June		July		August		September		October		November		December	
	Lal Bazaar	Tadong	Lal Bazaar	Tadong	Lal Bazaar	Tadong	Lal Bazaar	Tadong	Lal Bazaar	Tadong	Lal Bazaar	Tadong	Lal Bazaar	Tadong	Lal Bazaar	Tadong	**Lal Bazaar**	**Tadong**
Chayote	25	27	20	25	32	38	38	43	35	37	33	37	30	37	30	37	25	20
Chayote leaf tendril	50	60	60	64	54	54	54	60	54	60	54	56	45	50	40	45	N/A	N/A
Chayote root tuber	N/A	N/A	N/A	N/A	N/A	N/A	N/A	N/A	N/A	N/A	N/A	N/A	200	180	180	160	80	70
Pumpkin leaf tendril	N/A	N/A	60	60	60	40	N/A	N/A	40	40	40	35	40	44	44	44	N/A	N/A
Fern(Chiple Nigro)	30	44	34	40	25	35	30	34	40	40	40	45	40	54	40	50	N/A	N/A
Fern(Sawane Nigro)	34	40	40	45	35	44	40	44	50	50	50	50	50	60	45	55	N/A	N/A
Simrayo leaf	44	54	44	50	50	34	50	35	50	40	60	55	60	70	50	50	60	60
Fenugreek leaf	60	45	60	40	44	40	N/A	N/A	N/A	N/A	N/A	N/A	N/A	N/A	N/A	N/A	60	60
Bethu leaf	N/A	N/A	20	20	40	40	N/A	N/A	N/A	N/A	N/A	N/A	N/A	N/A	N/A	N/A	60	60
Latte leaf	35	45	50	40	50	40	40	40	40	40	50	54	N/A	N/A	N/A	N/A	60	60
Palak leaf	40	44	50	40	50	40	50	40	60	50	60	64	60	64	60	60	50	40
Sisno	30	30	34	25	34	30	40	40	40	40	40	45	40	50	40	44	40	40
Brassica **leaf**	60	70	60	70	50	65	50	65	50	60	50	54	60	60	60	60	60	60
Nakima	N/A	N/A	N/A	N/A	N/A	N/A	N/A	N/A	N/A	N/A	N/A	N/A	225	140	140	140	N/A	N/A

TABLE 4.3 (Continued)

Vegetables	April		May		June		July		August		September		October		November		December	
	Lal Bazaar	Tadong	Lal Bazaar	Tadong	Lal Bazaar	Tadong	Lal Bazaar	Tadong	Lal Bazaar	Tadong	Lal Bazaar	Tadong	Lal Bazaar	Tadong	Lal Bazaar	Tadong	**Lal Bazaar**	**Tadong**
Ghue Bori	N/A	N/A	110	100	62	68	40	46	42	45	95	105	80	90	66	65	N/A	N/A
Bihi	N/A	N/A	60	40	66	74	80	80	80	80	55	60	60	66	80	80	N/A	N/A
Cherry pepper	N/A	N/A	N/A	N/A	N/A	N/A	N/A	N/A	N/A	N/A	N/A	N/A	N/A	N/A	240	250	100	90
Chuche karela	N/A	N/A	N/A	N/A	N/A	N/A	N/A	N/A	N/A	N/A	N/A	N/A	N/A	N/A	30	40	30	40
Chamsur leaf	N/A	N/A	N/A	N/A	N/A	N/A	N/A	N/A	N/A	N/A	N/A	N/A	N/A	N/A	60	80	40	40

Note: N/A means not applicable.

Chayote, *nakima*, *ghue bori*, chayote root tuber, and *bihi* sold at Rs/kg

Leafy vegetable, fern sold in *bita* (bundle), and two *bitas* (bundle) are equivalent to 1 kg. 2 *bitas* = 1 kg (converted to kg from *bita* of leafy vegetable, Table 4.3)

TABLE 4.4 Price of Commercial Vegetable from April to December 2015

Vegetables	April		May		June		July		August		September		October		November		December	
	Lal Bazaar	Tadong	Lal Bazaar	Tadong	Lal Bazaar	Tadong	Lal Bazaar	Tadong	Lal Bazaar	Tadong	Lal Bazaar	Tadong	Lal Bazaar	Tadong	Lal Bazaar	Tadong	Lal **Bazaar**	**Tadong**
Cauliflower	30	30	50	60	80	60	80	80	80	80	100	100	100	100	60	90	40	30
Stuffing cucumber	40	40	60	60	60	60	60	60	60	80	N/A	N/A	N/A	N/A	N/A	N/A	N/A	N/A
Palak/kg	30	30	30	30	40	30	40	30	40	30	40	30	30	30	40	40	40	30
Chili	60	60	60	60	120	80	120	80	120	80	120	80	80	100	60	100	90	90
Latee leaf	40	20	40	30	40	40	40	40	40	40	40	30	N/A	N/A	N/A	N/A	N/A	N/A
Snake Beans	40	40	50	60	60	70	70	70	70	70	60	70	60	80	70	80	80	80

and their various medicinal aspects. Markets in Lal Bazaar and Tadong sell both local vegetables and commercial ones. But, based on study of the seller and perception of consumers, the consumers purchase the local ones. The price of local ones is also cheaper than that of the commercial ones. Chayote root tuber, *nakima*, and cherry pepper fetch a higher price, and the market of Lal Bazaar show lesser price than that of Tadong Bazaar. The increase in price of the local vegetable fluctuates with the month, season, and availability in the market.

Based on the above results, it is evident that the suitability associated with this local landrace productivity can not only improve the economic condition of the local people but also improve the sustainability of the farmers. Hence, policymakers have to pay attention critically for the constraints faced by the farmers by improving marketing channels, crop insurance, availability of suitable input, and proper storage condition.

KEYWORDS

- **availability**
- **indigenous**
- **market**
- **price**

REFERENCES

Pradhan, S., & Tamang, J. P., (2014). Ethnobiology of wild leafy vegetables of Sikkim. *Indian Journal.of Technical Knowledge*, *14*(2), 290–297.

Sharma, L. S., Pradhan, A., Bhutia, K. D., Tamang, D. L., & Bhuita, S. G., (2015). Local crop germplasm and landraces diversity in West Sikkim. *Journal of Agricultural Technology, 2*(1&2), 57–59.

Sikkim Tourism Development Corporation. http://www.sikkimstdc.com (accessed Jun 23, 2018).

Shodhanga. http://www.shodhanga.inflibnet.ac.in (accessed Jun 23, 2018).

PART II

PRODUCTION AND MARKETING OF HORTICULTURE CROPS

CHAPTER 5

HORTICULTURE FOR SUSTAINABLE DEVELOPMENT IN INDIA

K. L. CHADHA*

Adjunct Professor, IARI and President, The Horticultural Society of India, New Delhi, E-mail: klchadha@gmail.com

CONTENTS

ABSTRACT

Since the independence of the country, India has made rapid strides in agriculture by achieving green revolution during the mid-1960s that enhanced cereal production by five times, white revolution during the 1970s that enhanced milk production by seven times, and blue revolution during the

*Chief Guest on Inaugural Session of the International Symposium on Sustainable Horticulture, 14–16th March 2016, Mizoram University, Aizawl, India.

mid-1970s that enhanced fish production by nine times. We have been moving steadily for achieving horticulture revolution since the last three decades by enhancing horticultural production to the tune of 11 times. Economic Survey 2015–2016 of Government of India had strongly recognized this growth in the horticulture sector.

Therefore, this article emphasized on the role of horticulture R&D in achieving sustainable development and thus will be dwelling on the drivers for horticulture R&D growth, besides technological advancements and their impact on horticulture production, and productivity resulting in creating more livelihood opportunities and promoting income of small farmers.

5.1 DRIVERS OF HORTICULTURE GROWTH

The role of horticulture crops in health and nutrition has been known globally from times immemorial. However, the role of these crops through crop diversification as a remunerative, viable, and sustainable alternative to the rice-wheat based cropping system and the comparative advantage of high value horticulture crops in livelihood and their increasing role in health, nutrition, nutraceuticals, and exports were realized only about three decades back. As a result, policy attention to horticulture was started practically in the seventh Five Year Plan resulting in several initiatives, namely:

- Increasing financial allocation for R&D programs;
- Establishment of a sound R&D infrastructure;
- Launching of several flagship development programs;
- Pro-active policies encouraging international participation/joint ventures in the PPP (public–private partnership) mode.

The Government of India gave a clear signal of its intent and purpose of supporting horticulture, and as a consequence, the budget allocation to horticulture research and development has increased from Rs. 3.5 and Rs. 25 Crores in eighth Five Year Plan to Rs. 1,050 and Rs. 15,946 Crores in ninth Five Year Plan, respectively (Anon. 1997). The demand of horticulture produce and products has also been increasing due to increasing population and fast-growing cities, rising exports (9,961 crores in 2011–2012), and increasing need of the processing industry as a consequence of modernization of the post-harvest infrastructure, including roads, highways, airports, AEZs (Agri-Export Zone), and food parks.

5.2 MAJOR TECHNOLOGIES DEVELOPED

5.2.1 PLANTING MATERIAL

For propagation of high-quality planting material, vegetative propagation techniques have been developed in most fruit crops. Technology for micropropagation is now available in over 200 horticulture crops. It provides scope for throughout the year multiplication of disease-free planting material. Micropropagation has been commercially adopted in propagation of many horticultural crops such as banana, papaya, strawberries, date palm, cardamom, vanilla, large cardamom, gerbera, chrysanthemum, carnations, anthurium, orchids, syngoniums, and ferns. In tuber and root crops such as potato, ginger, and turmeric, this technology is very efficient in the production of large-scale disease-free clonal planting material of elite genotypes. Meristem culture in combination with thermotherapy or chemotherapy has been effectively used for the production of virus-free plants. Tissue culture standards have been set and guidelines for potato, banana, apple, citrus, bamboo, vanilla, and black pepper have already been notified. Shoot-tip grafting is being used in citrus propagation, while a salt-tolerant rootstock has been successfully used in grape. Standards for accreditation of nurseries have been developed and are being enforced by the National Horticulture Board.

In vegetables, techniques for mass production of seedlings in green house have been developed and commercialized. Grafting in cucurbitaceous vegetables is now possible and is being commercially used. CMS lines have been developed for the production of hybrids seeds in several vegetables, namely, chili, brinjal, tomato, and carrot, besides seed production for onion. Vegetable seedling nurseries have been established in many states.

In potato, the seed plot technique has been developed to enable seed production in north Indian plains. True potato seed (TPS) technology has also impacted the production of potato in some states, particularly Tripura. In tuber crops, a Minisett technique has been developed as a viable technology to produce large-scale planting material and has enhanced the multiplication ratio in tuber crops.

In coconut, plantations have been made with dwarf and tall varieties to produce D × T and T × D hybrid plants. In oil palm, a large number of nurseries have been established with elite material to produce seeds for extension of area under the crop.

Through diagnostics and management technology, the quality of seeds and planting material can be ensured. PCR-based diagnostic protocols have been developed for rapid detection of citrus greening and Tristeza virus. Diagnostics technologies like ELISA and ISEM have been developed for early detection of several diseases of asexually propagated plant.

5.2.2 VARIETIES DEVELOPED

In fruit crops, 146 varieties have been developed in 24 fruits by various institutions of the National Agricultural Research System. These include highly colored, dwarf, and regular bearing varieties of mango; gynodioecious and dwarf varieties of papaya; soft-seeded and colored flesh guava and pomegranate; and low chilling varieties of temperate fruits, namely pear, peach and plum. Several clonal selections and varieties have also been developed in grape, sapota, passion fruit, lime, *bael*, jackfruit, *jamun*, and walnut. High-yielding varieties have also been developed in *aonla* and *ber* and high pulp content variety in custard apple.

In vegetable crops, as many as 485 pure line selections and hybrids have been developed. These include F, hybrids in several crops, namely ash bitter and bottle gourds; cabbage and cauliflower; brinjal, capsicum, and tomato; and okra and watermelon. Besides a number of varieties tolerant or resistant to major diseases and pests, e.g., bacterial wilt in brinjal, chilly, sweet pepper and tomato; bacterial blight in cowpea and French bean, powdery mildew in chili, pea, and watermelon; downy mildew in muskmelon and watermelon; yellow vein mosaic in okra; and fruit and shoot borer in tomato. Chili cultivars resistant to aphid, mite and thrips and a tomato variety resistant to root knot nematode are also available. Varieties tolerant to climatic stresses, e.g. high/low temperature, high humidity and salt tolerance have also been developed in cauliflower, cabbage, tomato, carrot, radish, and turnip.

In potato more than 48 varieties have been developed for different agroclimatic regions, including varieties resistant to late blight and several varieties suitable for processing. Nearly 87 varieties have been released in tropical tuber crops, including sweet potato (37), cassava (16), and different yams.

Varieties developed in flower crops cover Bougainvillea, China aster, chrysanthemum, crossandra, marigold, rose, and tuberose.

Varieties developed in plantation crops include those for high fruit, copra, and oil yield besides their use as tender coconut. A number of improved varieties with high yield potential have also been released in Arecanut. A total of 42 superior varieties, 29 through selection and 13 hybrids, have been released in cashew, many of which are now under commercial cultivation. Further, 106 varieties of spices in 15 different crops have been released both by institutes and AICRP centers. Similarly, 124 varieties in medicinal plants, 8 in aromatic crops, and 8 in mushrooms have been developed for commercial production.

5.3 BIOTECHNOLOGICAL INTERVENTIONS

In Indian horticulture, biotechnology has played an important role particularly in the area of micropropagation, genetic modification, genome sequencing, and molecular breeding.

For instance, the tissue culture used in the multiplication of disease-free planting material has also been utilized for breeding of seedless varieties or making interspecific hybrids in citrus, grape, etc., through embryo rescue. Development of haploids and double haploids has provided scope for understanding inheritance patterns and hastening breeding programs. Plant regeneration and development of haploids and double haploids from anther and microspore cultures are reported in many crop species like Annona, papaya, citrus, litchi, apple, pomegranate, grape, chili, tomato, carrot, potato, brinjal, lily, petunia, and tulips. Similarly, successful isolation and culture of protoplasts were reported in citrus, tomato, sweet potato, brinjal, carrot, cabbage, mint, potato, cardamom, ginger, turmeric, vanilla, capsicum, nutmeg, garlic, fennel, fenugreek, peppermint, and saffron. But both the haploidy and protoplast culture are yet to be exploited to their potential.

The molecular breeding technique has created scope for developing crop varieties tolerant/resistant to biotic and abiotic stresses and yield enhancement. Marker-assisted breeding is helping to hasten and increase the efficiency of breeding and reducing the breeding cycle. Many markers are now available for precise and rapid transfer of desired traits. Several useful markers identified at IARI (Indian Agricultural Research Institute, New Delhi), IIHR (Indian Institute of Horticultural Research, Bangalore), IIVR (Indian Institute of Vegetable Research, Varanasi), and CPRI (Central Potato Research Institute, Shimla) include fertility restoration gene (Rf Pointed gourd: RAPD maker (0PC07) associated with identification of female sex; p-carotene gene (Or) in cauliflower and gynoecious (gy1) in bitter gourd; screening of gender of *Trichosanthes* at the seedling stage; and molecular basis of sex determination in dioecious species. In okra, of the SSR markers developed and commercialized, 55 markers have been sold to six private firms.

The use of biotechnology for the biosynthesis of secondary metabolites particularly in plants of pharmaceutical significance holds an interesting alternative to the production of plant constituents. Proliferation of stigma of saffron in vitro and chemical analysis of metabolites produced through tissue cultures have been reported in saffron, chili, anise, lavender, mint, rosemary, etc. Though the feasibility of in vitro production of spice principles has been demonstrated, the methodology for scaling up and reproducibility needs to be developed before it can reach commercial levels. Once scaled up, this technology has tremendous potential in industrial production of important high value compounds.

The whole genomes of many model plant species have been completely sequenced as a result of development of next generation sequencing (NGS) platforms. Genome sequencing in horticultural crops is underway in most horticultural crops. The sequencing of genome has already been completed for many crops including potato, tomato, chili, and grape, while draft genome sequence has been prepared for watermelon, apple, mango, etc., and is underway through international and national consortia.

Transgenic development is horticultural crops is underway for many traits, including resistance to pests, diseases, stressful environmental conditions, resistance to herbicides, improving the nutrient profile and shelf life, etc. Since the first development of the famous "FlavrSavr" tomato, new vistas opened in the genetic transformation of horticultural crops. The most recent success has been the release of the transgenic papaya variety "Rainbow," which is the first commercialized transgenic fruit crop and covers an estimated 71% of total papaya variety in Hawaii. In India, efforts are being made to develop genotypes resistant to fungal, viral, and insects and for drought and salinity tolerance. Significant progress has been made in potato, tomato, watermelon, papaya, brinjal, onion, and chili. Transgenic watermelon cv Arka Manik resistant to WBNV (Watermelon Bud Necrosis Virus) is under development, and T3 plants are being evaluated. Transgenic tomato Arka Saurabh, Arka Vikas, Arka Meghali for resistance to TLCV, Arka Saurabh, Arka Vikas, Arka Meghali for resistance to Tomato tospo virus, and for combined resistance to TLCV and Tospo virus have been generated, and they are at the evaluation stage. Transgenic tomato cv Arka Vikas with the chitinase gene, T1 stage, showed increased resistance to Alternaria solani compared to control. Bt transgenic brinjal and tomato have been developed, and the transformants are in different stages of testing and multiplication. Transgenic tomato and chili plants with the DREB 1A gene for abiotic stress resistance have been developed, which are at T2 stage of evaluation. The transgenics where field trials are in progress are potato, tomato, and brinjal in public sector and potato, tomato, brinjal, okra, and cauliflower in private sector. The genes used are cryIAa, cryIAc, cryIBa, cryIAabc, cryICa, cryIEc, rice chitinase, tobacco osmotin, antisense replicase gene of tomato leaf curl virus, etc. Around 17 crops are permitted for field trails in India recently, even though there is yet no clear decision on the release of GMOs.

5.4 HI-TECH INTERVENTIONS

In the last three decades, Indian horticulture has introduced, standardized, and exploited several hi-tech technologies, namely, micropropagation, high density planting (HDP), canopy management, soil- and nutrient-based fertilizer management, microirrigation, and fertigation, in situ water conservation, use of chemicals for high productivity, mechanization, use of plastics and protected/greenhouse cultivation for improving productivity and quality.

HDP technology has been standardized in several fruit crops to provide high yield, high net economic returns per unit area, more efficient use of natural resources (land, water, and solar energy) and inputs, resulting in higher yields. Successful but selective adoption of HDP is already in practice in mango, guava, banana, citrus, pine apple, pomegranate, papaya, cashew, and coconut. Technology for meadow orcharding through HDP and canopy management has been standardized in guava. Coconut and arecanut-based high density multispecies cropping systems have also been developed for better stability of income.

Canopy management is important for maintaining proper plant health, increasing productivity (per unit area), sustaining high yields, optimum fruit quality, restoration of regularity in bearing, and reducing cost of production. The technology for canopy management has been developed in mango, guava, grape, etc.

A large number of plantations are also becoming old and senile, thus requiring rejuvenation and replanting. Rejuvenation technology has therefore been developed for crops like mango, guava, *aonla*, litchi, apple, etc. Such plants refruit and give commercial crops within 3 to 4 years of top working. The best time of rejuvenation pruning is December–January. A project on apple and coconut replanting has also been launched in Kerala and Himachal Pradesh.

Effective water management and enhancing water and nutrient use efficiency have been successfully exploited in a number of horticultural crops through microirrigation and fertigation, resulting in saving of 25–60% water with 10–60% increase in yield over conventional irrigation. Fertigation (application of 100% soluble fertilizer through irrigation) requirement in several fruit, vegetable, and plantation crops have been standardized. Sampling techniques and optimum leaf nutrient standards have been developed for several crops to enable better fertilizer management.

Several plant growth regulators and chemicals are now commercially employed in production and quality improvement of horticultural crops. These include IBA (indole butyric acid) for rooting in cuttings, dormex for hastening bud break in grape, paclobutrazol for flower induction in mango, ethrel for flowering in pineapple, urea sprays for crop regulation in guava, NAA (naphthaleneacetic acid), and 2,4 D (2,4-dichlorophenoxyacetic acid) for control of fruit drop in mango and citrus, and gibberellic acid in improving berry elongation and quality of grapes. Maleic hydrazide is also being used as a sprout suppressant in potato and onion for prolonging storage life.

Seventeen Precision Farming Development Centres have been established under different agro-climatic conditions for developing protected and precision farming technologies. Use of plastics in green house construction, mulching, shade nets, and plastic tunnels are now very popular. Polygreen houses are being effectively utilized for cultivation of high-value crops, though adoption of protected cultivation is slow than desired. The major crops grown under protected cultivation include strawberry, tomato, cucumber, capsicum, chilies, rose, carnation, chrysanthemum, etc. Low cost polyhouses have significantly improved cultivation of leafy vegetables in the dry temperate areas of Leh and Ladakh in Jammu and Kashmir. Throughout the year production of many vegetables crops is now a reality.

Effective management practices have been developed for several diseases like fruit flies in mango and guava; diamond black moth in cabbage; and late blight, bacterial wilt, apical leaf curl, and stem necrosis in potato. Disease management for cassava mosaic disease has been accomplished by multiplication in vector-free zones.

5.5 MECHANIZATION

While India has 3.2 million tractors (largest number in the world), most operations are carried out manually or with animal power. Some new implements developed include tractor drawn implements for potato cultivation from sowing to harvesting and fruit harvesters for mango, guava, sapota, pomegranate, and citrus including manually portable ladders and tractor drawn hydraulically operated platforms for harvesting these crops, tree climbers have also been developed for coconut harvesting. In addition, seed and fertilizer drill for vegetable and potato, graders and washers developed by different Institutes have become popular among the growers. Automatic grafting machines available in advanced countries have also been introduced for adoption. Self-propelled pruners for tree crops and vine yards are available. For nurseries, media sievers and plastic bag fillers are now available.

5.6 POST-HARVEST MANAGEMENT

Maturity standards have been developed in several important crops. Pre-harvest treatments to reduce post-harvest losses have been developed in mango, citrus, and grape. Controlled atmospheric (CA) storage technology has been adopted in apple. Low cost eco-friendly cool chambers have been developed for on-farm storage of some commodities (onion, potato) besides standardization of storage temperature for most of the horticultural crops.

Several new products and technologies have been developed and commercially exploited. The commercialized technologies include raisin and wines (grape), chips and fingers (potato), concentrate (banana), gherkin in brine, pouched coconut water, snow ball tender coconut, etc. Several dehydrated and other miscellaneous products, namely juices from crops like passion fruit, Leh berry, pomegranate, and noni, are now commercialized. Consumer friendly products like frozen green peas, ready-to-use salad mixes, vegetable sprouts, ready-to-cook fresh cut vegetables, and ready-to-use cooked vegetables are now major retail items.

5.7 IMPACT OF RESEARCH AND DEVELOPMENT

The R&D efforts made in the last three decades have impacted horticulture in many ways:

- Significant developments in planting material production and registration of nurseries have resulted in availability of large quality, disease-free, planting material of high-yielding varieties. Commercial use of micro-propagated banana plants has resulted in significant yield increase.

- Several new fruit crops, namely kiwi, passion fruit, sea buckthorn, oil palm, noni, cherry tomato, gherkins, and baby corn, have been introduced and commercialized in different states. *Aonla*, *ber*, pomegranate, lime and lemon, papaya, and low chilling temperate fruits, namely pear, peach, and plum, have assumed commercial proportions.
- Hybrid vegetables have revolutionized vegetable cultivation. There has been virtual revolution in potato production and processing.
- There has been a significant increase in the production of floriculture crops, including cut flowers besides different species of mushroom.
- Hi-tech horticulture, particularly HDP, canopy management, rejuvenation, and promotion of beehives for pollination, have resulted in enhanced productivity of a number of crops. Microirrigation is now virtually a rule in new horticultural plantations. Fertigation and use of liquid fertilizers are picking up. Protected cultivation has become quite popular for commercial cultivation of a number of vegetables and cut flowers.
- Organic farming and conservation horticulture have emerged as viable alternatives in areas with depleted natural resources.
- Postharvest management of mango through vapor heat treatment has made mango exports to USA and Japan a reality. Irradiation facility for potato and onion for long duration storage has become a reality.
- Storage capacity of horticultural produce has increased with the establishment of a large number of cold stores. CA storage facilities are now available in several towns.
- Improvement in packing and packaging technology is evident from the range available in supermarkets.
- The availability of a wide variety of processed products has improved significantly over the years due to changing lifestyles and emphasis on value addition. Pigment production from flowers especially marigold and calendula has assumed commercial scale.
- Farmers and consumers have started realizing the importance of quality for both enhanced retailing and consumer satisfaction. Exports of horticulture produce and products have significantly increased.
- The above developments have brought about the following changes in the horticulture sector of the country:

- There is now a large R&D network in the country with several institutions and with number of programs.
- The importance of horticulture is being realized by one and all, be it a small farmer, the corporate sector, and policy makers at all levels.
- The area, production, productivity, and quality of different commodities have significantly improved.
- Globally, India has emerged as a major player in horticulture crop production and is the largest producer of fruits and vegetables after China, the largest producer of mango and banana, and the second largest producer of potato, tomato, and onion.
- The horticulture produce has percolated even to villages in the country, and its acceptability and use has increased considerably in every household.
- There is significant improvement in per capita availability and consumption of horticulture commodities.
- Varied fresh and processed horticultural products are now available in every nook and corner of the country.
- Horticulture has come out of village confines to the urban area and corporate sector.
- Indian horticulture has been gradually penetrating in the international market with significant increase in fruits, vegetables, flowers, and cashew exports.
- A massive infusion of private investment is being made in postharvest management and retail marketing of horticultural produce.

The impact of the above initiatives has become quite visible, and their role in development of this sector has been recognized in our country. The following statements made by the honorable Minister of Finance, Government of India, during his speech on unraveling Economic Survey 2015–2016 (Anon., 2016) in the parliament reveals the achievements made in the sector in the country.

- Over the last decade, the area under horticulture grew by about 2.7%, resulting in a total area of 24.2 million hectares under horticulture crops.
- During 2013–2014, the production of horticulture crops was about 283.5 million tons.

- Highest annual growth rate of 9.5% in fruit production was achieved during 2013–2014.
- The production of vegetables has increased from 58,532 thousand tons to 1,67,058 thousand tons between 1991–1992 and 2014–2015 and the annual production of total horticulture crops increased by 7%.
- Exports of horticultural produce rose to 536.42% in quantity and 2007.80% in value since 1991–1992 to 2012–2013.
- Share of horticulture output in agriculture is more than 33%.
- Thus, horticulture is emerging as the growth engine of Indian agriculture and future of sustainable agriculture in India.

KEYWORDS

- **hi-tech interventions**
- **horticulture**
- **mechanization**
- **post-harvest management**
- **rejuvenation**
- **varieties**

REFERENCES

Anonymous (1997). Ninth Five Year Plan. Planning Commission. Government of India.

Anonymous (2016). Economic Survey 2015–16. Ministry of Commerce & Industry, Government of India.

CHAPTER 6

ORGANIC FRUIT PRODUCTION IN INDIA: PRESENT SITUATION AND FUTURE PROSPECTS

S. K. MITRA,[1] HIDANGMAYUN L. DEVI,[2] and K. S. THINGREINGAM IRENAEUS[3]

[1]*Chair, Section Tropical and Subtropical Fruits, International Society for Horticultural Science, Belgium*

[2]*ICAR Research Complex for NEH Region, Tripura Centre, Lembuchera, West Tripura, 799210, India*

[3]*Department of Horticulture, College of Agriculture, Tripura, India*

CONTENTS

6.1 INTRODUCTION

Concern for food and environmental safety with less resource degrading inputs and chemical residues is increasing worldwide. The 2015 United Nations Climate Change Conference, COP 21 or CMP 11, was held in Paris, France, from November 30 to December 12, 2015. It was the 21st yearly session of the Conference of the Parties to the 1992 United Nations Framework Convention on Climate Change (UNFCCC) and the 11th session of the Meeting of the Parties to the 1997 Kyoto Protocol. Climate change is globally more frequently discussed under UN, and the recent one in France focused on combating global climate change by way of reduction in GHGs and adoption of green and resources conserving eco-friendly technologies. Representatives of 195 nations reached a landmark accord that will, for the first time, commit nearly every country to lower planet-warming greenhouse gas emissions to help stave off the most drastic effects of climate change. The result of the conference is an agreement (to be signed between April 22, 2016 and April 21, 2017) to set a goal of limiting global warming to less than 2 degrees Celsius (°C) compared to pre-industrial levels. The agreement calls for zero net anthropogenic greenhouse gas emissions to be reached during the second half of the 21st century.

In agriculture, organic agriculture is the only holistic agricultural production system in the world based on international standards and guided by the principles of health (i.e., sustaining ecosystem health and human health), ecology (i.e., enhancing living ecological systems, ecological balances, and cycles), fairness (i.e., ensuring equity, respect, justice, and stewardship of the shared world), and care (i.e., taking precaution and responsibility) (Jimenez, 2006). Food security is not simply a function of production or supply, but of availability, accessibility, stability of supply, affordability, and the provision of adequate quantity and quality of food and safe nutritious food at all times (Boon, 2007; IAASTD, 2009). Thus, improving food security requires an economically viable, environmentally friendly, socially acceptable, and appropriate agricultural system based on local people's needs. In a world where consumers are increasingly becoming aware of what they eat and consume, quality becomes a key issue and a unique selling proposition.

Though India has achieved the green revolution from the begging bowl status with greater use of synthetic agrochemicals like fertilizers and pesticides, adoption of nutrient-responsive high-yielding varieties of crops, greater exploitation of irrigation potentials, etc., the continuous use of these high energy inputs indiscriminately has led to deterioration of soil health and environments, and food safety has become a major concern. Apprehending the bad effects of chemical use, the Government of India has declared organic farming as the national program to promote organic agriculture in India by providing various incentives to give momentum to this noble movement. Our soils have stopped responding to the use of chemicals in the absence of requisite humus, which can be only provided by way of adopting organic farming. This is the only way to revive lost fertility of the soil. Climate change is also influencing the environment to the extent that organic farming appears to be the only alternative to combat these situations.

In India, with the phenomenal growth in area under organic management and growing demand for wild harvest products, India has emerged as the single largest country in the world with highest arable cultivated land under organic management. India has also achieved the status being the single largest country in the world in terms of total area under certified organic wild harvest collection. Development of organic agriculture is now being embraced by the mainstream and shows great promise commercially, socially, and environmentally. While there is a continuum of thought from earlier days to the present, the modern organic movement is radically different from its original form. It now has environmental sustainability at its core in addition to the founders' concerns for healthy soil, healthy food, and healthy people (Mitra and Devi, 2015).

6.2 WHY ORGANIC FRUITS?

Fruits have historically held a place in dietary guidance because of their concentrations of vitamins, especially vitamins C and A; minerals, especially electrolytes; and more recently phytochemicals, especially antioxidants. Additionally, fruits are recommended as a source of dietary fiber. Fruits have greater potential for export and are valued much as organic food among horticultural crops. Organic fruits are free of chemical residues, firm, juicy, and tastier with longer shelf life and are preferred over conventionally grown fruits. There is an increasing demand of organic fruits in the consumer market. Environmental pollution affecting soil, water, and air has curtailed and biodiversity has improved through organic means. It facilitates judicious use of farm resources, intercropping, and mulching. In some cases, resistance against insect pests and diseases in the inorganic system has been developed, as seen in apple.

6.3 CONVENTIONAL VERSUS ORGANIC FARMING

Input into organics is small compared to the conventional and comparisons between and organic farming and the former are actually not on a level-playing field. Organic research tends to be more diffuse, farm-base participatory, drawing on local knowledge and tradition. It also focuses on public goods, resources, and tools that are not readily patentable. This explains why organic farming attracts little investment from private sources as compared to conventional and biotechnological approaches (Parrot and Marsden, 2002).

The relationship between soil organic matter and crop productivity has been cited in literatures and can be concluded that crop response to applied fertilizer depends on soil organic matter availability. The problems inherent to tropical soils are soil acidity, excessive aluminum, deficient calcium, and low organic matter (Hue, 1992). Organic matter additions are the only means of making some soils economically productive. Therefore, the use of organic amendments is synonymous to soil productivity (Reichardt et al., 2000). Increasing soil organic matter has added benefit of improving soil quality and thereby enhancing the long-term sustainability of agriculture. Codex Alimentarius (1999) viewed organic farming as holistic production management systems (for crops and livestock), emphasizing the use of management practices in preference to the use of off-farm inputs.

Model estimates indicate that organic methods could produce enough food on a global per capita basis to sustain the current human population, and potentially an even larger population, without increasing the agricultural land base (Badgley et al., 2006). Based on available data in their review, it was suggested that in temperate and tropical agroecosystems, leguminous cover crops could fix enough nitrogen to replace the amount of synthetic fertilizer currently in use; this indicates that organic agriculture has the potential to contribute quite substantially to the global food supply, while reducing the detrimental environmental impacts of conventional agriculture. In a comparative study of yields of organic versus conventional or low-intensive food production for a global dataset of 293 examples, the average yield ratio (organic:nonorganic) was slightly <1.0 for studies in the developed world and >1.0 for studies in the developing world for different food categories.

Consumers are now turning to organic food because they believe it to be tastier as well as healthier, both for themselves and environment. Of the seven studies comparing taste of crop produced with organic management versus conventional management, there was positive impact of organic management in two crops of apples in Australia (Velimirou et al., 1995) and potatoes in Finland (Varies et al., 1996), though no differences were observed in the other five crops. In respect of pesticide residues, lower levels can be expected in both vegetables and fruits from organic production. Conventionally cultivated or minerally fertilized vegetables normally have higher nitrate content than organically produced vegetables (Woese et al., 1997). In the case of vegetables, in particular leafy vegetables, a higher dry matter concentration can be observed in organically grown or organically fertilized products than in conventionally grown products. On the other hand, it was inconclusive with regard to all other desirable nutritional values, as either the case that no major differences were observed in physico-chemical analyses between the products from different production forms or contradictory findings did not permit any clear statements.

In India, trials on organic production showed that papaya cv. Pusa Delicious was more responsive to farmyard manure and poultry manure than neem cake (Ray et al., 2008). Similarly, fruit yield was highest (70.2 kg/tree) with 15 g N/tree as farmyard manure in CO1 cultivar, although 15 g N/tree as groundnut cake proved better for tree growth (Patil et al., 1997) than other organic and inorganic nutrients in the form of urea. However, in the former study, the yield of organic amendments was lower than that of inorganic fertilizers with lower cost-benefit ratios (Ray et al., 2008). Organic matter content in soil was significantly influenced by treatments with different organic manures. Soils under different organic modules had significantly higher microbial population (bacteria, fungi, and actinomycetes) and activities of urease, phosphatase, dehydrogenase, and cellulases as compared to that under recommended dose of fertilizers. Application of FYM (farm yard manure) 20 kg/plant was the best organic module with regard to higher microbial populations and enzyme activities in soil (Ravishankar et al., 2010). In pineapple cv. Mauritius, 250 g poultry manure, *Azospirillum*, and phosphobacteria at 650 mg each, along with N, P_2O_5, and K_2O at 8:4:8 g/plant recorded highest values in terms of growth of plants, juice percent, and quality of the fruits (Devdas et al., 2005).

6.4 GLOBAL SCENARIO OF ORGANIC AGRICULTURE

In the world, nearly 43.1 million ha land is being certified as organic in 170 countries, constituting 1% of the total agricultural land of the countries under study on certified organic agriculture worldwide, according to the latest FiBL-IFOAM survey (Helga and Lernoud, 2015). The regions with the largest areas of organic agricultural land are Oceania (17.3 million ha) and Europe (11.5 million ha). Latin America has 6.6 million ha followed by Asia (3.4 million ha), North America (3 million ha), and Africa (1.2 million ha). Australia (17.2 million ha), Argentina (3.2 million ha), and the United States (2.2 million ha) are reported to be the countries with the most organic agricultural land. Apart from the agricultural land, there are further organic areas, most of these being areas for aquaculture, forests, and grazing areas on non-agricultural land. The areas of non-agricultural land constitute more than 35 million hectares. In total, 78 million hectares (agricultural and non-agricultural areas) are organic.

Of the total of almost 2 million producers globally as of 2013, more than 80% (1.7 million) of the producers with about a quarter of the world's agricultural land (11.7 million ha) are in developing countries and emerging markets. Thirty-six percent of the world's organic producers were in Asia, followed by Africa (29%) and Europe (17%). The countries with the most producers are India (650,000), Uganda (189,610), and Mexico (169,703). Land use details were available for almost 90% of the organic agricultural land. Unfortunately, some countries with very large organic areas, such as Australia, Brazil, and India, had little or no information on their land use. With a total of at least 7.7 million hectares, arable land constitutes almost 20% of the organic agricultural land. An increase of almost 3% over 2012 was reported.

6.5 PRESENT STATUS AND RECENT DEVELOPMENTS IN ORGANIC AGRICULTURE IN INDIA

Like in other Western countries, there is a growing realization and interest for organic foods in India. Concerted efforts are being made by various civil society organizations to join the movement for its potential in sustaining soil health, preventing contamination in surface and ground water aquifers, and ensuring safe and healthy food. To support the export prospects, the Ministry of Commerce launched the "National Programme on Organic Production" (NPOP) defining the National Standards for Organic Production (NSOP) and the procedure for accreditation and certification in 2000. India now has 30 accredited certification agencies for facilitating the certification to growers. For area expansion and technology transfer, the Ministry of Agriculture launched a national project on promotion of organic farming (NPOF-DAC) and earmarked funds for setting up of organic and biological input production units and vermicompost production units and for organic adoption and certification under various schemes such as NHM (National Horticulture Mission) (now MIDH – Mission for Integrated Development of Horticulture), NMSA (National Mission for Sustainable Agriculture), and RKVY (Rashtriya Krishi Vikas Yojana). To empower farmers through participation in certification process and to make the certification affordable for domestic and local markets, the Ministry of Agriculture has also launched a farmer group centric organic guarantee system under PGS-India program.

A network project on organic farming (NPOF-ICAR) under Project Directorate of Farming System Research with 13 collaborating centers across the country was launched by ICAR to augment organic research, and the package of practice for some important crops has been developed under the project. India has brought 4.72 million ha area under organic certification process, which includes 0.6 million ha of cultivated agricultural land and 4.12 million ha of wild harvest collection area in forests, as of March 2014. Growth of area under organic farming during different years is presented in Figures 6.1a, 6.1b, and 6.1c.

Mizoram passed the Mizoram Organic Farming Act, 2004, on July 12, 2004 and has declared itself as totally organic. Sikkim has taken up the task of converting the entire state into organic and has already brought more than 65,000 ha are under organic certification process. Among all the states, Madhya Pradesh has covered the largest area under organic certification, followed by Himachal Pradesh and Rajasthan. It has launched a noble initiative under

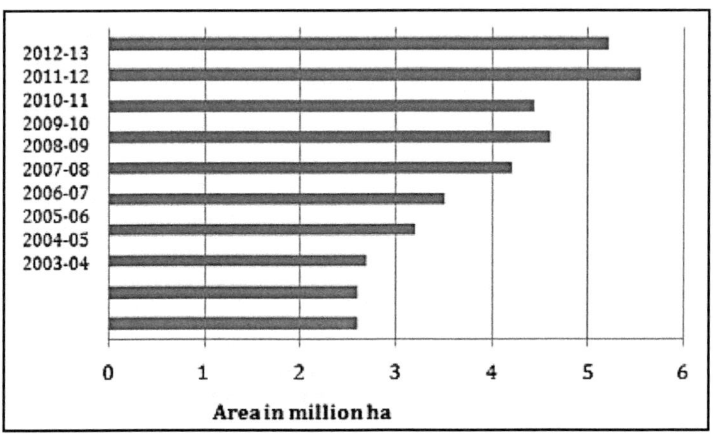

FIGURE 6.1A (See color insert.) Total area under certification (cultivated + wild harvest).

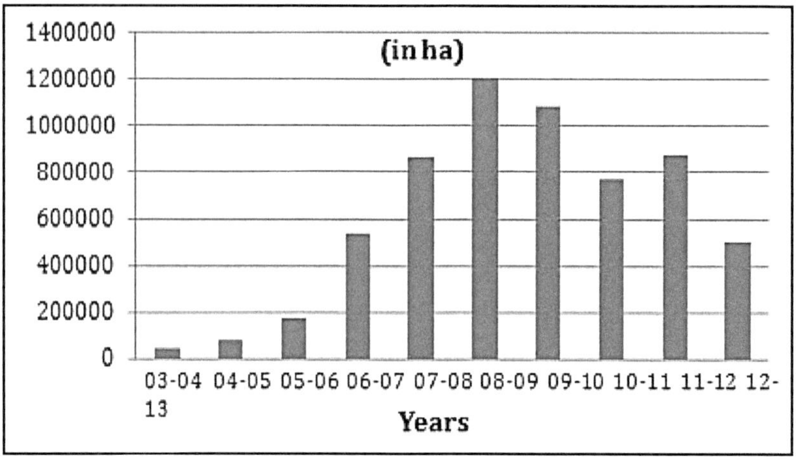

FIGURE 6.1B **(See color insert.)** Cultivated area under organic certification (in ha).

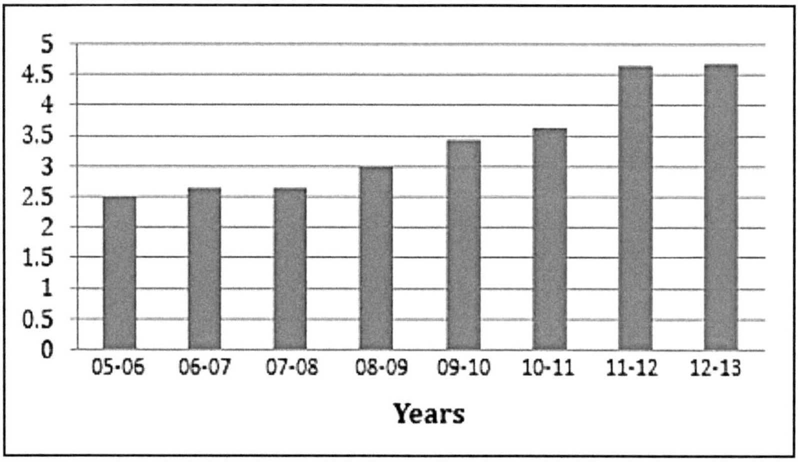

FIGURE 6.1C **(See color insert.)** Wild harvest collection area under organic certification (million ha).

which all the 313 blocks of the state have been turned into Bio Village. Uttaranchal also leads in organic agriculture movement. Karnataka and Andhra Pradesh formulated organic policy. Maharashtra, Tamil Nadu, Kerala, and West Bengal supported public-private partnership for promoting organic farming.

6.6 ORGANIC FRUIT PRODUCTION IN INDIA: PROSPECTS AND OPPORTUNITIES

India is one of the leading fruit producers in the world, producing about 10% of the world's fruit production. India produced 88.97 million metric tons of fruits from an area 7.22 million hectares in 2013–14 (Indian Horticulture Database, 2014). India ranks first in the production of bananas (22.04%), papayas (40.74%), and mangoes (including mangosteens and guavas) (32.65%). India ranks 10th among the top 10 countries in terms of cultivable land under organic certification. Global retail organic food sales are currently valued worth USD 31 billion and are growing at over 20% per annum. The same is estimated to increase to USD102 billion by 2020. The target for tropical and subtropical fruits and its products is 0.14 billion USD. Major organic fruits produced in India are mango, cashew, banana, pineapple, passion fruit, and orange. The total organic production of fruits and vegetables was estimated at 8227.74 MT and dry fruits at 521.46 MT in 2011–2012.

Associates of International Federation of Organic Agriculture Movement (IFOAM), Germany, have over 600 members from 120 countries, and its members spearhead the organic agriculture movement in India. Its member in India is All India Federation of Organic Farming (AIFOM), having in its fold a spectrum of NGOs, farmers' organizations promotional bodies, corporate units, and institutions. Still many small and marginal farmers, because of many reasons, have not fully adopted the conventional farming, and they follow more or less the traditional environment friendly system. They use local or own farm-derived renewable resources and manage self-regulated ecological and biological processes.

Since 2004, many states embraced organic farming and drafted policies. So far 11 states, namely Andhra Pradesh, Karnataka, Kerala, Uttarakhand, Maharashtra, Madhya Pradesh, Tamil Nadu, Himachal Pradesh, Sikkim, Nagaland, and Mizoram, have drafted the organic agriculture promotion policies.

6.7 OPPORTUNITIES FOR ADOPTION OF ORGANIC FRUIT PRODUCTION

- Fruit trees are perennial in nature and possess huge plant canopy with sufficient amount of leaves, and these leaves shed during winter, which contribute a lot to the soil organic matter content. Among horticultural crops, fruits that have greater scope for export are valued much as organic food. Of the total area under organic agriculture, globally, permanent crops account for 7% of the organic agricultural land, amounting to 3.2 million ha (Helga and Lernoud, 2015). The most important permanent crops are coffee (with more than 0.7 million ha, constituting almost one-quarter of the organic permanent cropland), followed by olives (0.6 million ha), nuts and grapes (0.3 million ha each), and cocoa (0.2 million ha).

- There is a growing interest and demand in the global organic food supply because of its safety and richer nutritional perspectives. After many years of consumers having to hunt around for their organic produce from several suppliers, perhaps directly from the farmer, the task is now a lot easier with specialist food shops and organic shelf space in supermarkets. Global links have been forged in all continents as organic agriculture has been seen to be an effective rural development option.
- Market opportunities for organic tropical and subtropical fruits and products from low cost-producing countries of the southeast Asian region. India's best prospects for organic exports are in a range of primarily tropical products where it has a well-established export market and/or a growing supply base of organic production (including "default" organic supply).
- In India, because of the nature of fragmented and marginal land holdings, plant nutritional requirement can be provided by application of locally available low-cost inputs that could be managed by the family and conversion to organic system is thus easier. In the country, 65% of the country's cropped area is organic by default and farmers still practices traditional method of production system. This default status coupled with India's inherent advantages such as varied agro-climatic regions, local self-sustaining agri-systems, sizeable number of progressive farmers and ready availability of inexpensive manpower translate into the potential to cultivate a vast basket of products organically.
- Organic agriculture is one of the fastest growing agribusiness sectors in the world. The concept of organic farming has strong marketing appeal, growth forecasts are almost all positive and it has been suggested that the "movement" is now an industry.
- Fruit orchards offer an option for intercropping crop, which is not only remunerative but act also as cover crops or a source of organic mat-

ter and nitrogenous nutrient requirement with the use of leguminous crops.

- A NPOP containing the standards for the organic products has been launched in association with four export organizations, viz., APEDA, Tea Board, Coffee Board, and Spices Board.
- Northeast hilly region of India has also been declared as "Organic Farming Zone," and the government has identified crops like pineapple (Tripura), passion fruit (Manipur), and ginger (Sikkim/Meghalaya) for organic production in the NEH region.
- APEDA has launched a program for the development of model organic farms for crops including fruits like passion fruit and pineapple in the northeast region.
- A "National Centre for Organic Farming" has been established at Ghaziabad to provide institutional support and to facilitate farmers to move into organic crop production.
- National Commission on Farmers has identified organic farming as a tool for second green revolution in the rain-fed and hilly areas of the country.
- National Horticulture Board has been promoting commercial horticulture through production and postharvest management. It has made eligible for capital investment subsidies for two important items, viz., organic farming and biotechnology, which also covers organic food production as a major component.
- A large amount of agricultural wastes is available in the country (5 tons of organic manure/hectare arable land/year), which is equivalent to about 100 kg NPK/ha/yr but it is not properly used. Technologies have been developed to produce large quantities of nutrient-rich manure/compost and several alternatives for supply of soil nutrients from organic sources like vermicompost and biofertilizers are now available.

6.8 MARKET AND EXPORT POTENTIAL

The main market for exported products is the European Union (Helga and Lernoud, 2015). Another growing market is the USA. During 2012–2013, India exported 165,262 MT of organic products belonging to 135 commodities valuing at US$ 312 million (approximately INR 1900 crore). Domestic market is also growing at an annual growth rate of 15–25%. According to the survey conducted by ICCOA, Bangalore, the domestic market during the year 2012–2013 was worth INR 600 crore (Figure 6.2).

With the growing worldwide market, and scope for organic farming, India holds significant potential in export of organic tropical and subtropical fresh fruit and processed products. India with its tradition of organic farming and growing consumer awareness has an edge to capitalize on the growing organic market. The vast production base offers India tremendous opportunities for export. During 2014–2015, India exported fruits worth Rs. 2771.32 crores. Mangoes, walnuts, grapes, bananas, and pomegranates account for larger portion of fruits exported from the country. The major destinations for Indian fruits are UAE, Bangladesh, Malaysia, UK, Netherland, Pakistan, Saudi Arabia, Sri Lanka, and Nepal. Presently, India's exports of organic produce are worth 1 to 2 crores only, mainly of tea, cotton, and spices. There is tremendous potential to increase India's share in international trade in organic food by including commodities such as fruits, vegetables, aromatic and medicinal plants, coffee, pulses, durum wheat, basmati rice and sugar, etc. India has competitive advantages in the world markets due to low production costs and availability of diverse climates to grow a large number of crops round the year. India is the largest mango producer in the world; however, a negligible amount of fresh and processed mangoes are

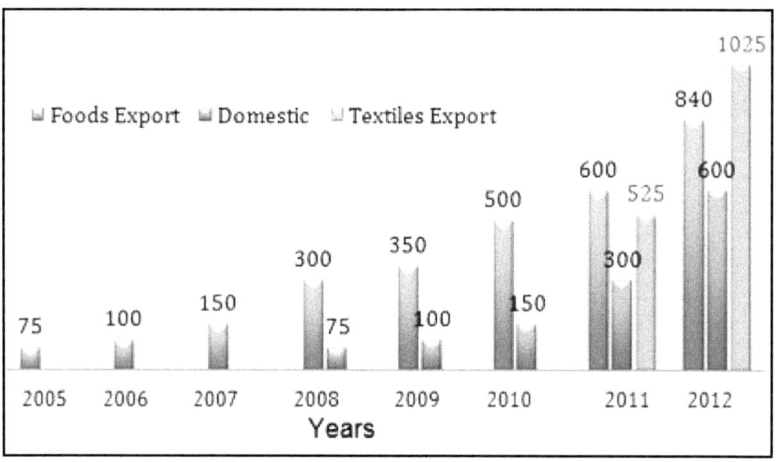

FIGURE 6.2 **(See color insert.)** Export and domestic market for organic products (Rs crores).

being exported due to high domestic demand. EU member states, especially UK, Netherlands, and Germany, have a good demand for organic mangoes, which can be exploited by India (Mitra, 2013).

Indian organic banana exports (dry banana and banana puree) were negligible as compared to world trade. India needs to focus on increasing organic banana exports; firstly, it should target the processed organic banana market (banana pulp, purees, and concentrates). Secondly, the focus should be on the geographically closer Japanese market and the EU. India is the fourth largest producer of pineapple, accounting for about 11% of the total world output. Less than 1% of the Indian pineapple production is exported to world markets (Mitra, 2013). India holds a good potential for export of organic pineapples to the three major pineapple importing markets (USA, EU, and Japan).

Passion fruit is popular in the Western countries because of its lemony tart flavor and consumed as fruit juice or as an ingredient in multivitamin drinks, which is the fastest growing segment in the juice market. The fruit grown in all the northeastern states is cultivated organically. India's share in the total world trade is negligible due to the lack of world-class processing industry in the region, and thus, export potential can be exploited.

Grapes are India's second largest exported fruit after mango. European Union (EU), especially the UK and Netherlands, are the main target destination market for Indian organic grapes, which can be exploited. A consumer trend these days is increasingly favoring organic wine, which further increases the demand of organic grapes (Mitra, 2013).

6.9 CONSTRAINTS IN ADOPTION OF ORGANIC SYSTEM

Marketability at a premium over the conventional produce has to be assured. Inability to obtain a premium price, at least during the period required to achieve the productivity levels of the conventional crop, will be a setback. High rainfall and humidity in the tropical and sub-tropical region creates favorable environment for disease and pest development, and pest control exclusively in the organic way becomes difficult. Nonavailability of quality planting materials and seeds further complicates the issue. The planting material itself has to be purely of organic origin too.

Inadequate certifying agencies with adequate verification facilities leading to certification and inability of the farmers to reach the certifying agencies is a major practical problem. In the northeastern and other hilly parts of the country, most of the land holdings are marginal and fragmented, and individual farmers are not able to go for large-scale commercial production. Thus, there are practical problems in adoption of organic farming for commercial purpose and this also makes certification process difficult. This can be managed by adoption of contract farming system among group of farmers. The practice of *jhum* cultivation results in heavy loss of organic matter from the soil and makes it susceptible to erosion hazards.

The costs of the organic inputs are increasing and higher than those of inputs used in the conventional farming system and other chemicals making them unaffordable to the small cultivators. There is a lack of an efficient quality control mechanism and organized marketing system for organically produced commodities. There is also a lack of awareness about organic food production standards, use of biofertilizers, biological control of pest and diseases, and organic food market.

Sufficient infrastructure for organic production, storage, transportation, certification, trading, and absence of large-scale processing units, post-harvest and value addition facilities is yet to be established especially in regions like northeast India where maximum potential for organic agriculture in the country exists.

Another important factor against adoption of organic production system is the extensive role of marketing personals and companies actively engaged in producing and promoting use of chemicals among the farmers. There is a lack of sufficient biofertilizers and biopesticides to control different kinds of pests and diseases, and improper marketing and distribution network for them because of lower margins of profit.

6.10 CHALLENGES

- Keeping up with the requirement of the growing population is the most feared and debated topic among all other problems. Thus, production and productivity have to be ensured through organic means by adoption of various technologies.
- Fair trade, supply chain development, and global harmonization for standards need to be developed.
- Dealing with crop-specific problems without using substances prohibited by organic standards.
- Considering the nutrient demand of perennial fruit crops over time, adequate methods are needed to maintain soil fertility, especially lev-

els of phosphorus, potassium, and micronutrients, in such a way that yields can be sustained.

- Conversion to organic agriculture requires innovative solutions that guarantee purity of seeds, planting material and provide assurance against the risk of pest and disease spread.
- Alternative approaches are needed, for instance, to control ripening of pineapples without chemicals or to control banana or citrus pests and diseases.
- Organic horticulture makes optimum use of crop residues, green manure, and biological nitrogen fixation, but this does not replace all nutrients removed by crops.
- Nitrogen supply needs to be supplemented by the provision of other essential plant nutrients that usually cannot be replenished sufficiently by organic matter in tropical acid soils.

KEYWORDS

- **certified**
- **climate change**
- **fruits**
- **manure**
- **microbial**
- **organic**

REFERENCES

Badgley, C., Moghtader, J., Quintero, E., Zakem, E., Chappell, J. M., Aviles-Vazquez, K., Samulon, A., & Perfecto, I., (2006). Organic agriculture and the global food supply. *Renewable Agriculture and Food Systems*, *22*(2), 86–108.

Boon, E. K., (2007). *Food Security in Africa: Challenges and Prospects*. Encyclopedia of Life Support Systems (EOLSS), Oxford, pp. 31.

Codex Alimentarius, (1999). *Guidelines for the Production, Processing, Labeling and Marketing of Organic Produced Foods*, Commission, CAL/GL, 32.

Devadas, V. S., & Kuriakose, K. P., (2005). Evaluation of different organic manures on yield and quality of pineapple var. Mauritius. *Acta. Hort.*, *666*, 185–189.

Helga, W., & Lernoud, J., (2015). *The World of Organic Agriculture – Statistics and Emerging Trends*. FiBL-IFOAM report. Research Institute of Organic Agriculture (FiBL), Frick, and IFOAM-Organics International, Bonn.

Hue, N. V., (1992). Increasing soil productivity for the humid tropics through organic matter management. In: Overcash, M. R., (ed.). *Tropical and Sub Tropical Agricultural Research, Progress and Achievements.* The Pacific Basin Administrative Group. Livestock waste Management, vol. I, CRC press, Boca Raton, FLa.

Indian Horticulture Database, (2014). National Horticulture Board, Ministry of Agriculture, Government of India 85, institutional area, Sector-18, Gurgaon-122-015. www.nhb.gov.in.

International Assessment of Agricultural Knowledge, (2009). *Science and Technology for Development- IAASTD.* Agriculture at a cross road. Global report. Island press, 1718 Connecticut Avenue, NW, Suite 300, Washington, DC, pp. 590.

Jimenez, J. J., (2006). *Organic Agriculture and the Millennium Development Goals*, Dossier, IFOAM, Bonn, Germany, pp. 27.

Mitra, S. K., & Devi, L. S., (2015). Organic horticulture in India. *Horticulture, Proc. QMOH 2015. First Int. Symp. Quality Management of Organic Horticultural Produce.* Ubon Ratchatani, Thailand, pp.1–6.

Mitra, S. K., (2013). Organic tropical and subtropical fruit production in India-prospects and challenges. *Acta. Hort., 975*, 303–307.

Parrot, N., & Marsden, T., (2002). *The Real Green Revolution: Organic and Agro Ecological Farming in the London,* Green peace environment trust. pp. 1–6.

Patil, K. B., Patil, B. B., & Patil, M. T., (1997). Studies on manurial requirement of papaya. *J. Soils and Crops*, *7*(2), 123–127.

Ravishankar, H., Karunakaran, G., & Hazarika, S., (2010). Nutrient availability and biochemical properties in soil as influenced by organic farming of papaya under Coorg region of Karnataka. *Acta. Hort., 851*, 419–424.

Ray, P. K., Singh, A. K., & Arun K., (2008). Performance of Pusa delicious papaya under organic farming. *Indian J. Hort.*, *65*(1), 100–101.

Reganold, J. P., Glover, J. D., Andrews, P. K., & Hinman, J. R., (2001). Sustainability of three apple production. *Nature, 410*(19), 926–930.

Reichardt, W., Inubushi, K., & Tidje, J., (2000). Microbial processes in C and N dynamics. In: Kirk, G. J. D., OIK, D. C. (Eds). *Carbon and Nitrogen Dynamics in Flooded Soils*, International Rice Research Institute, Makati city, Philippines, pp. 101–146.

Varis, E., Pietila, L., & Kolkkalaiknen, K., (1996). Comparison of conventional, integrated and organic potato production in field experiments in Finland. *Acta. Agriculture Scandinavia. Section B, Soil and Plant Science*, *46*(1), 41–48.

Velimirov, A., Plochberger, K., Schott, W., & Walz, V., (1995). Neue UtersuchungenZur Qualitat Unter schiedlich angebauter Apfel-Nilchtalles, was goldenist, 1st auch delicious! Das Bio skop. Faschzeitchrift fur Bioland bauund okologie, *6*, 4–8.

Woese, K., Lange, D., Boess, C., & Bogl, K. W., (1997). A Comparison of organically and conventionally grown foods – results of a review of the relevant literature. *J. Sci. Food Agric.*, *74*, 281–293.

CHAPTER 7

TROPICAL FRUIT TREE ARCHITECTURAL MANAGEMENT IN THAILAND

NARONG CHOMCHALOW,[1] PREM NA SONGKHLA,[2] and LOP PHAVAPHUTANON

[1]*Chairman, Thailand Network for the Conservation and Development of Landraces of Cultivated Plants, Bangkok, Thailand*

[2]*Director, Kehakaset Academic Institute of Agriculture, Bangkok, Thailand*

[3]*Department of Horticulture, Faculty of Agriculture at Kamphaeng Saen, Kasetsart University, Thailand*

CONTENTS

7.1 INTRODUCTION

AEC or ASEAN (Association of Southeast Asian Nations) Economic Community has been effective since December 31, 2015. It creates major impact on agricultural industry as well as on other industries in all 10-member countries of ASEAN. A total of 650 million people of the 10 ASEAN countries need more foods and other agricultural produces. The unification of the ASEAN countries also results in the competition of agricultural products from other AEC-producing countries as many of them are agricultural-product producing countries. Hand-in-hand with agricultural production is the need for increase management cost, health and sanitation, and environment. Tropical fruits are items of great importance for both the growers and consumers. One of the main factors involved in the production of tropical fruits are their agricultural management, which will help the growers to obtain both the quality and quantity of their produces.

7.2 TROPICAL FRUIT TREE ARCHITECTURE

Ideal architecture of tropical fruit trees involved the following subsections.

7.2.1 CONTROLLING SIZE AND SHAPE OF TROPICAL FRUIT TREES

Factors involved in controlling size and shape of tropical fruit trees are as follows:

1. Selecting appropriate cultivars: Cultivars of tropical fruit trees play a significant role in controlling size and shape of tropical fruit trees.
2. Training and pruning: Suitable size and shape of tropical fruit trees can be achieved by proper training and pruning.
3. Root pruning: Root pruning is an important practice to control the size and shape of tropical fruit trees.
4. The use of growth regulators: Growth regulators can be used to control the growth of tropical fruit trees.

7.2.2 TROPICAL FRUIT TREE CYCLE

The stage of growth of tropical fruit trees can be illustrated by the following diagram:

Dormancy → Flowering → Fruiting
Sprouting → Pruning → Harvesting

7.2.3 PATTERN OF TRAINING

Five patterns of training have been developed, namely:

1. **Central-shoot or pyramid pattern:** The shoot and main branch are allowed to grow vertically away from the ground, while sub-branches emerged around the trunk and would not overlap each other, forming a pyramid-shaped canopy. Sunlight is allowed to reach to the bottom.
2. **Open-center pattern:** Cut off top or leader branch when the tree reaches appropriate height. Choose the branches between each other and train them to have the same size. Cut off new shoots to obtain the shape of a vase. Examples: orange, custard apple, mango, rambutan, and star fruit.
3. **Modified open-center pattern:** This pattern is a combination of pyramid and open center patterns. The fruit tree should not be too high or too low, but with strong branches and high yield. Examples: jackfruit, rose apple tamarind, durian, mango, and citrus.
4. **Various shaped patterns:** T-shaped, V-shaped, or Y-shaped patterns are employed to obtain high yield and quality with smaller sized tree. Examples: grape vines, guava, coffee, tamarind, and rose apple.
5. **Other patterns of canopy:** Many other patterns are applied to fruit trees grown naturally with very minimal pruning and training. Only unwanted branches are cut off. Examples: Burmese grape, jackfruit, gandaria, pummelo, sapodilla, and tamarind.

7.2.4 HEALING AFTER PRUNING

This practice can be achieved by fungicide spraying, other substitutions to fungicide, and lime application.

7.3 TROPICAL FRUIT TREE ARCHITECTURAL MANAGEMENT

Maintaining proper tropical fruit tree architecture involved the following subsections.

7.3.1 TYPES OF TROPICAL FRUIT TREES

The following types of tropical fruit trees exist:

1. **Constant shape:** The tree grows only vertically, while the branches do not spread out. Thus, training would not be needed. Examples: pineapple, banana, papaya, and coconut.
2. **Small size:** The tree grows naturally to the height of 2–3 m with many branches. Planting space should be considered. Examples: pomegranate, coffee, and star gooseberry.
3. **Medium size:** The tree is not taller than 10 m. The growth rate of the canopy is slow. When the tree gets bigger as it gets older, the yield is lower. Therefore, training is needed. Examples: citrus, custard apple, guava, cacao, jujube, and sapodilla.
4. **Big size:** The branch spread out very fast, both vertically and horizontally, with maximum height of 10 m or more. Therefore, training is very important for growers to manage their orchards. Examples: rambutan, mangosteen, santol, jackfruit, and tamarind.
5. **Creeping or climber:** The tree needs support in order to grow upwards. Training is needed in order for the canopy to obtain direct sunlight. Examples: grape vine, passion fruit, and dragon fruit.

7.3.2 TYPES OF FLOWERING OF TROPICAL FRUIT TREES

Various types of flowering are envisaged, viz.:

1. **Terminal bud of young shoot:** The young shoot bears flower buds. If there are no shoots, the chance of having flowers is very slim. Flowers would bloom at the second and third positions of the terminal bud. Examples: guava, custard apple, sweet orange, santol, grapevine, tamarind, and jujube.
2. **Full grown shoot:** Flowering develops from bud of full-grown shoot. Examples: mango, gandaria, rambutan, lychee, longan, and mangosteen.
3. **Branch or trunk:** Occurs only when the tree is more than one-season old or from old branch. Examples: durian, jack fruit, rose apple, coffee, star fruit, and Burmese grape.
4. **Direction of the row:** To obtain optimum sunlight, row direction should be perpendicular to, or along with, sunlight direction. Planting space depends on tree's height, machine used, and transportation in the orchard.

7.3.3 INTENSITY OF SUNLIGHT

Quantity and quality of the fruit yield depend on sunlight intensity perceived by the leaves. Training allows the light to reach into the crown as well as facilitating chemical spray and harvesting. Planting space affects the pruning because the trees are easily influenced by sunlight intensity. Some are grown in a high humidity climate, examples: durian, rose apple, and mangosteen. Some are tolerant to intense sunlight; thus, it is necessary to change the time of pruning to the rainy season with low light intensity and leave some leaves and branches after pruning to protect parts of the branch from sunburn.

7.4 INNOVATION OF TROPICAL FRUIT TREE ARCHITECTURAL MANAGEMENT

7.4.1 TRAINING AND PRUNING TECHNIQUES

Training is a practice to balance tree branches and leaves in order to perceive the sunlight and gain the best orchard management efficiency. Pruning is cutting part of unwanted branches out to have a good tree shape to obtain the maximum yield. Unwanted branches such as those that are overlapping, weak, and infected are pruned off. The main objective is to allow more sunlight into the canopy. The followings are involved in training and pruning:

7.4.1.1 Controlling the Size and Shape of the Fruit Trees

This involved the choice of cultivars, training and pruning, root pruning, and using growth regulators to control the growth of fruit trees.

7.4.1.2 Patterns of Training

The following patterns of training are available: (i) central shoot or pyramid pattern, (ii) open-center pattern, (iii) modified open-center pattern, and (iv) other patterns of canopy training.

7.4.1.3 Healing after Pruning

This practice allows the shoot or main branch to grow vertically away from the ground, while sub-branches are twisted around the trunk and would not overlap one another, allowing sunlight to reach in. Examples: durian and mangosteen.

7.4.2 PROPAGATION TECHNIQUES

Cleft grafting was adopted for mango cultivation using introduced cultivars such as "Red skinned" from Florida.

7.4.3 CHEMICAL CONTROL OF FLOWERING

The use of chemical control of flowering has gained acceptance by the farmers as it is cost-effective. Several chemicals have been used, e.g. potassium chlorate, paclobutazol, etc.

7.4.4 OFF-SEASON PRODUCTION

Following the chemical control of flowering, the same chemical can be used to produce off-season crops that give much higher profit for the farmers as the produces are available at the time when no other products are available.

7.5 EXAMPLES OF VITAL TROPICAL FRUIT TREE ARCHITECTURE

7.5.1 MANGO

7.5.1.1 Mango Blooming Data

Flowers are borne from terminal buds of the branch. Techniques of pruning vary depending on cultivars, environment, and age of the tree.

7.5.1.2 Architectural Patterns of Mango Tree

The following architectural patterns are recognized:

1. **Vase-shaped or open-center pattern:** Pruning can begin with mid-branch cutting at a height of 1 m. When new branches grow, maintain 3-4 branches that create an angle of 30° to the trunk.
2. **Modified pyramid pattern:** Allow the tree to grow for 1–3 years. Then, cut off the shoots and leave 3-4 large sub-branches on the side.
3. **Close-shaped pattern:** A close-space of 4 × 4 m with open-center pruning. Prune the leader branch at the height of 50–60 cm. Select 3–4 newborn branches, each forming 30° angle with the tree trunk. Cut off other branches and prune again, leaving 2–3 branches, and then repeat the process as desired.
4. **Deep-cutting off or rejuvenating pattern:** (i) *For healthy tree but too tall to harvest:* Prune around strong branches and leave some branches to protect the tree from sunlight. (ii) *For old tree with poor structure:* Prune heavily to have tree trunk of 1–2 m high. Keep some branches with leaves as the source of nutrient and cut off unhealthy ones to create new structure.

7.5.2 LONGAN

7.5.2.1 Longan Blooming Data

Bloom at terminal buds of the branch. Flowers may not bloom at the same time. Natural blooming period is from December to February, depending on the cultivars, environment, and climatic condition.

7.5.2.2 Architectural Pattern of Longan Tree

The following architectural patterns are recognized:

1. **Standard or half-circle shape:** Allow sunlight to penetrate through all tree structures. Growers can control tree height by pruning 1–2 upper branches to reduce tree height and allow the sunlight to shine through the canopy. Outward and unwanted branches can also be pruned.
2. **Short-tree shape:** The principle is to prune the mid-bush and leave the outward branches of 1–2 layers for young trees. For old trees, outward branches should be left between 3 and 4 layers to leave some room for flowers to bloom. For upward bush pruning, leave the length of about 10 cm, and the canopy height should be between 1.3 and 1.5 m, which is suitable for close-space planting of less than 4 × 6 m tall trees.

3. **Square shape:** Prune off outward side branches by 30 cm. Reduce the size of the tree from 2.5 m to 1.7 m. Suitable for close-spaced trees that need high and outward side branches to be pruned. This practice provides highest area for flowering and fruiting. It is suitable for machinery use.
4. **Open-center shape:** Cut off 2–5 mid-branches to reduce the height of the tree to allow sunlight to penetrate into the canopy. Small branches where sunlight cannot reach should also be pruned. It helps to reduce the height of the tree with spacing of more than 4 × 4 m.
5. **Old longan tree**
   a. **To reduce the size of wide structure and height:** Similar to standard pruning process, but there are more branches to be pruned, especially the overlapped ones. The main branches should be kept while conducting major pruning as it will help to reduce burning from sunlight.
   b. **To encourage new shoots to develop:** Pruning is required during the first year and 3–4 main branches should be kept. It will create a better architecture for the tree. However, there are some disadvantages, namely: (i) there is no yield during the first year; and (ii) during the second year of the harvest period, the risk of the branches with high yield to be torn is high.

7.5.3 RAMBUTAN

7.5.3.1 Rambutan Blooming Data

Full-grown rambutan trees can be 10 m tall. Shorter trees are preferred for easy management because rambutan tree flowers at terminal buds of sub-branches.

7.5.3.2 Architectural Pattern of Rambutan Tree

7.5.3.2.1 Open Center

About 2–3 main branches should be left. Ropes are used to pull down the branches. New branches will not be tidy. Pruning will be needed again to create a tree balance.

7.5.3.2.2 *Medium Pruning*

It is good for old trees with the problem of fungal diseases. Thus, the trees should be cut down to make them shorter so that the canopy receives more light to allow fungicide and pesticide spraying as well as sunlight to reach inside of the canopy.

7.5.4 LYCHEE

7.5.4.1 Lychee Blooming Data

Lychee is a perennial plant similar to rambutan and longan. It blooms at terminal bud of the full-grown branches.

7.5.4.2 Architectural Pattern of Lychee Tree

Top branch should be pruned so that axillary buds would develop. Open-center training technique should be initiated. Pruning can be done when axillary buds are 25 cm long. Choose 3–4 lateral branches with a wide angle. The lead branch should be pruned off, leaving a 15-cm long branch. Other branches should grow to 25 cm, and then, choose the ones that are parallel with 3–4 main branches that were chosen, and then prune them to 15 cm.

7.5.5 MANGOSTEEN

7.5.5.1 Mangosteen Blooming Data

Mangosteen tree has large canopy with thick shiny leaves on both sides. The trunk is straight with a pyramid shape with the height of around 10–15 m. Sub-branches would sprout from all directions. Flowers are borne at terminal of the shoot. Fruits of good quality are those borne inside the canopy. Light pruning is needed on all sides and at the top of the canopy so that the fruits are able to receive sunlight, which results in high-quality fruits.

7.5.5.2 Architectural Pattern of Mangosteen Tree

Naturally grown canopy of mangosteen tree has a pyramid shape. Only one strong leader branch should be left. Prune off terminal branches. Where sub-branches overlap, cut off infected and unhealthy branches, leaving only sub-branches in the inner canopy for flowering. Any branches that bend down to the ground should also be cut off.

7.5.6 DURIAN

7.5.6.1 Durian Blooming Data

The tree has a long straight trunk, 25–50 m high. It has 3–30 flowers/cluster, borne from main or small branches or trunk, 5–7 cm. Flower is bell-shaped, blooms only once or twice a year. Fruits are borne when the tree is 4–5 years old.

7.5.6.2 Architectural Pattern of Durian Tree

7.5.6.2.1 Pyramid or Central Leader

Prune out newborn branches and leave the main branches to protect the tree from sunburn. In the next year, pruning should be done on the side to maintain new shape of the fruit tree and to keep the tree at a stable height, which would be easy to manage and the tree can hold the fruits better.

7.5.6.2.2 Modified Pyramid Shape

The tree should have 15 or more main branches. If the tree is higher than expected, pruning should be done after the harvest.

7.5.6.2.3 Heavy Pruning or Rejuvenating Technique

This is used for old durian trees. The top main branches can be pruned off step by step, from year-to-year until the tree reaches a satisfactory height and is easy to work on. Next year, side pruning should be done. During the third year, the tree's shape will be steady, resulting in quality fruits.

7.5.7 ROSE APPLE

7.5.7.1 Rose Apple Blooming Data

Rose apple tree has newborn branches all year round. Flowers are borne from the branches inside the canopy. The branches would be more than one season old. Cut off parts of lateral branches and branches that are pointing upward to allow the light to shine through the canopy. Newborn branches should also be cut off continuously. Prune off newborn branches and leave main branches to protect the tree from sunburn. In the next year, pruning should be done on the side to maintain new shape of the fruit tree and to keep the tree at a stable height, which would be easy to manage and the tree can hold the fruits better.

7.5.7.2 Architectural Pattern of Rose Apple Tree

7.5.7.2.1 Open Center

Prune the mid-branch when the tree is still young at the height of 60–100 cm. Leave out 4–5 main branches with the space of 20–30 cm between each other. There would be no mid-branch, but only side main branches.

7.5.7.2.2 Modified Open Center

The leader branch remains, while the mid-branch would be left until it reaches 150–180 cm. Then, the tip would be pruned, leaving the side branches to spread. The lowest branch is 60–75 cm and the tallest is 150–180 cm.

7.5.8 PUMMELO

7.5.8.1 Pummelo Blooming Data

Flower blooms at terminal bud of new shoot. Once the flowers bloom, the tree will take approximately 7–8 months to harvest, depending on the cultivars. Direct sunlight is needed for newborn branches as well as around the canopy. Pruning should be done annually.

7.5.8.2 Architectural Pattern of Pummelo Tree

Based on modified open center technique of pruning, pummelo trees with good height allow easy management. Plenty of fruits are borne.

7.5.9 GUAVA

7.5.9.1 Guava Blooming Data

Flowers are borne from axillary buds of newborn branches. The technique to induce flowering is to cut the edge of the green branch by cutting out the fourth pair of leaves from the tip of the branches. Prune the branches near the main branches, leaving 3–4 buds for new flowers to emerge. Pruning is the best practice to stimulate new branches and new flower buds to form and to obtain quality fruits.

7.5.9.2 Architectural Pattern of Guava Tree

Guava produces fruits when its branch is not more than one-year old. Pruning is required to create 3–4 main branches. The first main branch should be 50–80 cm above the ground; pruning is required for the branches to spread out equally. Pruning is required to stimulate new branches and new flower buds to form and to obtain fruits. When the tree is fully grown, pruning is done to control the shape of the canopy to receive maximum sunlight. For faster flowering, bending the branches as well as pruning is required for the new branches to form.

7.6 DISCUSSION

Fruit tree architectural pattern involves pruning and training – an art and science in which the growers must understand the basic concept and physiology of the fruit trees. How pruning and training techniques are used depends on factors such as type of tree, time, environment, soil, weather, and purpose of the growers. Consequently, growers should be careful in selecting appropriate techniques of pruning and training in their orchards.

KEYWORDS

- **architecture**
- **fruit**
- **pruning**
- **training**
- **tree**
- **tropical**

REFERENCES

Carlson, R. F., (1982). Fruit tree training and pruning. *Compact Fruit Tree*, *15*, 96–98.

Chomchalow, N., & Na Songkhla, P., (2014). *Tropical Fruit Tree Architectural Management in Thailand.* Kehakaset Magazine, Bangkok, Thailand (in Thai).

Domoto, P. A., (1991). *Pruning and Training Fruit Trees.* Iowas State Univ. Extension, Aimes, Iowa, USA.

Setpakdee, R., (1997). *Fruit Tree Architecture.* Kehakaset Printing, Bangkok, Thailand (in Thai).

Somerville, W., (1996). *Pruning and Training Fruit Trees.* Inkata Press, Melbourne, Victoria, Australia.

Wangnai, W., & Poonsri, P., (1999). *Pruning and Training.* 5th Thailand Encyclopedia for Youth, Bangkok, Thailand (in Thai).

Wanitchkul, K., (2003). *Training and Pruning Fruit Tree Architecture.* Kasetsart University Publ., Bangkok, Thailand (in Thai).

CHAPTER 8

ORGANIZING MARKET LED EXTENSION IN HORTICULTURE FOR LESS FAVORED AREAS (NEH —NORTH EASTERN HILL REGION): CHALLENGES AND OPPORTUNITIES

V. K. JAYARAGHAVENDRA RAO

ICAR-National Academy of Agricultural Research Management (NAARM), Hyderabad–500030, India

CONTENTS

North Eastern Region (NER) is home to some niche crops like Assam lemon, *Joha* rice, medicinal rice, and passion fruits that have high market demands. It accounts for 45% of total pineapple production in India, and an Agri-Export Zone (AEZ) is already set up in Tripura. Sikkim is the largest producer of large cardamom (54% share) in the world. It is the fourth largest producer of oranges in India. Best quality ginger (low fiber content) is produced in this region, and an AEZ for ginger is established in Sikkim. The extent of chemical consumption in farming is far less than the national average. Approximately 18 lakh ha of land in NER can be classified as "Organic by Default." There is thin population density per square kilometer (13–340 compared to 324 at the national level), and mid- and high-altitude farmers are dependent on within farm renewable resources.

Markets are of fundamental importance in the livelihood strategy of most rural households, rich and poor alike. Markets are where, as producers, they buy their inputs and sell their products, and where, as consumers, they spend their income from the sale of crops or from their nonagricultural activities to buy their food requirements and other consumption goods. Because of this, rural poor people in many parts of the world often indicate that one reason they cannot improve their living standards is that they face serious difficulties in accessing markets. Low population densities in rural areas, remote location, and high transport costs present real physical difficulties in accessing markets. The rural poor are also often constrained by their lack of understanding of the markets, their limited business and negotiating skills, and their lack of an organization that could give them the bargaining power they require to interact on equal terms with other larger and stronger market intermediaries. Furthermore, rural producers from developing countries face significant impediments in accessing rich countries' markets. To help the rural poor access efficient and more equitable markets, there are three types of interventions: field operations, development and sharing of knowledge, and policy advocacy. In field operations, we seek to reduce the transaction costs between poor rural producers and private-sector intermediaries. This includes supporting the establishment of commercially oriented producer organizations (groups, associations, and cooperatives), helping and training producers to identify new markets, linking farmers with traders and processors, constructing and improving rural roads, building market information systems, etc. Developing the fund's knowledge in the area of market linkages requires improving the process of learning from its own projects. It also requires establishing effective monitoring and evaluation systems, working closely with cooperating institutions to improve impact assessment and supervision, and strengthening partnerships with a range of different players. We must also be active at the national, regional, and international levels, promoting a global policy environment that increases market access for the rural poor. The crucial role of market linkages for rural poverty reduction has only recently received the attention it deserves in the development arena. More needs to be done, especially on the implementation side. This paper is committed to the objective of improving the rural poor's access to markets, and in this context, is seeking ways to: effectively increase the market share of the rural poor and improve the terms in which they participate in markets; achieve greater market access and market development for the rural poor; and effectively improve at the national, regional, and international levels the rules of trade in favor of the rural poor.

8.1 HOW TO GO ABOUT IT?

Our control theme of extension requires change from production oriented to market oriented. This is because many times, only production-led extension and support system cannot fulfill the farmers need of income as well. The results of several studies show that only increase in production cannot be a suitable solution for farmers' economy. On the other hand, farmers who plan their production according to the market got better returns and profit. In this way, currently, the acceptance and requirement of market-led extension over conventional extension is higher. The basic thought behind the market-led extension is quality improvement with appropriate concern over price. The equation may be increase quality and reduce cost/price major aspects of market-led extension through:

8.1.1 AGRICULTURE POLICY AND ACTS

Minimum support price of major agricultural commodities was fixed by the central government through its agriculture cost and price commission. Other important acts are passed by the concerned state government wherein each state has a State Agriculture Marketing Board, which is the nodal agency for agriculture product trading.

8.1.2 SUPPLY CHAIN MANAGEMENT

Effective supply chain management (SCM) results in lower transaction costs and increased margins for which supply chain parties work together with interdependence with customers, retailers, wholesalers/distributors, manufacturers, and raw material suppliers.

8.1.3 MARKET INFORMATION SERVICES

Market information service is a continuing and interacting structure of people, equipment, and procedures together, which involves sorting, analyses, evaluation, and distribution of pertinent and accurate information for use by marketing decision makers to improve their marketing planning, implementation, and control.

8.1.4 MARKET INTELLIGENCE

Allows to remain competitive by improving strategic decision and leads to better performance against competitors through market research. Websites for market information: 1. Agrisurf, 2. NETVET, 3. Agriwatch, 4. Commodity India, 5. Agfind, 6. Agmarknet, 7. Hortiindia, 8. APEDA, 9. NCDEX, 10. e-CHOUPAL, and 11. Commodity Board.

8.1.5 CROP INSURANCE

Several plans are made available like (i) crop revenue coverage, (ii) revenue assurance, (iii) group revenue insurance policy, (iv) income protection, and (v) group risk plan.

8.1.6 CONTRACT FARMING

A contact is an agreement between two stake holders through a (i) centralized model, (ii) nucleus estate model, (iii) multiparty model, (iv) informal model, and (v) intermediary model.

Success stories of contract farming:

- Potato, tomato, groundnut, and chili in Punjab;
- Safflower in Madhya Pradesh;
- Oil palm in Andhra Pradesh;
- Seed production contracts for hybrid seed companies in Karnataka;
- Cotton in Tamil Nadu and Maharashtra;
- Bengal potato growers.

8.1.7 PROCESSING AND VALUE ADDITION

Converting crop produce into semi- and final products like juice, jelly, and jam.

8.1.8 POSTHARVEST MANAGEMENT

Management for reduction in post harvest losses.

8.1.9 COMMODITY MARKETING AND FUTURE TRADING

for price risk management, forward contracts have become the primary trading instruments in commodity markets. The commodities are traded in commodity exchanges, keeping a future date of harvest in mind at a predetermined price, it is just like hedging in mango. In futures contracts, the buyer and the seller stipulate product, grade, quantity, location, and leaving price as the only variables.

8.1.10 SWOT ANALYSIS OF MARKET

Strengths (demand, high marketability, good price, etc.), weaknesses (the reverse of the above), opportunities (export to other places, appropriate time of selling, etc.), and threats (imports and perishability of the products, etc.) need to be analyzed about the markets. Accordingly, the farmers need to be made aware of this analysis for planning production and marketing.

8.1.11 INTERNATIONAL TRADING AND IMPLICATIONS OF WTO REGIME ON AGRICULTURE, ETC

Agreement on agriculture (AoA), improved market access through tariffication, improved domestic support, and reduced export subsidies.

8.2 ROLE OF EXTENSION PERSONNEL IN LIGHT OF MLE (Maximum Likelihood Estimation)

1. Doing SWOT analysis.
2. Organization of farmer interest group (FIG).
3. Enhancing the capabilities of local established groups.
4. Enhancing the communication skills of farmers.
5. Establishing market linkages between farmers groups, markets, and processors.

6. Helping in production and marketing plan.
7. Educating farmers to establish agriculture as enterprise.
8. Educating farmers about direct marketing.
9. Capacity building of FIGs
10. Expose the farmers to acquire complete market intelligence.

8.3 CASE STUDY

8.3.1 ELEGANT ANALYSIS OF THE IMPACT OF THE CELL PHONE ON FISH MARKETING IN KERALA, SOUTHWESTERN INDIA, BY JENSON

With cell phone calls possible, off-shore fishermen call different beach auctions to decide the best market on that day. The result is that one-third of fishermen take fish to more distant auctions. If farmer/fisherman takes risk and sell it in distant markets where he gets more price, he can increase his income, this requires management skills.

The result was that fishermen and consumer were winners, as the market became more efficient and rational.

8.3.2 UTTAM BANDHAN

- Serve entire value chain through FIGs and FOs, including soil testing for micronutrients, balance use of fertilizers, advisory, input supply and marketing
- 50:50 cost sharing between ATMA and Chambal for the services
- Developing an organic linkage through Uttam Krishi Sevak (UKS) via continuous feedback
- Dissemination and extension of agricultural services using ICTs at FIAC established by ATMA and operated by UKS.

8.4 CONCLUSION

The focus of the extension functionaries need to be extended beyond production. Farmers should be sensitized on various aspects on quality, consumer preferences, market intelligence, processing and value addition, and other marketing information. This would certainly promote and motivate farming community toward profitable agriculture.

KEYWORDS

- **extension**
- **income**
- **market**
- **production**
- **promotion**
- **risk**

REFERENCES

Gauraha, A. K., Lakpale, N., & Hulas, P., (2012). *Training Manual on Model Training Course*. On market led extension. Organized at directorate of extension, services, IGKV, and Raipur from, pp. 156.

Indira Gandhi Krishi Vishwavidyalaya. http://igau.edu.in/pdf/pubdes18.pdf (accessed Jun 23, 2018).

International Fund for Agricultural Development. http://www.ifad.org/gbdocs/gc/26/e/markets.pdf (accessed Jun 23, 2018).

Jayaraghavendra Rao, V. K., (2013). In: AEM 201 "Market led extension" PGDAEM course material published by National Institute of Agricultural Extension Management, Rajendra Nagar, Hyderabad–500 030, Andhra Pradesh, India. pp. 237.

National Centre for Agricultural Economics and Policy Research. http://www.ncap.res.in/upload_files/new.pdf (accessed Jun 23, 2018).

CHAPTER 9

TRANSFORMING MARKETING LINKAGES IN HORTICULTURE THROUGH EFFECTIVE INSTITUTIONS AND INNOVATIVE AGRI-BUSINESS MODELS

VASANT P. GANDHI

Centre for Management in Agriculture Indian Institute of Management, Ahmedabad, India

CONTENTS

ABSTRACT

India is among the world's largest producers of horticultural crops, but marketing linkages in horticulture remain extremely weak. Huge differences between what the consumers pay and what the farmers receive are very common, and margins are often found to be as high as 80–90%. This reflects not only unreasonable profits but also marketing inefficiencies because of which both the farmers and consumers suffer. Transformation of the marketing linkages remains a major concern and the challenges for solutions to this include: how to organize sustained production and procurement from large numbers of small farmers, how to ensure adoption of the right technology and practices by farmers to generate quantity and quality output at a reasonable cost, how to ensure investment in the state-of-the-art processing technology and meet working capital need, how to deliver a strong marketing effort to reach consumers and open nascent markets, and how to bring inclusive management and control to ensure win-win across stakeholders including farmers, consumers, and investors. To address these challenges, effective institutions and agri-business models are a must, and it is heartening to see that a number of promising models have emerged in India. These include the HPMC model, the AMUL model, the Pepsi model, the E-choupal model, the McCain model, the Heritage model, the Suguna model, the Desai Cold Storage model, etc. This chapter uses the available literature and data to study and compare these models to see how and to what extent they meet the above mentioned challenges. The lessons and findings are found to be very useful in indicating ways for transforming the marketing linkages of horticultural crops and for designing supportive policies and practices that would lead to sustainable horticulture development.

9.1 INTRODUCTION

Agribusinesses and agri-value chains have been growing at a fast pace in India in the recent years. They have been given high priority by the government due to their significant potential for bringing value addition to agricultural output, and enhancing small farmer income and rural employment. The priority can be traced to the vision of the father of the nation, Mahatma Gandhi, who as early as 1920s, saw agro-industries and agri-value chains as extremely important for India's development and the independence movement, see Table 9.1. Even today, they are given substantial importance (India Planning Commission, 2008) due to various national priorities including enhancing value addition to agricultural output, rural employment, income, and food availability, and alleviating hunger and poverty. The sector, however, faces numerous difficulties including sourcing of quality raw materials, rural market imperfections, supply-chain inefficiencies, financial constraints, and product marketing challenges (Srivastava and Patel, 1989; CII-Mckinsey, 1997; Gandhi, Kumar, and Marsh, 2001; Goyal, 1994).

9.2 PHASES OF DEVELOPMENT

The development of agribusinesses and agri-value chains has gone through three important phases in India as shown in Table 9.1. The first phase was Mahatma Gandhi's village-based agro-industries approach, founded mainly on a strong social and political ideology (Goyal, 1994). This "*swadeshi*" (our own country's) concept saw a significant role for them in connecting the rural masses (90% of the population at that time) to the mainstream of the Indian economy as well as to the independence movement. It also sought to reduce the dependence on imports from Britain. After playing a major role, it later failed mainly because it became a blanket basis for nationalists to favor traditional industries over modern industry despite less efficiency and incompatibility with market demand (Gandhi and Jain, 2011).

In the second phase, spanning from India's independence to the early 1980s, the policy was largely dominated by the ideas of Prime Minister Nehru and his central planner Mahanalobis, who argued that India must give high priority to the capital goods sector consisting of large-scale public industries, while the consumer goods sector should be confined to small-scale and agro/rural industries that required less capital and more labor. Regulations and

TABLE 9.1 Phases of Development of Agro-Processing and Agribusiness in India

Phases	Features	Shortcomings
Pre-Independence: Mahatma Gandhi-*Swadeshi* Phase – upto 1947/1950	Encourage the use of own rural output	Less efficient techniques of production
	Discourage imports	Incompatibility with market demand
	Generate rural employment and incomes,	Basis for opposing modern industry
	Connect and uplift rural masses to mainstream of economy, independence movement	
	Fight against colonial rule	
After-Independence: Nehru-Mahanalobis Phase: 1950-1984	Industrialization strategy	Limited scale
	Capital goods reserved for large scale public industries	Inefficient technologies
	Consumer goods reserved for agroindustries/small scale	Inability to meet mounting demand for quality goods of growing population with rising incomes
	Logic of acute capital scarcity – agro/small scale industries need less capital	Supply shortages and lack of competitiveness
	Agro/small scale labour intensive, generate more employment	
Modernization Phase: 1984-onwards (mainly 1991- onwards)	Economic liberalization	Large scale
	Focus on efficiency – modernization, competitiveness	Private, capital-intensive
	Attract foreign investment	Weak value chain linkages with rural areas and small farmers
	Focus on quality – use best technology	Low contribution to rural employment
	Focus on meeting consumer demand	Weak development linkage – for which agro-industries were given priority

policies were accordingly designed, and this was consistent with acute scarcity of capital at that time and the need to increase employment. However, such small-scale industries, due to their limited scale and inefficient technologies, were unable to meet the mounting demand for quality consumer goods as population grew rapidly and incomes increased, making India a country of shortages and also one that lacked competitiveness in industry.

This forced the third phase from the mid 1980s and particularly after economic liberalization in early 1990s, in which the emphasis shifted to modernization, deregulation, and opening out to competition. The industry transformed toward better technology, meeting market demand for quality and quantity, efficient management, and competitiveness. However, this trend is often seen to lead toward large, private, capital-intensive enterprises often with weak value chain linkages with rural areas and small farmers. The result is a negative outcome for rural employment, and a weakening of the development linkage for which agro-industries have been given high priority in the first place in India.

9.3 OVERVIEW OF THE HORTICULTURAL ECONOMY OF INDIA

India now ranks first in the world in the combined production of fruits and vegetables. Out of 654 million tons of fruit production in the world, India accounts for 89 million tons. Of the 1159 million tons of vegetables produced in the world, India produces 162 million tons and giving it a share of 14%. The horticultural crops in the country presently cover 24.2 million hectares of land, i.e., 5% of the gross cropped area and contribute 33–36% of the gross value of India's agricultural output. India is the largest producer of mango, guava, banana, and papaya in the world and has third position in the production of orange, fifth in the production of apple, sixth in production of pineapple, and ninth in the production of grapes (Table 9.2). Among major vegetables, India occupies the first position in okra and brinjal production and second in tomato, onion, potato, and cauliflower.

9.4 FEATURES OF AGRI-VALUE CHAINS AND AGRO-INDUSTRIES IN INDIA

CII-McKinsey (1997) indicated that there is a tremendous scope and potential for the development of agri-value chains and food industries in India.

TABLE 9.2 Production of Major Fruits and Vegetables: India's Position in the World, 2013-14

Sr. No.	Fruits/ Vegetables	Production (000' MT)		India's Share	India's Rank
		India	World		
	Fruits				
1	Mango and Guava	22099	48989	45.1	1
2	Banana	29725	106848	27.8	1
3	Apple	2498	76673	3.3	5
4	Pineapple	1737	23615	7.4	6
5	Papaya	5639	12890	43.7	1
6	Orange	3886	67110	5.8	3
7	Grapes	2585	68412	3.8	9
	Total Fruits	88977	654449	13.6	2
	Vegetables				
1	Tomato	18736	163030	11.5	2
2	Onion	22600	85944	26.3	2
3	Brinjal	13558	49782	27.2	1
4	Okra	6346	8706	72.9	1
5	Potato	41555	365438	11.4	2
6	Cabbage	9039	70644	12.8	2
7	Cauliflower & Broccoli	8573	22840	37.5	2
	Total Vegetables	162897	1159890	14.0	2

Source: Horticultural Statistics 2015 & National Horticultural Database 2015, Department of Agriculture & Cooperation, Ministry of Agriculture, GOI.

However, there are numerous constraints to its growth and development. These have been brought out by Boer and Pandey (1997), Gulati et al. (1994), Kejriwal (1989), Srivastava and Patel (1989), Gandhi and Jain (2012), and Gandhi, Kumar, and Marsh (2001). These include the following:

- Raw material supply constraints
- Poor quality, inappropriate varieties, residues
- Short period of availability – seasonality
- Small producers, scattered supplies, perishability
- Competing markets – large market for fresh products
- Constraints in processing

- Old technology – poor efficiency, quality
- Poor capacity utilization due to seasonality
- Unsuitable for export or high value markets
- Constraints in marketing
- Limited market size/nascent markets, changing customer preferences
- High product and brand development costs
- Long inefficient supply chains, small retail stores
- Financial constraints
- Requires more working capital hard to get, higher interest rates
- High investment requirements for latest technology
- Government policy
- Processed/packaged foods considered as luxuries taxed heavily – affects the economics
- Many special regulations – e.g., MMPO: Milk and Milk Product Order
- Squeeze between input price support and output price control
- Ad hoc export and import controls

9.5 INEFFICIENCIES IN HORTICULTURE MARKETS AND VALUE CHAINS

Findings of a study of fruit and vegetable marketing in the city of Ahmedabad as well as in Ahmedabad, Chennai, and Kolkata (Gandhi and Namboodiri, 2006) provide an analysis of the farmer to consumer price difference, the marketing cost and the implicit profit margins for fruits and vegetables. The analysis shown in the table below indicates that for vegetables, the cost frequently amounts only about to about 10–20% of the price difference. The profit margin, on the other hand, comes out very high and is frequently 80–90% of the price difference. This is indicative of possible large trader profits and relatively poor marketing efficiency including substantial wastage and spoilage (Table 9.3).

Similar results for fruits are given in the table below. The results indicate that the costs amount frequently to only about 20% of the price difference, with the exception of apple where it amounts to only 6–7%. The profits margin seems to be very high and amounts frequently to 80% of the price difference, and in the case of apple to 93%. This is indicative of high profits and relatively poor market efficiency.

TABLE 9.3 Vegetables: Farmer-Consumer Price Difference, Percentage Marketing Cost and Profit

	Farmer-Consumer Price Difference Rs./ unit		Marketing Cost Rs./ unit		Cost Over Price Difference %		Profit Margin Over Price Difference %	
	Min	Max	Min	Max	Min	Max	Min	Max
Potato (G)	311.85	382.13	71.00	78.74	22.77	20.61	77.23	79.39
Onion (OG)	246.06	265.73	92.27	99.49	37.50	37.44	62.50	62.56
Tomato (OG)	873.20	1297.82	153.55	179.51	17.58	13.83	82.42	86.17
Cabbage (G)	411.71	563.77	83.33	100.40	20.24	17.81	79.76	82.19
Cabbage (OG)	432.24	624.25	106.21	122.17	24.57	19.57	75.43	80.43
Cauli flower (G)	1001.94	1211.40	113.94	129.89	11.37	10.72	88.63	89.28
Cauli flower (OG)	1052.82	1277.46	144.78	168.54	13.75	13.19	86.25	86.81
Brinjal (G)	486.58	712.42	91.14	102.38	18.73	14.37	81.27	85.63
Green pea (OG)	592.67	1050.17	219.20	272.33	36.99	25.93	63.01	74.07
Lady's finger(G)	746.65	885.40	126.22	160.34	16.90	18.11	83.10	81.89

Note: Min is at the lowest price reported and Max is at the highest price reported, and reflects quality difference.

(Reprinted with permission from Gandhi, V. P., & Namboodiri, N. V., (2006). "Fruit and vegetable marketing and its efficiency in India: A study of wholesale markets in the Ahmadabad area," CMA monograph, Indian Institute of Management, Ahmadabad. 2016.)

The study finds that the extent of contact between farmers and commission agents is low and needs considerable improvement. It also shows that the adoption of open auctions in the markets is very low and much potential for gain in market efficiency has not been realized. The study finds that the share of the farmer in the consumer rupee works out to only 48% for vegetables and 37% for fruits. Further, the explicit marketing costs work out to only a very small percentage of the price difference between the farmer and the consumer, and the profit margin works out frequently to 80–90% of the price difference. These figures are indicative of relatively poor efficiency of the marketing system despite the presence of the APMC and the regulated markets.

9.6 CHALLENGES AND MODELS FOR AGRI-VALUE CHAINS AND AGRIBUSINESSES

The challenges and complexities arising from all these constraints faced by the sector as well as its peculiarities such as variability, seasonality, and perishability, and the need to meet different conflicting goals such as profitability and development, raise the need for innovative approaches and institutional models for the organization of agri-value chains and agribusiness activity in India (see Gandhi and Jain, 2011; Gandhi, Kumar, and Marsh, 2001).

Based on the literature and the experiences in India, a set of key success factors have been identified that appear critical for sustainable success of integrated agribusiness enterprises and the sector in India. These are as follows:

1) Strong procurement system: The produce is generally cheapest at the farm level. The success in reaching large numbers of small farmers and organizing production and procurement from them (direct sourcing) keeping transaction costs low is critical, with further advantages of obtaining the right quality, and achieving economic competitiveness, with high impact on rural incomes and employment.
2) Transforming agricultural technology and practice: The ability to transform farming by bringing adoption of new technology/practices by the farmers plays a major role in the production of abundant quantity and required quality of raw material at a reasonable cost, also creating a win-win situation for raising farmers' income.

3) Capital resources and investment in the best processing technology and operation: The ability to invest in state-of-the-art processing and handling technology to deliver high-quality food/agro products, meeting its high fixed and working capital needs given the characteristics of perishability, variability, and seasonality, is very important.
4) Delivering a strong marketing effort: The ability to deliver a strong marketing effort to reach large numbers of consumers with perpetual need, address consumer preferences, create and open nascent markets especially for processed foods and compete with traditional systems, and competitors.
5) Share benefits across the chain: Sharing benefits or profits across the often long food value chain is found critical for sustaining success. Building a strong committed value chain that consistently delivers benefits and performance across all stakeholders including farmers, supply-chain members, investors, consumers, and the economy is important. This often requires appropriate representative management, control, ownership, and institutional structures.

These features can be used to examine alternative models and approaches. They may seem a tall order. But a number of successful agribusiness models have emerged in India that meet many of these conditions and show their relevance and the way. The models are examined and broadly evaluated below to derive lessons for food value chains and agribusiness development.

9.7 GOVERNMENT ORGANIZATION MODEL: HPMC

In this model, the government or a government-owned corporation is the major player. One of the well-known examples of this is the Himachal Pradesh Fruit Processing and Marketing Corporation (HPMC). The corporation is fully owned by the government and is managed by government staff. The corporation sets up a network of infrastructure and processing facilities including produce collection centers, warehouses, cold storages, and processing plants. The produce is purchased from the farmers at the announced prices. It is then stored, processed, and marketed nationally by the government corporation. The marketing of fresh produce is typically up to the wholesalers' level only, and trade by wholesalers and retailers remains private. The HPMC has set up two collection centers, three warehouses, and five cold storages in the state of Himachal Pradesh, principally for apples.

It has also set up cold storages in the metropolitan cities of Delhi, Mumbai, and Chennai.

Even though the HPMC was fairly successful at one time, reports indicate that it has not been able to sustain this high performance (Vaidya, 1996). Lately, it has been unable to neither attract enough farmer suppliers nor expand distribution beyond its own outlets. While government-owned agro-industries are well funded for investment in infrastructure and technology and have government support, they depend on bureaucrats for management often with limited business skills. Managers are frequently transferred at the whims of changing governments and are accountable primarily to their superiors and neither to farmers nor consumers for their performance. They demonstrate little commitment to procure from small farmers, on the one hand, and to meet dynamic marketing demand, on the other hand, thereby blocking the long-term success of the enterprise.

9.8 THE AMUL COOPERATIVE MODEL

A model that has been very successful in the dairy agribusiness in India is the AMUL cooperative model. This has evolved out of a successful dairy cooperative initiative of 1946 in the Kaira district of Gujarat state. The model and its methods were perfected under the leadership of its enlightened chairman, Tribhuvandas Patel, and its competent professional manager, Dr Varghese Kurien. It has grown enormously over the years, spawning many cooperatives and a state cooperative federation that now markets milk and milk products across the whole country.

In this model, the ownership rests with the farmers on a cooperative basis. It typically has a three-tier organizational structure, with primary cooperatives at the village level, cooperative unions at the district level, and a cooperative federation at the state level. The milk is produced by the farmers/milk producers in the villages, and the village cooperatives procure the milk from them. The district union collects the milk from the village cooperatives and transports it to its owned dairy plant and undertakes its processing. The state federation undertakes the marketing of the milk and milk products of all the unions nationally. The organizations are governed at the top by farmer-elected rotating boards/managing committees who confine themselves only to periodic strategic and policy decisions. The operational management is entrusted entirely to professional managers/staff who are largely independent and highly empowered. Apart from the milk business, the cooperative also engages substantially in providing technical services and inputs to the farmers such as veterinary and breeding services, nutritional feeds, and extension services. These enhance cohesion and commitment in the organization and help its long-term growth and development. The profits are shared as bonus or reinvested in the business. Due to its impact, the model is supported by the government and international donors.

The base of the model is the village cooperative society that consists of milk producer members–shareholders, and an elected managing committee consisting of 9 to 12 voluntary representatives with an elected chairperson. The managing committee appoints a paid secretary and staff for day-to-day operations. The cooperative society collects milk from the milk producers and makes payments at district union-fixed prices based on transparent measurement of the quantity and quality of milk. It also provides a few services to the members such as veterinary first aid, artificial insemination (AI) breeding service, and sale of nutritious cattle feed. The village societies are members of the district-level cooperative milk union, represented by their chairpersons. The union is governed by an elected board of directors consisting of 9 to 18 representatives from village society chairpersons and an elected board chairperson. The board appoints a professional managing director and staff. The union collects the milk from village societies, sometimes chills it, and transports it to its own modern dairy processing plant. Here, it is pasteurized, stored, packaged, or processed into milk products. The union is also proactive in new initiation, training, and supervision of the village societies, and a number of important services including veterinary doctor services, AI breeding services, cattle feed supply and vaccination. The district unions are members of the state-level cooperative milk federation represented by their chairpersons. The federation is governed by a board of directors elected from among the union chairpersons, and an elected federation chairperson. The board appoints a professional managing director and staff. The federation undertakes and coordinates the marketing of the milk and milk products of the milk unions. The Amul structure is outlined in Figure 9.1.

The state federation, the Gujarat Cooperative Milk Marketing Federation (GCMMF), markets the milk and milk products under the popular brand names "AMUL" and "SAGAR" (Kurien, 2003) and has developed a massive network covering over 3500 dealers and 500,000 outlets. There are 47

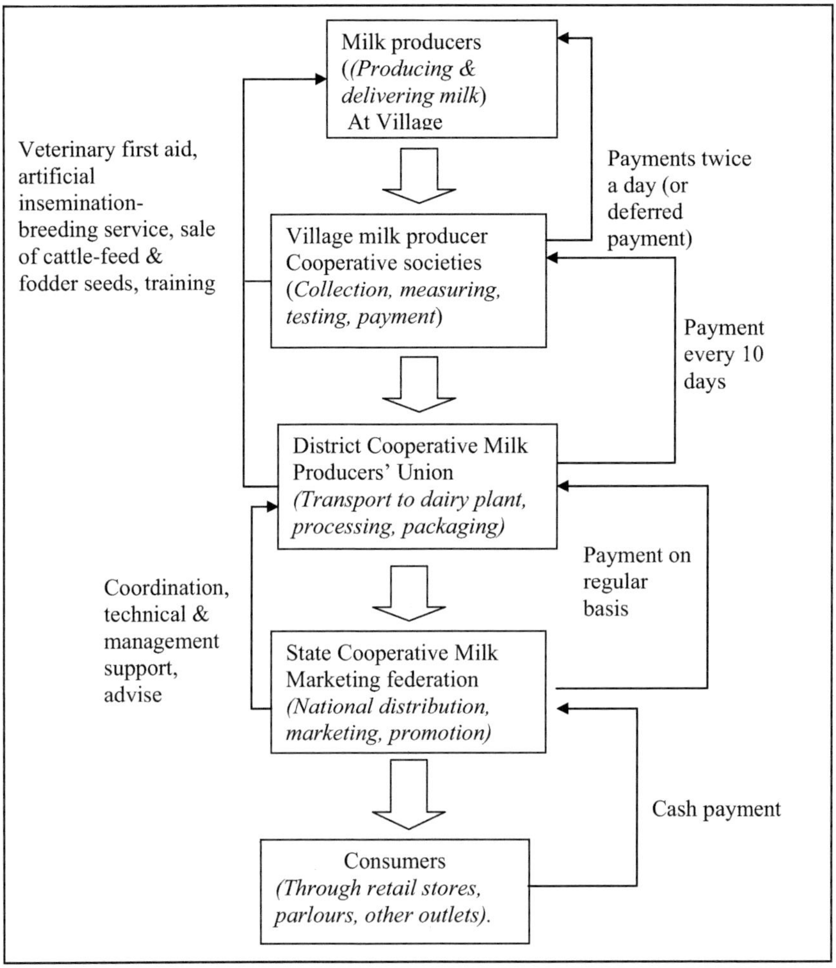

FIGURE 9.1 Outline of the AMUL model. (Reprinted with permission from Gandhi, V. P., (2014). "Growth and transformation of the agribusiness sector: Drivers, models, and challenges," Indian Journal of Agricultural Economics, 69[1].)

depots with dry and cold warehouses to carry the inventory. The distribution network comprises 300 stock-keeping units, 46 sales offices, 3000 distributors, 100,000 retailers with refrigerators, a 18,000-strong cold chain, and 500,000 non-refrigerated retail outlets. Products marketed include fresh milk, UHT milk, beverage milk drinks, infant milk, milk powders, sweetened condensed milk, butter, cheese, ghee, yogurt/curd, breadspreads, pizza, *mithaee* (ethnic sweets), ice creams, chocolate, and confectionery.

AMUL represents a model of an agribusiness enterprise that has ensured a high level of governance and business effectiveness. The model benefits from commitment of the farmers and cost-efficiency in raw material production as well as procurement, which have become its major competitive advantage. It also extensively engages with small farmers as well as the landless rural poor who may keep even 1–2 animals, contributing significantly to rural incomes and employment. However, some of its drawbacks include the need for enlightened and committed leadership, and of capable management, which is sometimes difficult to ensure. The board is elected and could become politicized and detract from sound business practices. Further, antiquated laws governing cooperatives often lead to government interference, and inability to use financial markets for raising cheap capital useful for expansion and growth.

9.9 THE NESTLÉ MODEL

Nestlé is one of the largest private food and beverages companies in the world. The company uses the milk district model for its agribusiness activity in India. Nestlé milk processing factory in the Moga district of Punjab produces milk powder, infant products, and condensed milk. In 2008, it covered about 100,000 farmers and had a procurement of 1.25 million liters milk/day. A milk district setup involves negotiating agreements with farmers for twice-daily collection of milk, establishing collection centers and chilling centers at larger community collection points or adapting existing collection infrastructure, arranging transportation from collection centers to the district's factory, and implementing a program to improve milk quality. Each of the six districts from which Nestlé sources raw milk are referred to as "Moga Milk Districts."

In the Nestlé or "Moga model," the job of sourcing milk from farmers is carried out by a private commission agent appointed by the company. Nestlé operates a network of 1100 agents who receive a commission on the value of the milk supplied to the dairy. Dairy farmers supply milk under contract, and the company maintains their records. The company has stringent quality specifications. Nestlé staff members regularly monitor milk quality and performance *vis-à-vis* contractual obligations, and the farmers obtain feedback on milk quality at the collection points. Company technologists determine quality in laboratories with samples being taken in the presence both of the farmers and the company representatives. Nestlé is not obliged to collect milk that does not meet the quality standards specified in the contract. The contract also allows the technologists to penalize the producer with a 30-day ban. If antibiotics are found, the price of milk is reduced by 15%. Repetition of any discrepancy is considered a serious breach of contract. Farmers have the right to complain through registers located at each collection point if they believe there is a problem. The system works because it provides an assured market for the farmers at remunerative prices for the milk.

9.9.1 COMPARISON OF THE NESTLÉ MODEL WITH THE AMUL MODEL

In terms of scale and reach, Nestlé's milk procurement pales in comparison with that of AMUL. During 2000–2001, AMUL's unions procured an average of 4.58 million kg of milk per day from over 2 million farmer members in Gujarat (Business Line, 2001), see Table 9.4. Every third liter leaving a milch animal's udders in the state was collected by the societies affiliated to AMUL. Nestlé's operations are much smaller and confined to districts around Moga. Nestlé's average procurement of 0.65 million kg per day covers barely 3% of Punjab's annual milk output. The average Nestlé farmer supplies about 7.25 kg of milk per day, whereas the figure for AMUL is about 2 kg per day, indicating AMUL's reach extends substantially to small/marginal farmers and landless farm laborer who may own only 1–2 milch animals.

With respect to price, Nestlé in 2000–2001 paid an average price of Rs 9.84 per kg, lower than the Rs 13–14 per kg that AMUL paid to its farmers. However, adjusting for the fat content, there is little difference between the farm gate prices paid by Nestlé and AMUL (Table 9.5). In 2000–2001, Nestlé's payments to Moga's farmers for milk as well as development inputs amounts to almost 47% of the value of the company's sales of milk products. In comparison, this proportion for AMUL and its unions is over 80%. Thus, a much larger share of the consumer rupee reaches the farmers in case of AMUL as compared to Nestlé. It must be noted that Nestlé is a company accountable to its shareholders and investors, while AMUL is an entity owned by and accountable to the farmers (Business Line, 2001).

TABLE 9.4 Fruits: Farmer-Consumer Price Difference, Percentage Marketing Cost and Profit

Fruits	Farmer-Consumer Price Difference Rs./ unit		Marketing Cost Rs./ unit		Cost Over Price Difference %		Profit Margin Over Price Difference %	
	Min	Max	Min	Max	Min	Max	Min	Max
Mango(OG)	899.33	1048.14	228.65	269.72	25.42	25.73	74.58	74.27
Apple(OG)	4548.81	6480.74	331.66	398.81	7.29	6.15	92.71	93.85
Sapota(G)	407.30	1028.94	175.55	223.16	43.10	21.69	56.90	78.31
Banana(G)	455.63	769.52	157.76	164.18	34.62	21.34	65.38	78.66
Sweet orange(OG)	37.51	43.09	7.91	8.25	21.09	19.15	78.91	80.85
Pine-apple(OG)	90.20	83.00	19.06	18.43	21.13	22.20	78.87	77.80
Pomagranate(OG)	1371.47	1242.08	211.21	294.79	15.40	23.73	84.60	76.27

Note: Min is at the lowest price reported and Max is at the highest price reported, and reflects quality difference.
(Reprinted with permission from Gandhi, V. P., & Namboodiri, N. V., (2006). "Fruit and vegetable marketing and its efficiency in India: A study of wholesale markets in the Ahmadabad area," CMA monograph, Indian Institute of Management, Ahmadabad. 2016.)

TABLE 9.5 Comparison Between Nestlé and Amul models

Feature	Nestle	Amul
Milk collection (mill kg/day)	0.65	4.58
No. of farmers covered (mill)	0.1	2.0
Percent of milk production collected	3	33
Avg. milk supply per farmer (kg/day)	7.25	2.0
Avg. price paid (Rs/kg)	9.84	13-14
Percent of consumer rupee reaching the farmer	47	80

Data source: *Business Line* 2001.

9.10 HERITAGE FOODS MODEL

The Heritage Group based in Andhra Pradesh was founded in 1992 by Chandra Babu Naidu, a former Chief Minister of Andhra Pradesh. It is a growing private enterprise with three business divisions: dairy, retail, and agri, under its flagship company Heritage Foods (India) Limited (HFIL). Heritage's milk products have a market presence in the states of Andhra Pradesh, Karnataka, Kerala, Tamil Nadu and Maharashtra, and it has retail stores in the cities of Hyderabad, Bangalore, and Chennai.

The company covers about 200,000 farmers and has the capacity to process 1.5 million liters of milk per day. The annual turnover reached Rs 34.7 million in 2006–2007. Heritage has established a supply chain that procures milk from farmers in rural areas, mainly in Andhra Pradesh and some parts of Karnataka, Maharashtra, and Tamil Nadu states. The Heritage model involves harnessing the current milk collection centers as well as rural retail shops to penetrate the rural market. Two-way or reverse logistics are used to transfer and sell goods from the urban markets to rural markets, and through this retail, milk procurement is also mobilized. This enables economies of scale in supply chain costs, serves both the rural customer and producer, and improves penetration in the rural areas. This also provides opportunities for Heritage to launch its private labels in rural markets. The company's rural retail network has increased to 1515 stores with 13 distribution centers. A typical rural store is about 10 square meters in size and is based on a franchise model to cater to villages with a population of less than 5000. The objective is to deliver popular fast-moving consumer goods (FMCG) products and quality groceries at affordable prices to interior villages across South India and leverage for the milk procurement network.

Apart from milk, vegetables and seasonal fruits are also procured through contract farmers and reach packhouses via collection centers strategically located in identified villages. The collection centers undertake washing, sorting, grading, and packing and dispatch to retail stores through distribution centers. Other features of the model include promotion of an annual crop calendar of sourcing that seeks to ensure regular supply and higher income per unit area, technical guidance – agri-advisory services, training of farmers, input supply and credit linkage, package of improved farm practices for better productivity and quality, an assured market at the doorstep, assured timely payments, and transparency in operations. The Heritage model provides an example of using the existing marketing points and chains for the purpose of agribusiness rather than building new/dedicated chains. This achieves faster roll-out and reach. It also provides an example of using two-way or reverse logistics for improving the efficiency and economics of the supply chain.

9.11 SUGUNA POULTRY MODEL

India has a rapidly growing poultry demand, and its size is now estimated to be around Rs 12 billion (Business Standard, July 2008). However, the poultry industry is highly fragmented and disorganized. In this disorganized sector, Suguna Poultry has emerged as one of the largest organized players and is believed to now rank among the top 10 poultry companies worldwide. The company is based in Coimbatore, Tamil Nadu, and has operations in 11 states in India, offering a range of poultry products and services. The company pioneered contract farming in the poultry industry in India and sources its products through 12,000 contract farmers across different states. Its fully integrated operations extend from broiler and layer farming to hatcheries, feed mills, processing plants, vaccination, and exports. Suguna sells live broiler chicken, eggs, and frozen chicken, and has set up a chain of retail outlets providing consumers with fresh, clean, and hygienic packed chicken. It has also implemented the Hazard Analysis and Critical Control Points (HACCP) system and has state-of-the-art processing plants.

In Suguna's business model, farmers who own land and have access to resources such as water, electricity, and labor can become growers of Suguna's Ross breed of chicks. Suguna takes the responsibility and provides all the other required inputs – day old chicks, feed, medicines as well as supervision to the farmers. Suguna also brings good management practices and technical know-how that lead to higher productivity. The method of growing the chicks is standardized and must conform to the exact standards laid down by the company; quality control checks are carried out by company staff to ensure the norms are being met. The broilers are procured by Suguna as long as they comply with established quality norms, and the farmer is paid a "growing" commission or charge. If a farmer does not comply with procedures as laid down or sells chickens to another party, this is considered as a breach of trust and the contract is unlikely to be renewed. Suguna also offers farmers a safety net: it bears production and market risks, taking responsibility for losses from a change in the market environment. A rise in the feed prices does not affect the farmers because they are supplied with feed directly by Suguna. Similarly, when the bird flu attacks occurred, Suguna absorbed the financial loss suffered by the farmers. Thus, farmers receive assured returns. Regardless of the market prices, the farmers receive the assured growing charge/cost and incentives.

The Suguna model offers fast scalability because the company does not have to buy or lease farms. It keeps costs low and offers economies of scale including in buying raw materials, feed, and medicines. Suguna has benefited large numbers of rural households, improving their lives with its innovative business model. Seeing the impact, other states such as Andhra Pradesh, West Bengal, Punjab and Jharkhand invited the company to set up operations in their states. Suguna has proved that every state in India is fit for poultry operations with its presence in 11 states. The model has also attracted visitors from abroad who are keen to learn from Suguna's initiatives and success and adopt the model in other countries.

9.12 PEPSICO MODEL

PepsiCo has been working with farmers in Punjab since the 1980s, starting with procuring tomatoes and producing tomato pulp. The operation has grown, and the model demonstrates effective backward integration by a private company already having strong marketing capabilities and established products and brands. Under this model, contracts are made with small farmers for the production and procurement of tomatoes but with an emphasis of building a win-win relationship of trust between the company and the farmers. The company brought in experts and promoted the use of appropriate varieties and farm technology, bringing to bear research and know-how available worldwide. Seedlings were provided to the farmers and planting was scheduled and programmed using computers. Tomatoes were procured by the company, and it used the best technology in processing and its strong marketing capabilities and networks in selling quality end-products.

More recently, a similar initiative has been launched for potato, see Figure 9.2. The product quality parameters put in place through the chain are driven by the specific needs of processing, and of buyer requirements. Processing requires potatoes with low sugar content (0%) and high solid content (between 15 and 20 percent). Because the company is HACCP and ISO

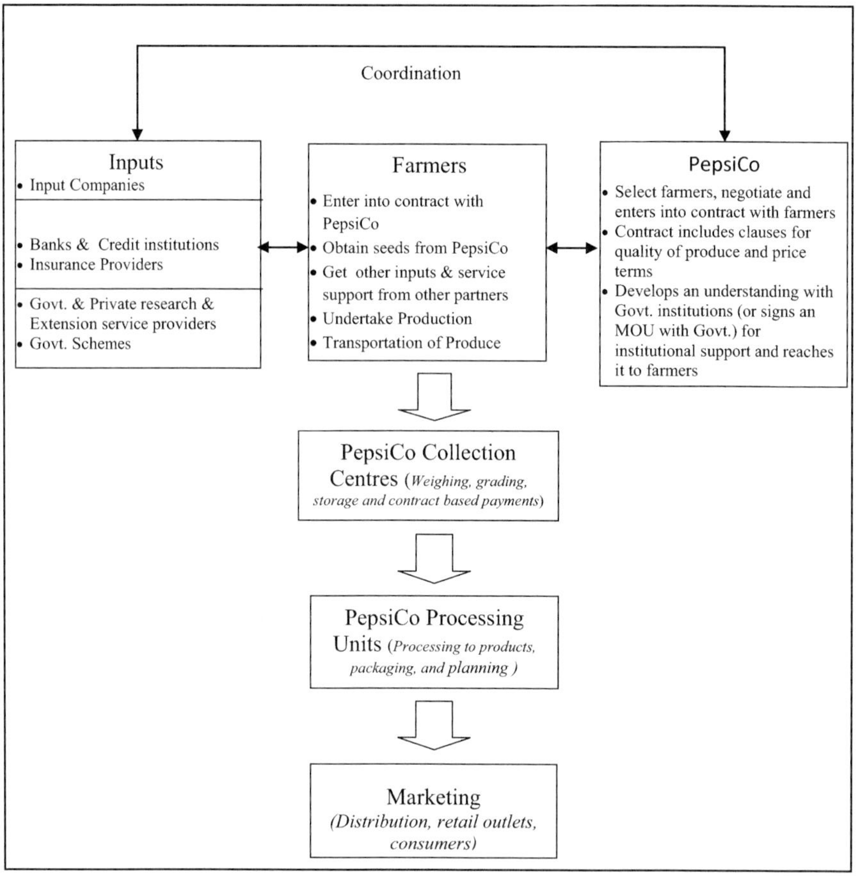

FIGURE 9.2 Tripartite model of PepsiCo India. (Reprinted with permission from Gandhi, V. P., (2014). "Growth and transformation of the agribusiness sector: Drivers, models, and challenges," Indian Journal of Agricultural Economics, 69[1].)

certified, stringent quality control is required at all levels in the chain. The requirements are met by ensuring quality compliance at every stage: farming, storing, processing, and packaging (Punjabi, 2008). The company has set up a 27-acre research and demonstration farm in Punjab to conduct trials for new varieties of tomato, potato, and other crops.

Extensive trials are undertaken before introducing the varieties to farmers, and a package of agronomic practices suitable to the local agro-climatic conditions is developed in collaboration with Central Potato Research Institute (CPRI). This includes specific fertilizer recommendations and spraying schedules. Seed potatoes of the specific varieties are provided by the company. The company ensures that farmers have availability of all the required inputs at the right time. The costs of inputs if provided are deducted during buy back of potatoes. The company had also introduced crop insurance and weather insurance, and PepsiCo created an institutional framework roping in the Central Potato Research Institute (CPRI), agrochemical company Du Pont, Agricultural Insurance Company (AIC), and ICICI Lombard General Insurance company (Punjabi, 2008).

Teams of agricultural graduates employed by the company work with the farmers to provide technical advice and monitor production. One technical expert deals with approximately 100 farmers. As a result, the use of chemicals and fertilizers is timely and effective (Punjabi, 2008). The agronomists regularly monitor the fields including at planting, spraying, and harvesting. If an outbreak of any disease or pest is seen or expected, farmers are advised for timely spraying. Major problems are attended to in consultation with the company researchers if necessary (Punjabi, 2008). After harvest, the selected procured potatoes are taken to the hi-tech processing plant. At the plant, they are washed, peeled, and inspected for physical damage and discoloration. Then, they are run through rotating slicers, deep fried, mixed with spices, and packed. The plant has a well-equipped quality testing lab. The new tomato varieties are said to have brought a yield increase from 16 tons to 54 tons per hectare (Punjabi, 2008). The introduced high-yielding potato varieties have increased farmer yields and incomes and enabled PepsiCo to procure world class chip-grade potatoes. The company has partnered with more than 10,000 farmers working over 10,000 acres of potato across the states of Punjab, Uttar Pradesh, Karnataka, Jharkhand West Bengal, Kashmir, and Maharashtra.

This model is more than simple procurement or contract farming and entails substantial company involvement in developing a mutually beneficial partnership between the agribusiness and the farmers. The model can result in very good benefits to small farmers in a limited area, but it requires a long-term view and commitment from the company and a willingness to absorb substantial start-up costs and initial losses (Gandhi, Kumar, and Marsh, 2001). Singh and Bhagat (2004) conclude that the PepsiCo model is a better model of contract farming as compared to others such as HLL and Nijjer, though there are some operational problems. As the acreage under the crop increases, the production increases and the open market prices may fall. The company may then base its contract price on this low open market prices and even fail to honor the contract. Singh and Bhagat (2004) indicate that it is necessary to learn from the experience of HLL that contract farming without building mutual trust might be problematic for the company itself. It should treat farmers as partners and share the benefits and risks with them, thereby creating a long-term sustainable business relationship and a win–win situation for both the farmers and corporates.

9.13 McCAIN INDIA MODEL

The McCain Foods subsidiary, Canada-McCain Foods India (Private) Ltd, was established in 1997, and its potato processing plant in Gujarat (Mehsana) became operational in 2007, with a processing capacity of 30 000 MT per annum or 4 MT per hour. The plant aims to produce French fries and other potato products such as flakes, patties, mashed potatoes, *aloo tikki*, and wedges for retail and food service businesses across the Indian subcontinent. It will employ 100 and 125 people at full capacity. It will also create indirect employment in storage, supply chains and outsourced services.

McCain has been buying its potato stock for processing via contracts, of which 85% were in Gujarat and 15% in Punjab. McCain has a choice of three types of contracts with the farmers: fixed, flexible, and open. In fixed contracts, the price is fixed and the transaction at harvest at that price is guaranteed, irrespective of the market price. In a flexible contract, a range is fixed within which the deal will be settled finally. If prices at the time of harvesting are more than the upper boundary then the farmer is bound by the upper price. In case, prices are below the lower boundary, the company is obliged to pay the lower price. In the open contract, both parties are free to transact or not too. Another type of contract may involve a mix of two of the three kinds described above, in some agreed monetary proportion.

In the McCain system, the farmer receives seed potato from the company. Fifty percent of the potato seeds are offered on credit and the cost is settled at the time of procurement. McCain also provides planters, diggers, and agri-inputs, including drip and sprinkler irrigation equipment. Loans are also made available through various banks for high quality potato production. Company agronomists visit the farmers every week and provide guidance on crop management. After harvest, the farmer takes the potatoes to the procurement-cum-storage hubs of McCain. His produce is graded and weighed, and quality testing is carried out. Then, the farmers are given an IOU for the amount the company will pay them; the payment is made within 15 days. The produce is kept in cold storage and when required, is transported to the McCain plant at Mehsana for processing. Quality is a very important criterion for the selection of growers. Generally, the growers must have 4–6 years of experience in quality potato growing. If the quality or the size are not up to the mark, then the produce is rejected, and the farmer can sell it where he likes or sell to McCain at a lower price.

9.14 COMPARISON OF McCAIN AND PEPSICO (FRITO LAY)

Both McCain and Frito Lay contract directly from the growers, specifying clear quality parameters. Both shifted from acreage contracts to combination acreage/quantity contracts that specified the minimum quantity to be delivered per unit area. The firms require delivery at a specified place, with growers bearing the delivery costs. However, there are differences: McCain had a smaller area of operation and a more specialized market for its products. While Frito Lay paid market prices, McCain offered a range of pricing options to its growers.

Frito Lay contracts offered only one price for all rejected chip-grade potatoes and could reject produce at its own discretion, buying it at the lower price stipulated for rejected produce. The final quality tests were carried out at the factory at Channo and undersized potatoes were not returned to the farmer (Singh, 2007). The McCain contract specified that if there was a deviation of more than 2 percent in some quality parameters (size, machine damage, mixing of varieties, presence of solid matter), the company could reduce the prices paid by an unspecified amount. Thus, there was uncertainty and risk for growers resulting from information asymmetry and lower bargaining power.

Both company's contract documents were replete with various obligations on the part of the farmer as regards quality maintenance, quantity, cultivation practices, post-harvest care of the produce, etc., but with very little obligation on the part of the company. All production risks were borne by the farmers and neither firm provided any reprieve to a contract farmer in case of crop failure. Even having entered into a contract and followed all instructions, there was uncertainty about whether the producer would have a market, because neither firm's contract obliged the company to buy the produce (Singh, 2007).

The cost of production for McCain potato contract growers was found to be slightly lower than the McCain non-contract growers, but much higher than those selling at the APMC (Agricultural Product Market Committee) or at farm gate. The yield and marketing costs for McCain contract growers were higher than that for any other channel. Marketing costs were higher for growers involved with McCain than they were for the APMC alternative, while net income from McCain was higher than that from alternative channels. Variation in net income between growers in the same category was much lower in the case of contract growers because of set prices. The McCain growers found that the use of sprinklers rather than flood irrigation reduced their labor requirements, improved soil quality, increased potato yields and quality, and saved water and extraction costs (Singh, 2007).

In the case of Frito Lay, the cost of production was higher and transaction costs somewhat lower than for APMC and farm-gate growers, despite the fact that contract growers had to deliver to the factory. Gross and net income was lower than that of growers using other channels due to lower yields; contract prices for high-quality produce and rejected produce were lower than post-harvest and off-season prices. The trend was for farmers directly supplying the companies to have higher production costs, regardless of whether a contract was involved, than those selling to other market outlets.

9.15 ITC E-CHOUPAL MODEL

ITC, through its International Business Division (IBD), undertakes procurement, processing, and export of agricultural commodities such as soybean, wheat, shrimp, and coffee. ITC-IBD has developed a unique IT-enabled procurement, information, and marketing model in rural areas, through village centers called *e-choupals*.

The model was launched by ITC in the villages of Madhya Pradesh in the year 2000. ITC opened three soya processing and collection centers and then identified six nearby villages for establishing *e-choupals*. The company identified an educated farmer to head the *e-choupal* in each village. The person is called the *sanchalak* and is trained to operate and coordinate the activities of the *e-choupal*. To establish the *e-choupal,* a personal computer is installed at the house of the *sanchalak,* and the *sanchalak* is given training in using it. The computer is connected to the Internet via telephone as well as satellite and has back-up power. The *sanchalak* helps the farmers in using the system, guiding them to the specially created website of the company, and to see the prevailing prices and other related information on it. To initiate a sale, the farmer brings a sample of the produce to the *e-choupal.* The *sanchalak* inspects the produce and performs quality tests (including foreign matter and moisture content) to assess the quality in the presence of the farmer and explains if there are any deductions. He then obtains the benchmark price from the computer, makes the appropriate deductions, and conveys a conditional quote to the farmer. If the farmer chooses to sell to ITC, the *sanchalak* gives the farmer a note with his name, village name, particulars about the quality tests, approximate quantity and conditional price. The *sanchalaks* is paid 0.5 percent of the value of soya procured by ITC.

The farmer takes the note from the *sanchalak* and proceeds with his produce to the nearest ITC procurement hub. At the ITC procurement hub, a sample of the farmer's produce is taken and set aside for laboratory tests. A chemist visually inspects the soybean and verifies the assessment of the *sanchalak*. Deductions for the presence of foreign matter such as stones or hay are made based on visual comparison with other produce such as of his neighbor's and the farmer may accept the deductions and the final price. Laboratory testing for oil content is performed after the sale and does not alter the price. The farmer's produce is then weighed on an electronic weighbridge, following which the farmer can collect his payment in full at the payment counter. The farmer is also reimbursed for transporting his crop to the procurement hub. The process is accompanied by appropriate documentation. The farmer is given a copy of inspection reports, agreed rates, and receipts for his records. The system also has *samyojkas* (who were former commission agents) who are responsible for collecting the produce from villages that are located far away from the processing centers and bringing it to the ITC centers. The *samyojka* is paid 1% commission. At the end of the year, farmers can redeem accumulated bonus points through the *e-choupal* for farm inputs or insurance premiums. The ITC *e-choupal* model is shown in Figure 9.3.

By 2007, the *e-choupal* services reached over four million farmers in about 40,000 villages through over 6500 *e-choupals*. This extended across the states of Madhya Pradesh, Uttar Pradesh, Rajasthan, Maharashtra,

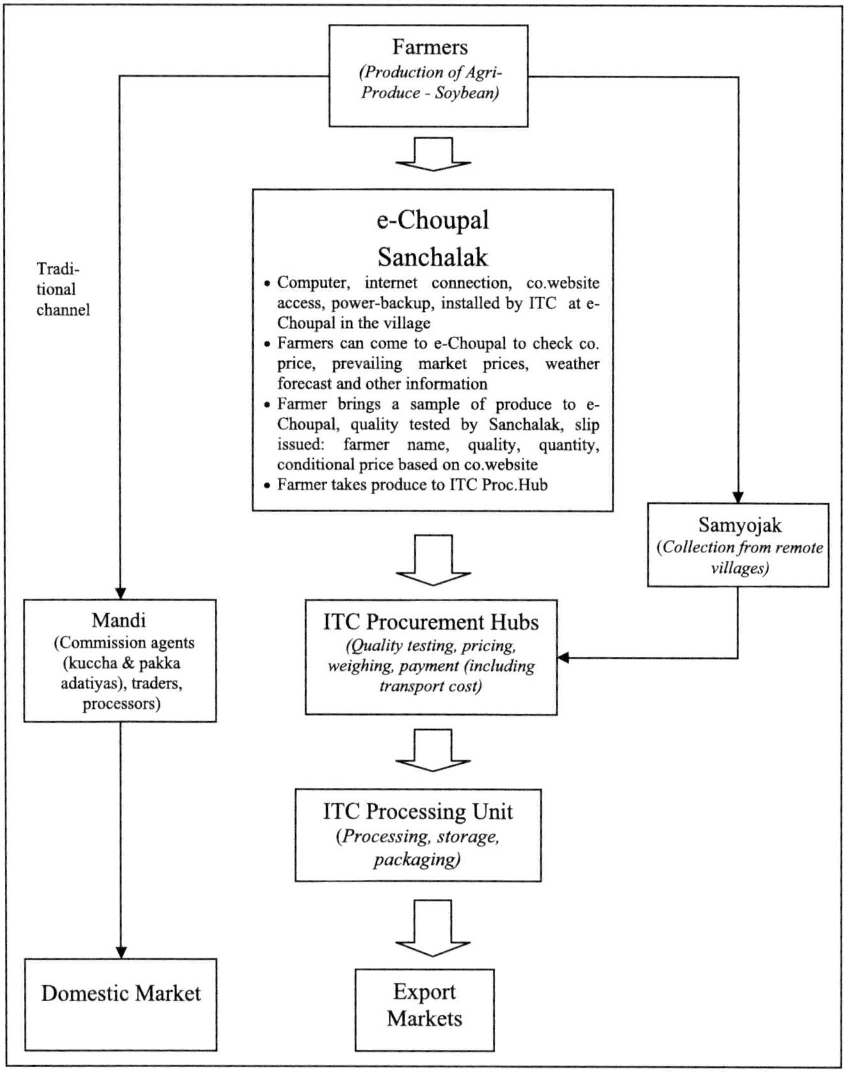

FIGURE 9.3 Outline of the ITC E-Choupal model. (Reprinted with permission from Gandhi, V. P., (2014). "Growth and transformation of the agribusiness sector: Drivers, models, and challenges," Indian Journal of Agricultural Economics, 69[1].)

Karnataka, Andhra Pradesh, and Kerala. ITC is extending its business model to other Indian States including West Bengal, Himachal Pradesh, Punjab, and Haryana. Some procurement hubs also have *Choupal Saagars*, which offer goods and services farmers may need including agri-equipment, agri-inputs, personal consumer products, insurance service, pharmacy and health center, agri-extension clinic, fuel station, and food court. Information and services provided by the *e-choupal* web site and e-commerce system include: weather information, information on scientific practices, guidance on how to improve crop quality and yield, and access to input supply (fertilizers, pesticides) along with recommendations and to soil testing service. The model has principally aimed at increasing the efficiency of procurement, resulting in value creation for both the company and the farmer. In addition, the model takes internet penetration to the villages, offering information and global commercial contact. The *e-choupal* allows the farmers daily access to information on prices of many *mandis* which helps them to make better decisions on when and where to sell the produce. Thus, *e-choupal* tries to provide farmers a better price. The incremental income from a more efficient marketing system is estimated to be about US$ 6 per ton on average or an increase of about 2.5 percent over the *mandi* system.

Singh and Bhagat (2004) report that many farmers did not agree that they are getting a better price for their produce, and that there are only minor benefits like de-bagging expenses, etc. One significant advantage, however, is correct weight, which is a major worry in traditional *mandi*. Even though there is a potential, the web portal does not have the enough richness to become an information and knowledge dissemination kiosk. The information on best practices, crop production, inputs, fertilizer, and seeds was not of high quality. The main information disseminated is of prices and weather conditions. ITC has not paid enough attention to input trading through its *e-choupals*, and proper partnerships with input companies are often not worked out. However, the model offers a quantum change and a huge potential for better service.

9.16 DESAI FRUITS AND VEGETABLES (DESAI COLD STORAGE)

Even small family firms show success in this activity. The substantial production of fruits and vegetables in the south Gujarat area gave two families the idea of starting a cold storage facility in 2001. There were two partners in this venture. Mr Ajitbhai Desai was the active partner, while the other partner was associated with a sugar mill in the area. The construction of the cold storage began in 2001 and was completed in 2002. In venturing into this business, the main intention was to help other traders and farmers to keep their produce for a longer time. But no suitable enquiry was carried out as to the extent the facility would be used; in the end, no such service was provided to any customer. The facility is now used only by the partners themselves, and they do not keep any of their products in cold storage for very long. Nevertheless, they have developed a highly successful business mainly trading in fruits and vegetables, not only locally but also nationally and internationally.

The major problem the partners encountered at the outset was convincing farmers to sell their produce to them. To overcome this, they devised a number of activities to create awareness among farmers in their area about the facilities they were offering. In 2001, Desai organized a number of face-to-face meetings with farmers in different villages. To attract farmers, they offered integrated pest management (IPM) without any charge. They also offered the harvesting and handling technology through which farmers could save labor and obtain higher quality produce. Some of the other benefits offered were:

- waver of any commission on the transactions (normal market charges amount to about 10%)
- assurance of 100% buying
- quality based pricing – higher than the prevailing market price
- direct buying from the farm gate
- assured minimum price. The minimum price would be calculated on the basis of the production cost and harvest labor cost, with a margin for the farmers added in. Through this mechanism, even if the market price crashed, the farmers would receive the predetermined minimum price agreed upon. If market prices increased, then the transactions would be carried out at the prevailing market price.

Desai began trading only in mangoes. During the first year of operation in 2002, with the help of traders from Canada, France, the Netherlands, and other countries, they exported 600 tons of the fruit; in 2003, this increased to 1200 tons, again with the help of merchant traders. In 2002, they also tried to sell in the domestic market through some super market chains in major cities and to the system run by NDDB Mother Dairy. But this experience was disappointing: the response from consumers was not as expected and the venture did not work. By 2004, they were exporting 17,000 tons of mangoes. The main destinations for these were the Middle East and the UK. They also continued to sell locally on a minor scale through merchants and Mother Dairy on a demand basis. Though the mangoes business is its major one, Desai also exports banana, papaya, and pomegranate. It has now succeeded in expanding its export destinations to include China and Australia; the required approvals/contracts have already been signed.

The mango season in India is spread over a period of about 8 months. It can start as early as January or February in Kerala, moving up to Andhra Pradesh and the Konkan region of Karnataka as the year progresses, then further north to Maharashtra and Valsad and the Saurashtra region in Gujarat, ending during August in Uttar Pradesh and Bihar. As such, Desai can expect to have a viable mango business for about 8 months per year. To meet the requirements of expanding the business into an eight-month operation, Desai are buying in local markets in Gujarat and are carrying out improvements and value additions to the produce at their present unit. In the future, the company plans to implement the procurement system it has already developed at Gujarat in other states and areas. Desai has already contacted farmers in some of the abovementioned regions, who have been guaranteed the benefits and facilities offered to producers in Gujarat. The state governments of Andhra Pradesh and Maharashtra have also started some processing units to help the farmers.

Desai has noticed that during the initial 2–3 months of the season, they will not have any competitors from any Asia region countries; hence their business can have a major competitive advantage during this period. Major competition comes from Pakistan but only toward the end of the season (July–August), when the *Chosa* varieties of mango from Uttar Pradesh and Bihar are available.

To improve and maintain quality and prevent deterioration in exports, certain processing and handling measures are followed stringently in the cold storage premises.

To expand into the banana business, Desai plans to import modern technology in all aspects of its operation, including farming, handling, and processing; the company plans to spend about Rs 5 million for this purpose. Large-scale banana farmers around Navsari and in nearby areas of Gujarat have been contacted and offered the same benefits and guarantees as the mango farmers. Contact meetings with these farmers have already taken place at different locations.

Since inception, Desai has also been dealing in green vegetables, the main ones being lady finger (okra), bitter gourd, bottle gourd, and other minor interests such as chili and lemon. During the first year of this operation (2002), the company bought this produce on the market and also purchased from farmers around the cold storage facility. As with the mangoes, vegetable exports were also initially distributed via merchant traders. Desai exported about 40% of its procured produce through these traders and sold the remainder in the domestic market. Here, Mother Dairy was the major purchaser.

In this initial period, the company convinced farmers to tie-up directly with them on the same lines as with the mango farmers. Slowly, direct input arrivals have picked up. Farmers are required to sort their produce into A, B, and C quality categories before delivery, and this is taken into consideration in deciding prices. The purchase prices of vegetables are fixed slightly higher than the Surat APMC daily market price. The payment is made to farmers the next day, after the exact market prices of Surat APMC have been ascertained. In 2003, Desai started direct export of vegetables to the UK and Middle Eastern countries. In addition to this, they are still keeping up their export business through merchant traders and in the domestic market through Mother Dairy.

To maintain the quality of the produce, Desai has devised a plan that also benefits the farmers significantly. Typically, small and marginal farmers had been exploited by retailers in the supply of inputs such as seeds, pesticides, and fertilizers. These farmers had also been exploited in selling their output to traders and commission agents. To liberate farmers who agreed to supply the company with produce, Desai devised an innovative idea. Inputs such as seeds, pesticides and fertilizers are procured in bulk and are supplied to farmers without any margin and on credit, free of any interest. The cost is then taken into account against payment for supply of the output.

In 2004, Desai distributed about 3000 kg of lady's finger seeds (okra). By supplying this to its farmers, as well as bottle gourd and bitter gourd, the company ensured that the output would be of the same good variety and quality. For lemon, chili, beans, and other vegetables, Desai did not provide seeds, but a bulk purchase agreement was signed. In chili alone, the 2004 business amounted to more than 500 tons, while the weekly procurement of lemon was as much as 15 tons.

CFI AG has made significant investments in Desai Fruits and Vegetables (DFV) – the new name for Desai Cold Storage since April 2006 – to enhance its financial, operational and management capabilities. DFV has since created multiple integrated pack houses for bananas and continues to invest in creating facilities at the farm and village levels. Since its inception, the company has invested in R&D for agricultural practices and in developing long-distance transport protocols. These initiatives have helped Desai emerge as one of the leading exporters of high-quality fruits and vegetables in India. It now supplies a wide range of high-quality tropical products: the portfolio includes over 10 types of fruits and 30 types of vegetables. However, the key focus over the years has been on four fruits – mangoes, bananas, pomegranates, and grapes. This flagship produce makes up over 90% of total sales.

Over 2500 small and medium-sized farmers supply to Desai. The company has a unique contract farming mechanism by which it controls the inorganic inputs, technology and work practices at the farm level to ensure that the product is absolutely natural and safe. The model has been so successful that Desai has a waiting list of over 800 farmers wishing to be included in its program. The company works with its customers to create specialized, customized solutions in order to deliver quality at a reasonable cost. It ensures the quality of every product that it supplies by controlling every step involved in the production process, from land preparation to logistics. The quality assurance system is based on the following principles:

1. *process control* – achieved through contract farming and technical control;
2. *farm approval system* – extensive checks are conducted for work practices, use of chemicals, and quality of products, before any farm is approved for procurement;
3. *traceability* – of the farm, inputs used, processing location and transportation;
4. *quality checks* – a Desai product passes through a stringent quality assurance process:
    - quality control by a supervisor at the point of procurement from farmer;
    - quality checks at the point of packing;
    - random checks of consignments;
    - random sampling at laboratories for residue analysis.

- Desai has been conscious of the importance of superior infrastructure from its inception. As a result, one of the most modern and state-of-the-art packhouses has been developed; innovative trailer designs have been created for smooth transportation; reefer vans have been bought; and pre-cooling, cold storage and processing facilities have been enhanced.

9.17 COMPARISON OF THE DIFFERENT MODELS

How do the models compare? The comparison can be done on the basis of the key success factors described above, and the table below provides a broad comparison and evaluation of these models (as well as a few other models – for models not described here, see Gandhi and Jain, 2011). As can be seen, the performance on the key success factors varies substantially across the models (Table 9.6). Whereas Amul and ITC e-choupal are strong in reach to small farmers, Suguna and Pepsi are strong in ensuring adoption of the right technology for quality and quantity. Nestle, Pepsi, and Amul are strong on investing in modern processing technology as well as at delivering a strong marketing effort to reach a huge food market. Amul and Suguna are strong on bringing commitment and benefits to all stakeholders, and Pepsi is reasonably good.

9.18 CONCLUDING OBSERVATIONS

Agri value-chains and agribusinesses have been growing quite rapidly in India in the recent years and have been given substantial priority due to their significant potential to contribute to income and development. The beginning can be traced to Mahatma Gandhi's emphasis on the need for encouraging of village-based agro-industries to uplift rural masses and connect them to the independence movement and the national economy. The sector still contributes significantly to employment and is very important to value addition and income generation in the rural areas. However, a number of difficulties are faced, and the success seems to depend on a set of key success factors built into the agribusiness model design. The key features include reaching large numbers of small farmers and economically procuring quantity; modernization farming through bringing adoption of right technology for quantity and quality; investing in modern processing; and meeting the capital requirement;

TABLE 9.6 Broad Comparisons of Different Models on Performance Parameters

Agribusiness Model	Procurement: Reaching large numbers of small farmers	Transforming agriculture: Bringing adoption of new technology by farmers	Capital investment in modern processing technology & operations	Delivering strong marketing effort	Sharing benefits across the value chain, management and control bringing commitment across stakeholders
AMUL	Strong	Reasonable	Strong	Strong	Strong
Nestlé	Limited	Reasonable	Strong	Strong	Limited
Heritage	Good	Limited	Good	Good	Limited
Suguna	Strong	Strong	Strong	Good	Strong
Pepsi	Reasonable	Strong	Strong	Strong	Reasonable
ITC e-Choupal	Strong	Limited	Strong	Strong	Limited
Other Models					
Nandini	Good	Limited	Limited	Reasonable	Good
Mother Dairy	Limited	Limited	Good	Good	Reasonable
Safal Market	Limited	Limited	Good	Limited	Limited
HPMC	Reasonable	Limited	Good	Poor	Poor
McCain	Reasonable	Strong	Strong	Strong	Limited
Desai Fruits & Vegetables	Reasonable	Good	Good	Strong	Reasonable

(Reprinted with permission from Gandhi, V. P., (2014). "Growth and transformation of the agribusiness sector: Drivers, models, and challenges," Indian Journal of Agricultural Economics, 69[1].)

delivering a strong marketing effort; and sharing benefits across stakeholders through appropriate management, control, and institutional structures so as to build a strong value chain.

A number agribusiness models with varying designs that have emerged in India were studied ranging from Amul to Suguna to ITC e-choupal. The comparison shows varying approaches and performance on the different key factors identified, with strong implications for the success of the models. The AMUL cooperative model seems an excellent farmer-owned design that brings commitment, contribution, and substantial benefits to the stakeholders including small farmers. A number of good private agribusiness models have also emerged often from strong industrial and marketing capabilities, which have created strong backward linkages to the farmers, improving technology and efficiency. They need commitment, investment, and a strong partnering approach. Even though the success factors remain important, the same approach may not be good for different agribusiness activities and regions, and it is critical that alternative models are experimented with. The models that can effectively address all the five key success factors and that can potentially transform and modernize the whole value chain deserve particular encouragement.

KEYWORDS

- **agri-business**
- **consumer**
- **horticulture**
- **investors**
- **linkages**
- **marketing**

REFERENCES

Boer, K. De, & Pandey, A., (1997). India's Sleeping Giant: Food. *The McKinsey Quarterly, No. 1.*

Bowonder, B., Gupta, V., & Singh, A., (2002). *Developing a Rural Market e-hub: The case study of e-Choupal Experience of ITC*. New Delhi, Government of India Planning

Commission (Available at: http://planningcommission.nic.in/reports/ sereport/ser/stdy_ict/4_e-choupal%20.pdf).

CII-McKinsey & Co. (1997). *Modernising the Indian Food Chain: Food and Agriculture Integrated Development Action Plan (FAIDA).* New Delhi, CII and McKinsey and Co.

Damodaran, H., (2001). A tale of two milk districts. In: *Business Line*, New Delhi Edition. (Available at http://www.thehindubusinessline.in/businessline/2001/12/09/stories/14090303.htm).

Gandhi, V. P., & Jain, D., (2011). "Institutional innovations and models in the development of agro-industries in India: Strengths, weaknesses and lessons", In: *Innovative Policies and Institutions to Support Agro-Industries Development*, Carlos A. da Silva (ed.), FAO, Rome, 203–257.

Gandhi, V. P., & Mani, G., (1994). Agro-processing for development and exports: The importance and pattern of value addition from food processing. In: *Indian Journal of Agricultural Economics*, *50*(3).

Gandhi, V. P., & Namboodiri, N. V., (2006). "Fruit and vegetable marketing and its efficiency in India: A study of wholesale markets in the Ahmadabad area" CMA monograph, Indian Institute of Management, Ahmadabad. *216*.

Gandhi, V. P., & Namboodiri, N. V., (2006). "Fruit and vegetable marketing in India: Consolidated study of wholesale markets in Ahmadabad, Chennai and Kolkata,. CMA monograph, Indian, Institute of Management, Ahmadabad, *221*.

Gandhi, V. P., Kumar, G., & Marsh, R., (2001). Agro-industry for rural and small farmer development: Issues and lessons from India. In: *International Food and Agribusiness Management Review*, *2*(3/4), pp. 331–344.

Gandhi, V. P., (2014). "Growth and transformation of the agribusiness sector: Drivers, models, and challenges", *Indian Journal of Agricultural Economics*, *69*(1).

Goldberg, R., (2006). *Nestlé's Milk District Model: Economic Development for a Value-Added Food Chain And Improved Nutrition.* (Available at: http://www.wcfcg.net/gcsr_proceedings/Professor%20Ray%20A%20Goldberg.pdf).

Goyal, S. K., (1994). Policies towards development of agro-industries in India. In: Bhalla, G. S. (Ed), *Economic Liberalization and Indian Agriculture, Chapter VII.* New Delhi, Institute for studies in industrial development, pp. 241–286.

Gulati, A., Anil, S., Kailash, S., Shipra, D., & Vandana, C., (1994). *Export Competitiveness of Selected Agricultural Commodities*, National Council of Applied Economic Research, New Delhi.

Heritage Foods Limited. http://www.heritagefoods.in/ (accessed Jun 23, 2018).

India: Ministry of Planning, (2013). Annual survey of industries (2011/12), *Ministry of Planning*, Government of India, New Delhi.

India: Planning Commission, (2008). Eleventh five year plan (2007–2012), *Inclusive Growth,* New Delhi, Government of India. vol. I.

India: Planning Commission, (2013). Twelfth five year plan draft (2012–2017), *Economic Sectors*, Government of India, New Delhi. vol. II.

Kejriwal, N. M., (1989). Performance and constraints in accelerating production and export of fruits and vegetables. In: Srivastava, U. K., & Vathsala, S., (eds) *Agro-Processing: Strategy for Acceleration and Exports.* New Delhi and Oxford, IBH Publishing Co. Pvt. Ltd.

Kurien, V. (2003). *Successful Corporatization: The AMUL Story.* Speech delivered at the Madras management association, Chennai.

Magfree.net. http://www.nposonline.net/ (accessed Jun 23, 2018).
Ministry of Food Processing Industries. http://mofpi.nic.in/ (accessed Jun 23, 2018).
Nestle. http://www.nestle.com/ (accessed Jun 23, 2018).
Outlook Business. http://www.business.outlookindia.com/ (accessed Jun 23, 2018).
Punjabi, M. (2008). *Supply Chain Analysis of Potato Chips: Case Study of PepsiCo's Frito Lay in India*. A draft submitted to United Nations Food and Agriculture Organization, New Delhi.
Singh, R., & Bhagat, K. (2004). Farms and corporate: New farm supply chain initiatives in Indian agriculture. In: *Indian Management*, pp. 76–77.
Singh, S. (2007). Leveraging contract farming for improving supply chain efficiency in India: *Some Innovative and Successful Models.* (Available at: http://www.globalfoodchain-partnerships.org/india/Presentations/ Sukhpal%20Singh.pdf).
Sridhar, G., & Ballabh, V. (2006). Indian agribusiness institutions for small farmers: Roles, issues and challenges. In: *Institutional Alternatives and Governance of Agriculture*, Indian Institute of Management, Kozhikode, pp. 1–18.
Srivastava, U. K. (1989). Agro-processing Industries: Potential, Constraints and Tasks Ahead. In: *Indian Journal of Agricultural Economics*, *44*(3), pp. 242–256.
Suguna Foods Private Limited. http://www.sugunafoods.co.in/ (accessed Jun 23, 2018).
Vaidya, C. S. (1996). Strategy for Development of Agriculture in Himachal Pradesh with Reference to Economic Liberalisation Policy, Agro-Economic Research Centre, Himachal Pradesh University, Shimla.

CHAPTER 10

STATUS AND STRATEGIC PLAN FOR THE SUSTAINABLE DEVELOPMENT OF HORTICULTURE SECTOR IN ANDAMAN & NICOBAR AND LAKSHADWEEP GROUPS OF ISLANDS, INDIA

R. K. GAUTAM, S. DAM ROY, AJIT ARUN WAMAN, POOJA BOHRA, I. JAISANKAR, V. DAMODARAN, K. SAKTHIVEL, and T. BHARATHIMEENA

ICAR–Central Island Agricultural Research Institute, Port Blair–744101, Andaman and Nicobar Islands, India, E-mail: ajit.hort595@gmail.com

CONTENTS

ABSTRACT

The Andaman and Nicobar Islands (ANI) and the Lakshadweep groups of islands (LDI) are the unique tropical ecosystems of India. Both the groups of islands constitute one of the 22 agro-biodiversity hotspots of the country, harboring variety of flora of horticultural importance. The geographical separation of these islands from the continental India has proved to be advantageous as well as disadvantageous at occasions. Being situated in the oceans, they are vulnerable to a number of climatic vagaries such as Tsunami, cyclones, and storms. In such fragile ecosystems, it is important to strike a balance between the development and ecological subsistence for achieving long-term sustainability of the horticulture production system. In this article, we analyze the historical and present status of the horticulture in these islands, various constraints encountered during the production, strategies to overcome them by developing locally suitable technologies, interventions made, gaps and issues to be addressed, and the role of various research and developmental agencies in the islands.

10.1 ANDAMAN AND NICOBAR ISLANDS

10.1.1 INTRODUCTION

The Andaman and Nicobar Islands (ANI) are located in the Bay of Bengal with a total geographical area of 8,249 sq. km., of which majority (about 4/5th) is in the Andaman groups. The archipelago is geographically divided into North Andaman, Middle Andaman, South Andaman, Car Nicobar, and Nan Cowry groups of islands, and the altitude ranges between 0 and 365 m. The island is characterized by equatorial warm humid tropical climate with mean monthly temperatures of 29–32°C (max) and 22–24°C (min). The islands receive a total of 3,100 mm rainfall spread over 8–9 months, while the remaining period receives very low quantity of rains, resulting into drought-like condition.

Topography of ANI is mostly undulating with both low and high lands found in the islands, which support cultivation of coconut, arecanut, and other horticultural crops. The total area available for agriculture in the ANI was 50,000 ha; however, the occurrence of *Tsunami* during 2004 has reduced this area to 43,339 ha. About half of the population is directly or indirectly dependent on agriculture and allied activities for livelihood. Though the agricultural activities initiated during 1858 AD (Kumar and Gangwar, 1985), the systematic cultivation began after interventions by the State Directorate of Agriculture (1945), and the ICAR institutions, which formed the ICAR-Central Island Agricultural Research Institute (erstwhile CARI) in 1978.

10.1.2 HORTICULTURAL CROPS IN THE ISLANDS

Plantation crops mainly, coconut and arecanut-based farming systems are largely found in the upland agriculture, while under lowlands paddy-vegetable system is popular in ANI. Of the total area under agriculture, the horticultural crops occupy about 70% area. Though coconut, arecanut, and spices are important from the economic security points of view, the vegetables, tuber crops, and underutilized fruits are important from nutritional points of view. The area and production of various horticulture crops are presented in Table 10.1. As the ANI is becoming a popular tourist destination, the pressure on these commodities is increasing. The tropical climate favors cultivation of roots, tubers, some vegetables, and fruit crops; however, only a portion of the demand is being met through local production. There is considerable scope for the productivity enhancement; however, thorough analysis of the constraints and implementation of appropriate strategies is the key for achieving the targets. The ANI and Lakshadweep groups of Islands (LDI) are among the 22 agro-biodiversity hotspots. The diversity present in these areas could be used to identify genes for crop improvement, alternative crops, and species suitable for use as rootstocks for cultivated crops (Gautam et al., 2015).

TABLE 10.1 Area and Production of Different Horticultural Crops in A&N Islands

Commodity	**Andaman and Nicobar Islands**	
	Area (000 ha)	**Production (000 ha)**
Coconut	21.90	89.45
Arecanut	4.23	5.88
Vegetables	6.89	51.79
Fruits	3.55	29.73
Spices	1.68	3.22
Flowers	0.13	0.29

NHB (2014).

10.1.3 TECHNOLOGICAL INTERVENTIONS MADE FOR DEVELOPMENT OF HORTICULTURE IN THE ANI

ICAR-CIARI, Port Blair being the only research institute working on horticultural crops in the ANI, has been involved in the need-based research for the benefit of the farmers. Besides this, line departments of ANI administration and other government agencies working in the islands have also contributed significantly for the overall development of horticulture sector in the Islands. An overview of these activities has been presented under the following subsections.

10.1.3.1 Varieties Developed by ICAR-CIARI

The systematic crop improvement program at ICAR-CIARI, Port Blair, has resulted in the development of 19 varieties of horticultural crops suitable for cultivation in A&N Islands (Table 10.2). Most of these varieties have been developed from the local germplasm through the selection process. A number of these varieties are being popularized amongst the farmers. apart from these, a number of improved varieties have been introduced from various parts of the country to identify types suitable for cultivation in the island ecosystem.

TABLE 10.2 Improved Varieties of Horticultural Crops Developed by ICAR-CIARI, Port Blair

Crops	Varieties/superior genotypes
Coconut (*Cocos nucifera* L.)	CARI Surya, CARI Anapurna, CARI Chandan, CARI Omkar
Arecanut (*Areca catechu* L.)	Samridhi
Noni (*Morinda citrifolia* L.)	CARI Sanjivini, CARI Samridhi, CARI Rakshak, CARI Sampada
Brinjal (*Solanum melongena*)	CARI Brinjal-1
Greater yam (*Dioscorea alata* L.)	CARI Yamini
Sweet potato (*Ipomoea batata* L.)	CARI SP-1, CARI SP-2
Culantro (*Eryngium foetidum* L.)	CARI Broad Dhaniya
Poi (*Basella alba* L. and *B. rubra* L.)	CARI Poi Selection, CIARI Shan
Amaranthus (*Amaranthus* spp.)	CIARI Harita and CIARI-Lal Marsha
Ground orchid (*Eulophia andamanesis*)	CARI Pretty Green Bay

10.1.3.2 Farmer Friendly Technologies Developed by ICAR-CIARI

It is a proven fact that a crop/variety will be able to express its full potential only when region-specific package of practices has been developed for its cultivation. The phenotypic response of a genotype will be dependent on its interaction with the growing environment, which is suitably modified using the recommended package of practices. In view of this, the institute has remarkably developed the following technologies for the A&N Islands.

(a) Package of practices for vegetable crops, noni, flower crops, etc.
(b) Horticulture-based cropping system models
(c) Rain shelter technology for cauliflower cultivation
(d) Protected cultivation of high-value leafy vegetables
(e) Development of structure for mites-free capsicum production
(f) Trench system for vegetable cultivation in problematic soils
(g) Local growing media for capsicum and tomato
(h) Coconut-based Silvi-pasture system in humid tropical climate of Andamans
(i) Gliricidia-alley cropping system in Andaman
(j) Fodder tree-based Silvi-pastoral system for Bay Island condition
(k) Raised bed technology with coconut husk burial for vegetables
(l) Black Pepper cultivation on hedge rows and *Gliricidia* standards
(m) Protected cultivation of cut flowers
(n) IPM (Integrated Pest Management) and IDM (Integrated Disease Management) modules for important crops
(o) ICAR-CIARI bio-consortia for the management of bacterial wilt of brinjal
(p) Rodent management in coconut
(q) Mushroom production technology

10.1.3.3 Technological Interventions Made by the Line Departments and Other Agencies

The Department of Agriculture, ANI, and other developmental agencies including the National Horticulture Mission, Coconut Development Board, AYUSH, NABARD, etc., have played a pivotal role in the development of agriculture in the islands. The Department of Agriculture, ANI, facilitates the procurement and distribution of the various inputs to the farmers. It also multiplies the planting material in large number for sale at subsidized rate.

There are a number of schemes, which are being implemented successfully in different parts of the islands. The major initiatives especially through the High Value Agriculture Development Agency include assistance for the local production as well as import of quality planting material of high value crops; setting up of new plant tissue culture unit in the islands; encouragement for adoption of drip irrigation; fertigation and mulching techniques; import of planting material of exotic fruits such as mangosteen, rambutan, sweet carambola, durian, etc., from the mainland India and supply to the farmers at subsidized prices; promotion of cultivation of perennial and nonperennial horticultural crops; protected cultivation; organic farming and certification; pollination support through bee-keeping; mechanized operations; and integrated postharvest management and technology dissemination.

10.1.4 CONTRIBUTION OF LOCAL PRODUCTION TO REQUIREMENT

The various technologies developed and the developmental activities undertaken have resulted into improved availability of the commodities in the islands. The contribution of local production in meeting the local requirement has increased from 1980 to 2011 to the tune of 40–80% in fruits and 20–80% in vegetables. As there is no increase in the land area, the increment in the production could be grossly attributed to the adoption of improved technologies in the ANI.

10.1.5 HORTICULTURAL CROPS: CONSTRAINTS AND STRATEGIES

Though the cultivation of horticultural crops is highly remunerative, a few challenges, both technical and technological, hamper the production to a great extent. The various constraints faced by the island farmers and pragmatic steps required to be taken up for solving these problems are mentioned hereunder.

10.1.5.1 Plantation Crops

Plantation crops are the important cash crops of the islands. Coconut and arecanut have been cultivated on commercial scale and are considered as the major sources of income, food, fuel, shelter, and other uses. Arecanut, apart from being used as stimulant, has multifaceted uses in the tribal communities. The leaves are used for preparing the floors and thatching of huts, while spathes are used as utensils.

10.1.5.1.1 Constraints

In case of coconut and arecanut, the existence of old and senile plantation is one of the major causes for the low productivity in the islands (Rethinam, 2009). Monocropping is a commonly followed practice, which reduces the per unit area productivity. Lack of trained work force for harvesting and de-husking operations in palms is one of the major constraints in all the areas of the islands. Bud rot (*Phytopthora palmivora/Phytopthora arecae*), yellow leaf disease (arecanut), and leaf blight/spot (*Pestalotia* spp.) are the major diseases hampering the productivity of palms. Rodents are pests of prime importance in the islands, causing up to 50–90% crop loss in severe cases. Rhinoceros beetle is another pest infesting the palms, thereby reducing the productivity.

10.1.5.1.2 Strategies

This involves replanting, rejuvenation of the old, senile, and unproductive palms in a systematic way with quality planting materials of high yielding varieties and hybrids (Rethinam, 2009). The planting material could also be produced through community nurseries. Emphasis need to be given for adoption of inter/mixed/multiple and multistoried cropping and mixed farming systems with high-value crops. Further, integrating animals, birds and fishes as well as recycling the residues to the crops on large scale must be promoted. Adoption of mechanization especially for climbing the palms for harvesting and de-husking could improve the efficiency of the available labor. For control of rodents, community-based management practices like mechanical trapping and use of zinc phosphide baits may be taken up. Use of *Oryctes baculovirus* (OBV), *Metarhizium* spp. formulations, and aggregation pheromone could be promoted for eco-friendly management of rhinoceros beetle. Coconut-based eco-tourism can be developed integrating many activities at one place. Organic farming needs to be focused in greater extent to fetch international marketing and hence technologies for organic production to be developed.

10.1.5.2 Vegetable Crops

Vegetables are integral component of agriculture and generate livelihood opportunity to more than 50% of island population, which depends on agricultural activities.

10.1.5.2.1 Constraints

Lack of inputs at right time for scientific cultivation is hampering the production of the majority of vegetable crops. The biotic and abiotic stresses are largely hammering the cultivation. Among the diseases, bacterial wilt (*Ralstonia solanacearum*), damping off (*Pythium* spp.), brinjal Phomopsis blight (*Phomopsis vexans*), chili fruit rot/leaf spot (*Colletotrichum capsici*), chili leaf curl, and mosaic virus are the major problems in solanaceous vegetables. Mosaic disease, leaf curl disease, and leaf spot disease are commonly seen in various cucurbitaceous crops, while anthracnose is a common problem in French bean. Yellow vein mosaic virus of okra is also noticed in some parts of islands. Brinjal shoot and fruit borer, okra shoot and fruit borer, and cucurbit fruit fly are amongst the major pests of vegetables in the islands. Limited area for expansion and heavy rains during monsoon are known to limit the cultivation of vegetable crops.

10.1.5.2.2 Strategies

Timely supply of quality inputs like seeds of high-yielding varieties, fertilizers, pesticides/fungicides, or organic inputs is required. Development of island-specific IPM and IDM modules for the management of major pests and diseases is required for successful cultivation of vegetables. Popularization of protected cultivation, viz., poly house, poly tunnel, shadenet house, rain shelter to promote vegetable cultivation in rainy season, is necessary.

Promotion of cold storage facilities would facilitate easy and safe transportation and storage of harvested vegetables. Promotion of farmers' cooperative societies could ensure better marketing of the vegetable produce to get remunerative price to the farming sector. Indigenous vegetables of the islands also need to be promoted (Singh et al., 2009).

10.1.5.3 Tuber Crops

In ANI, tubers and aroids are secondary foods for settlers but constitute important part of diets of native Nicobari, Jarawa, and Onge tribes. The major tuber crops in the islands are *Manihot esculenta* Crantz, *Ipomoea batatas* (L.) Lam., *Dioscorea alata* (L.), *Colocasia esculenta* (L.) Schott, *Xanthosoma sagittifolium*, and *Amorphophallus poeniifolius* (Dennst.) Nicholson (Medhi et al., 2001).

10.1.5.3.1 Constraints

Considerable diversity has been reported from the islands; however, the lack of quality planting material of superior types is one of the basic constraints in most of the tuber crop species. There are no island-suited varieties, which could be successfully incorporated in existing planting systems. Lack of awareness about the scientific production technologies are the cause for low productivity. Most of the produce is sold as such and little efforts have been made for value addition.

10.1.5.3.2 Strategies

Available wild indigenous germplasm of tuber crops need to be explored, conserved, documented, and evaluated. There is an urgent need to take appropriate steps for selection and multiplication of elite planting materials. Identification of short duration, high-yielding varieties suitable for intercropping in plantation crops could be beneficial. Considering the increasing market demand for tuber crops, efforts should be taken to create awareness about the scientific cultivation and management practices and popularization among the farmers and tribal communities through trainings and demonstrations. The technologies on value addition of root and tuber crops developed by the ICAR-CTCRI, Trivandrum, and SAUs need to be popularized in these islands.

10.1.5.4 Fruit Crops

The islands host a wide range of diversity of genetic resources in fruit crops, some of these being native to the region, while many are introduced species. Area-wise, important fruits in islands are banana (51.7%), papaya (10.0%), mango (9.1%), pineapple (8.7%), and sapota (5.0%), while other underutilized fruits share 6.2% of total recorded fruit coverage (3,241 ha).

10.1.5.4.1 Constraints

In case of banana, introduction of viral diseases from the mainland along with planting material has resulted in spread of devastating viral diseases, thereby reducing the yields. Leaf spot/ blight (*Cercospora musicola*) is an emerging problem in various parts of the islands. Pseudostem weevil is also causing severe yield loss in commercial banana fields. In case of underutilized and exotic fruit crops, most of the growers are not aware about the cultivation practices. Lack of idiotypic mother plants, non-availability of planting material of improved varieties, and limited area for expansion are restricting the large-scale adoption of these crops in the islands (Srivastava et al., 2009). In case of mango, anthracnose (*Colletotrichum gleosporoides*) and fruit fly are the major impediments in commercial scale cultivation. In papaya, fungal diseases mainly leaf spot and foot rot are affecting the cultivation. Further, the incidence of infestation with mealy bugs has resulted into complete crop losses in parts of the island.

10.1.5.4.2 Strategies

Mass awareness campaigns for the complete eradication of viral diseases is required. Procurement of planting material (especially tissue cultured plantlets of banana) only from certified agencies and nurseries could also restrict the spread in new areas. Quarantine certificates for plant movement should be implemented. Use of clean, pest free material and timely application of appropriate chemicals could control the spread of pseudostem weevil. For underutilized and exotic fruits, superior varieties need to be introduced, and mother blocks should be established for facilitating the multiplication of these crops under island condition, thereby reducing the dependency on mainland supply. Value addition in these crops should be taken up to popularize them among the masses. In case of mango, adoption of phytosanitary measures including collection and destruction of the infested fruits, parapheromones could help in tackling the pest and disease problems. Removal of weeds, which act as alternate hosts; need-based pesticide application; and introduction of encyrtid parasitoids are the key strategies for the eradication of mealy bugs of papaya.

10.1.5.5 Spice Crops

Being *low volume-high value* crops, spices offer commercial status to the agriculture. Black pepper, cinnamon, nutmeg, clove, ginger, and turmeric are the major spice crops grown in the ANI. Most of the species being shade-loving in nature, the interspaces between coconut and arecanut could be effectively utilized for the cultivation of spice crops.

10.1.5.5.1 Constraints

In the case of black pepper, adoption of obsolete varieties is one of the major factors for low productivity. Prevalence of water scarcity during dry season and water logging in wet season also reduces the yields by manifolds. A number of farmers are not aware about the improved production and post-harvest technologies, which are the cause of concern for the low productivity of the vines. Foot rot (*Phytopthora capsicii*) and pepper leaf blight (*Colletotrichum capsicii*) are the major diseases noticed in the farms. In the case of nutmeg, the use of seedling progeny increases the gestation period and the dioecious nature reduces the productivity largely. Non-availability of the planting material of improved varieties and seed propagation of the crop are also hampering the productivity. Nutmeg is sensitive to water stress condition and the prevalence of water scarcity during dry season results in lower productivity. In the case of cinnamon, the lack of awareness about scientific cultivation and harvesting practices is one of the major problems. Harvesting of bark requires skill and the shortage of trained workers during harvesting season deters the farmers from harvesting the plants. Seedling progenies in clove-like nutmeg increases the gestation period and makes harvesting difficult. Among ginger and turmeric, the availability of quality planting material is difficult. The production is also hampered due to the incidence of rhizome rot/leaf blight of ginger and turmeric (*Pythium*/*Fusarium* spp.).

10.1.5.5.2 Strategies

Introduction and evaluation of improved varieties with tolerance to biotic and abiotic stresses could help in improving the productivity. Subsequently, the mother blocks of promising varieties need to be established for large-scale multiplication and distribution to the farmers. Wild relatives of spices, which are endemic to the islands could be tested for their response to various biotic and abiotic stresses and their suitability for possible use as root-stocks for cultivated species could be explored. Awareness programs and trainings to the stakeholders about latest technologies including crop production, integrated pest and disease management, postharvest management, and value addition could improve the profitability of the spice cultivation in the Islands. Promotion of intercropping of spices under existing coconut and arecanut plantations could help to utilize the available area effectively. Bush pepper cultivation and grafting of pepper and nutmeg could solve the problems of biotic and abiotic stresses in these crops. Establishment of demonstration plots of improved technologies would help the farmers in adoption of new technologies. Development/identification of island-suited varieties in the perennial spices could boost the productivity to a great extent. Organic certification of the spices and promotion of "the islands" brand could help in harnessing the export potential.

10.1.5.6 Flower Crops

ANI being the hub of orchids and ferns, there is a considerable scope for sustainable development of this sector in the islands. Other crops like tuberose, jasmine, marigold, and specialty flowers, which are otherwise imported from the mainland, have commercial potential. The technology could be disseminated through trainings and other demonstrations. Biodiversity need to be explored for the identification of ornamental species with commercial potential. Being a tourist destination, landscape gardening is of great practical utility. This sector has been picking up gradually among the island farmers.

10.1.5.7 Noni

Full production technology has been developed for noni cultivation along with suitable varieties. However, timely and assured lifting of the produce, marketing and value addition could boost its success.

10.1.5.8 Mushroom

Being the places of tourist interest, there is ample scope for mushroom cultivation in the islands. The substrate like paddy straw is also available in plenty in the islands.

10.1.5.8.1 Constraints

Lack of awareness among the island farmers on scientific mushroom cultivation has limited its widespread adoption. Lack of availability of spawn (seed) materials also deters the farmers from taking up the mushroom cultivation.

10.1.5.8.2 Strategies

Regular awareness cum training programs on scientific mushroom cultivation during *kharif* season and training on spawn production technologies during *rabi* season needs to be given to the stakeholders (farmers/field officers/ extension workers/agricultural officers) by ICAR-CIARI, Port Blair, along with various organizations such as NABARD, State Department, CIPMC, NGOs, and SHGs. Timely and regular availability of quality spawn material need to be ensured by ICAR-CIARI, Port Blair, along with line departments. The private entrepreneurs on spawn production may be developed at farmer/ youth level through proper trainings.

10.1.6 GENERAL STRATEGIES FOR THE DEVELOPMENT OF HORTICULTURE SECTOR IN THE ISLANDS

The climatic conditions of the islands are favorable for growing a large number of horticultural crops, and developed technologies have shown promise in improving the productivity and off-season cultivation. However, various issues such as poor transport infrastructure, inadequate storage facilities, and a fragmented supply chain are eroding the islands' advantage as a low cost-producing region. Logistics delays cause wastage of fruits and vegetables and nullify the hard work of the farmers.

Considering the vast opportunities, a mission mode approach is required to achieve holistic development of horticulture in the islands. Farmers need to be educated and practically trained about the latest technologies for making the venture profitable. Demonstration plots of all potential crops and technologies need to be developed. Promotion of high value-export suited crops such as spices, exotic fruits, and ornamental crops through scientific production and postharvest management practices need to be done on priority basis. Multiple crops-based cropping systems, high-density orcharding, cooperative and contract farming, precision horticulture (including protected cultivation, advanced irrigation systems, mulching, mechanization, etc.), integrated pest and disease management, organic farming, value addition, and postharvest management would be the key strategies in the context of islands.

Reducing the dependence on mainland for the planting material by promoting the establishment of mother blocks of improved varieties and setting up of nurseries and plant tissue culture unit for high-value horticulture crops is the need of the hour. This would also check the entry of new pathogens and pests from the mainland. Though the progress of horticulture sector in the islands is satisfactory, it needs to be improved particularly to assure availability of quality produce at a reasonable price. A balance between the subsistence and commercial farming could help in achieving the profitability of the sector without affecting the precious biodiversity and environmental values.

10.1.7 GAPS AND ISSUES

10.1.7.1 Livelihood Options for Agriculture and Allied Activities in the Islands

Tourism and agriculture and allied activities are the livelihood opportunities in the ANI. However, the major tourist spots being centered in and around South Andaman, agriculture and allied sectors remain the only options for livelihood for people living in islands, which are located away from the mainstream. Thus, the development and dissemination of agriculture technology need to be focused more in these far-off areas.

10.1.7.2 Marketing in Major Horticulture Crops

The surplus production pockets of the islands for major crops like vegetables need great support for reducing postharvest losses through timely and assured lifting of their produce and movement to other places. In this regard, support from agencies like NABARD and government departments for forming and streamlining farmer producer organization will be beneficial. Similarly, timely movement of other perishable items should be facilitated from production to consumption centers.

10.1.7.3 Pest and Disease Management

10.1.7.3.1 Rodent Management

Rats and mice are a constant threat to cultivated crops in the ANI. Not only do they attack standing crops, but they also eat and spoil food commodities in storage. Around 18 species of rodents have been known to infest various crops including coconut and vegetable crops of which three are new species reported from the islands. The larger bandicoot rat, *Bandicota bengalensis*, a new entrant pest in the islands is supposed to have arrived along with cargo of food grains from the mainland. Zinc phosphide (2%) and bromodialone (0.005%) cakes are commonly recommended rodenticides to kill rats and mice. However, the animal is too shrewd for rat traps and also can learn to avoid poison baits. Research activities need to be initiated toward community-based approaches in rodent management, which can be the only solution to subdue the growing population and inter-island spread of *B. bengalensis* and other noxious rat species.

10.1.7.3.2 Fruit Fly Management

Fruit flies in cucurbits cause drastic yield reduction and most often farmers resort to pesticide application to suppress their population. Lack of awareness on male annihilation technique using methyl eugenol or feeding attraction lures is attributed to severe crop losses encountered. Moreover, the non-availability of lure traps in the island conditions is a bottle neck in effectively combating fruit fly menace in cucurbits.

10.1.7.3.3 Biocontrol Agents

The rhinoceros beetle *Oryctes rhinoceros* is the most dreaded pest of coconut in the islands. Adults bite through tender fronds in folded condition and softer portions of unopened inflorescence, while grubs are saprophagous and breed in manure pits. Repeated attack by the pest causes palm stunting, reduction in numbers and size of nuts, and may even kill young palms and seedlings outright. Jacob (1996) worked on the introduction and establishment of Oryctes Nudi virus for the control of the beetle in several parts of North and South Andamans and achieved more than 80% control of the beetle population. Hence, research initiatives should be take on re-introduction of Oryctes Nudi virus for the effective management of *O. rhinoceros*. Further, alternative biocontrol measures like use of entomopathogenic fungi such as *Metarhizium* formulations in manure pits for the control of grubs need to be explored.

The papaya mealy bug *Paracoccus marginatus*, an introduced pest in the islands, is wreaking havoc in papaya crop and also in many vegetable crops. Introduction and establishment of *Acerophagus papaya*, an encyrtid parasitoid, has given tremendous levels of successful control of the mealy bug in south India, especially Karnataka, Tamil Nadu, and Andhra Pradesh. Hence, initiatives to explore the presence of native parasitoids and introduction of *A. papaya* in the ANI needs to be taken up for the sustainable management of mealy bugs in papaya.

Borer pests cause critical yield losses in fruits and vegetables. Pesticidal sprays often prove ineffective because the damaging life stage of the insect pest is cryptic and well concealed within the plant. Egg parasitoids in this case might prove to be an effective solution so that the population build up is hampered well before the pest initiates crop damage. *Trichogamma* sp. parasitoids available as tricho cards from NBAIR, Bangalore, need to be evaluated under island conditions and their use should be promoted in long term.

Hence, there is a need of evolving and promoting suitable IPM and IDM modules for pest/disease management as well as for the assessment of pesticide residues. Strengthening of island-specific IPM program and biocontrol agents with proven efficacy along with continuous monitoring of pesticide residues is essential. In this regard, search and cultivation of suitable disease/pest-resistant varieties are very useful. New molecules should be tested for their effectiveness under island conditions. There should be check on the use of spurious pesticides.

10.1.7.4 Organic Production – Mandate and Feasibility

Further efforts are required to standardize and fine-tune the organic production technology. Organic farming has significance in the island ecosystem as the islands are reservoirs of biodiversity, and existing practice of inclusion of animal component in farming would facilitate the practice of organic farming (Dam Roy et al., 2015). However, organic agriculture/limited organic agriculture should be followed on need basis and in specific areas. Adequate provision for organic certification and marketing should be ensured for the economic and practical feasibility of the proposition. Brainstorming workshops on organic production for the islands should be organized. In this context, the utilization of host plant resistance against main diseases prevalent in the islands is also of great importance (Singh et al., 2015).

10.1.7.5 Protected Cultivation Technology for High-Value Vegetables and Flowers

The tourism has become a major economic activity, and in the next five years, high-end tourism will emerge as a major component. This will alter the requirement pattern of the vegetables and flowers. To make farmers a partner in this growth, it is essential that the production of vegetable and flowers be ensured throughout the year. This will require large-scale protected cultivation, which needs intensification of research on structural design as well as package of practices compatible to different requirements and farmers' socio-economic conditions.

10.1.7.6 Post-Harvest Technology and Value Addition

The production centers are distributed all over different small islands, while the consumption center is mainly Port Blair. Due to this, large-scale processing units are not possible. Thus, there is a need for local level processing technologies to enhance the shelf-life, reduce losses, and value addition for crop and horticultural products

10.1.7.7 Biosecurity and Quarantine

There is a likely introduction of new diseases and pests in the islands from the mainland and vice versa. The unregulated movement of these pernicious pathotypes/biotypes either through crops commodities or through seeds can cause havoc in terms of new diseases and pests on the hitherto pest-free islands. The chemical options for the control of such biotic menace are also limited in the view of organic agriculture advocated for the islands. Hence, ICAR-CIARI has initiated the establishment of composite biosecurity research facility at the institute. In addition, the proposition needs more proactive cooperation and control by the regulatory agencies and related departments.

10.1.7.8 Biodiversity Management for Sustainable Development

The islands, being one of the 22 agro-biodiversity hotspots of India, need careful attention and support for exploration, documentation, and harnessing their genetic resources. This should be done through farm conservation of native diversity, respecting IPR guidelines, farmers' rights, benefit sharing, and custodian farmers for future developments on sustainable basis.

10.2 LAKSHADWEEP GROUP OF ISLANDS

10.2.1 INTRODUCTION

Coconut is considered as the only major crop of the Lakshadweep islands as out of 3,228 ha of total geographical area, about 2,598 ha is under coconut cultivation (Table 10.3). Laccadive Ordinary, Laccadive Small, and Laccadive Micro are the popular ecotypes in Lakshadweep islands. The islands also boast to have the highest productivity of 22,310 nuts/ha. Considering this, emphasize has been given to develop integrated farming systems, which can support product diversification and value addition. Products such as *dweep halva* (prepared from coconut powder), coconut jaggery, vinegar and *neera* are the locally prepared value-added products. The Department of Agriculture of the UT provides various inputs at subsidized rates, and spraying of plant protection chemicals is carried out free of cost.

Since the last two decades, most of the coconut production in the Lakshadweep islands is organic. However, the complete benefits of organic production could be reaped only after obtaining the organic certification, which was done under National Horticulture Mission during 2010. The available information suggested that about 3,844 farmers have been registered with a total area of 920 ha in this certification program during the first phase.

10.2.2 CONSTRAINTS FACED IN COCONUT CULTIVATION

Nonpossibility of area expansion is the basic constraints in the LDI. Further, the lack of availability of fresh water for irrigation is the major reason for rainfed cultivation in the islands. The harvesting of coconut requires skilled persons, and there is always scarcity of trained climbers for harvesting. Mandari, bud rot, and stem bleeding have been reported to be the serious problems in the Lakshadweep islands. The UT agricultural department conducts regular awareness and training programs, and incentives are given for the removal of affected palms. Leaf spot have also been reported in the islands. Rats are the major pest of coconut in all the islands, causing about 20–30% damage to the produce. However, implementation of massive rodent control programs and timely management measures have brought down the damage to about 5–10%. Rhinoceros beetle is also causing significant damage; however, the adoption of integrated practices including baculovirus has reduced the incidence to a greater extent. The organic manure and composting sites are serving as breeding grounds for the beetles. Outbreaks of hairy caterpillars are a common phenomenon at the end of dry season. Due to the highly dense nature of coconut plantations, the Eriophyid mite has emerged as an alarming pest in some islands in the recent past. Apart from the organic certification mission, the other coconut development programs include:

- Coconut harvesting scheme for the farmers and department farms

TABLE 10.3 Area and Production of Main Horticultural Crops in Lakshadweep

Commodity	Lakshadweep islands	
	Area (000 ha)	**Production (000 ha)**
Coconut	2.60	48.80
Vegetables	0.25	0.33
Fruits	0.22	0.48

- Insurance and accident relief to coconut climbers
- *Neera* tapping program
- Land rent for established model organic farms
- Supply of organic inputs and farm implements
- Recycling of coconut waste, organic composting, fish composting, and vermicompost
- Coconut value-addition centers
- Compensation for bud rot/stem bleeding affected coconut palms
- Control of hairy caterpillar
- Maintenance of biocontrol lab

10.2.3 OTHER HORTICULTURAL CROPS

Most of the fruits and vegetables requirement of the islands is met by the supply from the mainland India. In order to reduce the dependency and make the islands self-sufficient, a number of developmental activities have been initiated. Banana, vegetable crops, and tuber crops are being promoted for cultivation. Distribution of various inputs such as quality planting material, implements, and fertilizers is done by the Department of Agriculture at subsidized rates, while pesticides are supplied free of cost. About 50 ha area has been brought under the organic vegetable cultivation in the islands. Mushroom cultivation is also being promoted in the islands.

10.2.3.1 Constraints

Sucking pests, fruit borers, and nematodes are commonly occurring pests in various intercrops grown in coconut plantations. Banana bunchy top, mosaic, and little leaf are the common viral diseases of banana and vegetables.

10.2.4 MAJOR DEVELOPMENTAL PROGRAMS AND INITIATIVES

Among the developmental programs being run by the administration, the major emphasis has been given to intensive cultivation of vegetables, fruit and tuber crops, procurement of seeds of high-yielding vegetable varieties, establishment and maintenance of herbal garden, maintenance of departmental agricultural farm, and maintenance of soil testing laboratory.

Apart from the efforts made by the Department of Agriculture, a number of activities have been taken up by the Krishi Vigyan Kendra for planning and conducting production-oriented need-based training courses, organizing field days, farm visits, farmers fair, radio talk, farm science clubs, etc. Demonstration plots have also been established by KVKs scientific lines for disseminating the latest technical know-how. Apart from the direct benefits to the farmers, practical trainings are also being imparted to the teachers and the students of vocational agriculture and home science and nutrition education for rural community, particularly for rural women.

10.3 CONCLUSION

This chapter has apparently dealt with the major issues plaguing the horticultural development in the tropical island ecosystem of India; initiatives taken up by the stakeholders, be it the administration or the research institutions located in these islands; and future strategy for the upliftment of farmers in these islands. Since the advent of agricultural activities in these islands, there has been a positive shift toward emerging technologies developed in the mainland India or elsewhere. However, the speed of adoption has not been satisfactory enough time due to various reasons. In the wake of rising concerns about vulnerability of islands to climate change, future interventions must focus on devising island specific technologies and products for sustainable development of these biodiversity hotspots.

KEYWORDS

- **biodiversity**
- **climate change**
- **island ecosystem**
- **nutritional security**

REFERENCES

Dam Roy, S., Velmurugan, A., Swarnam, T. P., & Gautam, R. K., (2015). Organic farming in Andaman and Nicobar Islands: *A Scientific Perspective for Policy Decisions*, CARI, Port Blair.

Gautam, R. K., Waman, A. A., Bohra, P., Singh, P. K., Singh, A. K., Singh, S., Baskaran, V., Abirami, K., Jaisankar, I., & Dam Roy, S., (2015). In Souvenir, National seminar on harmonizing biodiversity and climate change: *Challenges and Opportunity*.

Kumar, V., & Gangwar, B., (1985). Agriculture in the Andaman's-an overview. *J. Andaman Sci. Assoc.*, *1*, 18–27.

Medhi, R. P., Damodaran, T., & Shiva, K. N., (2001). Genetic resources of horticultural crops in Andaman and Nicobar Islands. *J. Andaman Sci. Assoc*, *19*, 77–79.

Rethinam, P., (2009). *In Proceedings of National Workshop Cum Seminar on Status and Future Strategies for Horticultural Development in Andaman and Nicobar Islands*, Port Blair, India, Jan 23–25, Srivastava, R. C., Singh, D. R., Sudha, R., Jaisankar, I., Singh, S., Damodaran, V., Pandey, V. B., & Singh, L. B., (eds.). CARI: Port Blair.

Singh, D. R., Singh, S., Pandey, V. B., & Srivastava, R. C., (2009). In: *Proceedings of National Workshop cum Seminar on Status and Future Strategies for Horticultural Development in Andaman and Nicobar Islands,* Port Blair, India, Jan 23–25, Srivastava, R. C., Singh, D. R., Sudha, R., Jaisankar, I., Singh, S., Damodaran, V., Pandey, V. B., & Singh, L. B., (eds)., CARI: Port Blair.

Singh, D. R., Singh, S., Sankaran, M., Damodaran, V., Sudha, R., & Dam Roy, S., (2012). *Horticultural Technologies for Andaman and Nicobar Islands*, CARI, Port Blair.

Singh, S., Gautam, R. K., Singh, D. R., Sharma, T. V. R. S., Sakthivel, K., & Dam Roy. S., (2015). Genetic approaches for mitigating losses caused by bacterial wilt of tomato in tropical islands. *European J. Plant Pathology*, *143*(2), 205–221.

Srivastava, R. C., Singh, D. R., & Singh, S., (2009). *In Proceedings of National Workshop cum Seminar on Status and Future Strategies for Horticultural Development in Andaman and Nicobar Islands*, Port Blair, India, Jan 23–25, Srivastava, R. C., Singh, D. R., Sudha, R., Jaisankar, I., Singh, S., Damodaran, V., Pandey, V. B., Singh, L. B., Eds. CARI: Port Blair.

CHAPTER 11

ORGANIC VEGETABLES FARMING AND ITS PROSPECTS IN PAKYONG EAST DISTRICT OF SIKKIM, INDIA

ANJANA PRADHAN, DEEKI LAMA TAMANG,
SANGAY GYAMPO BHUTIA, and PABITRA SUBBA

Sikkim University, 6th Mile, Tadong, Gangtok East Sikkim, India,
E-mail: yemilan@yahoo.co.in

CONTENTS

ABSTRACT

Organic vegetable farming in Sikkim has been a major scope for income raising and uplift of farmers. In Sikkim, farmers are mostly cultivating local germplasm that they consume as well as sell in the local market. A study was conducted in Pakyong district under four GPU among 45 respondents to evaluate the effects of organic vegetable farming done by farmers. Socio-economic condition of farmers involved in organic farming was analyzed based on the market value and cost of production of vegetables. Descriptive and analytical research design was used that involved questionnaire and interview method.

11.1 INTRODUCTION

Organic farming can generally be described as a method of production that utilizes non-synthetic inputs and emphasizes biological and ecological process to improve soil quality, manage soil fertility, and optimize pest management. The industrialization of agriculture in the 1940s served as a point of departure from conventional agriculture to shape a new production paradigm that avoided chemical inputs. As consumer demand for organic food increased, organic industry stakeholders requested the creation of United States federally regulated standards to facilitate national and international trade (Treadwell et al., 2003).

Today, organic production is a combination of new technology and traditional methods. As a result of recent research, there are many new tools for organic farmers to use including soil analysis, plant nutrient monitoring, and integrated pest management systems. Additionally, there are many new commercial organic fertilizers and pesticide products on the market that have made organic farming more user-friendly than ever (www.smallfarms.ifas.ufl.edu).

Organic vegetable farming in Sikkim has been conceptualized by the Hon'ble Chief Minister way back in 2003. The government of Sikkim stepped into organic mission process from 2003, the year when it stopped imports of chemical fertilizer in the state. Since then, the cultivable lands in Sikkim are practically organic and farmers in Sikkim are traditional users of organic manures (www.sikkim.gov.in).

There is a major scope for income raising and upliftment of farmers. In Sikkim, farmers are mostly cultivating local germplasm that they consume as well as sell in the local market. A study was conducted in Pakyong district under four GPU among 45 respondents to evaluate the effects of organic vegetable farming done by farmers. Socio-economic condition of farmers involved in organic farming was analyzed based on the market value and cost of production of vegetables. Descriptive and analytical research design was used that involved questionnaire and interview method.

11.2 METHODOLOGY

This study is based on information accumulated from the field survey conducted in four GPUs in east district of Pakyong, namely Karthok, Chalamthang, Pacheykhani, and Dugalakha. Forty-five households were randomly selected. The farmers were directly interviewed and consulted using prepared questionnaires for collecting the information on the overall activities, i.e., social aspect, economic aspect, and other relevant information.

11.3 RESULTS AND DISCUSSION

The survey that was conducted in different GPUs in east district of Pakyong yielded various results. From Table 11.1, we can see that the number of males in Karthok was more than that of the number of females; likewise, in the other GPUs like Chalamthang and Pacheykhani, the number of males were more than the females. Only in Dugalakha, the female outnumbered the males.

From Table 11.2, distributions of respondents by organic farming practiced at home show that all the people residing in the different GPUs had

TABLE 11.1 Distribution of Respondents by Age and Sex

GPU	Description		Total
	Male	**Female**	
Karthok	33	25	58
Chalamthang	20	18	38
Pachekani	30	28	58
Dugalakha	25	28	53
Total	**108**	**99**	**207**

TABLE 11.2 Distribution of Respondents by Organic Farming Practiced at Home

GPUs	Vermi-compost	Cow urine	E.M. compost	Compost pit of cow dung	Rural dry leaf compost
Karthok	-	15	15	15	9
Chalamthang	-	10	10	10	-
Pachekani	2	9	10	9	7
Dugalakha	-	10	10	10	-
Total	2	44	45	44	16

mostly used the different products made from the cow's waste. From the table, we can see that rural dry leaf compost had been used in fewer amounts. The respondents had used vermicompost in lesser amount as they informed us that they had lesser knowledge about the compost and the unavailability of the specific worms in the market had made them in using the lesser amount of the vermicompost in their practice of organic farming.

About the land holdings of the owner, they had more number of unirrigated lands than irrigated lands. This may be due to the lesser amount of rainfall and lesser availability of land, as a larger amount of land is needed for irrigation (Table 11.3).

Organic farming is an approach of farming without the use of chemical input (Singh, 2012). It increases the demand for green safe food. In the different GPUs surveyed, the trend of adopting organic farming has been increasing day by day, and it is the sign toward better living of human beings (Table 11.4).

Table 11.5 shows the income that they get from selling their organic vegetable to the local market or to the main market in Gangtok; all the people from the different places get an income of 15 to 2000 rupees per month, which they utilize for different purposes like paying the school fees, buying medicines, etc.

TABLE 11.3 Distribution of Respondents by Land Type

GPUs	Irrigated	Unirrigated	Both
Karthok	1	8	6
Chalamthang	3	5	2
Pachekani	4	5	1
Dugalakha	-	8	2
Total	**8**	**25**	**11**

TABLE 11.4 Trend of Consumption of Organic Farming

GPUs	Increasing	Decreasing	Constant
Karthok	Y	-	-
Chalamthang	Y	-	-
Pachekani	-	-	-
Dugalakha	-	-	-

TABLE 11.5 Income in Organic Vegetable Farming

GPUs	Income (Rs)/month			
	500–1000	1000–1500	1500–2000	2000–2500
Karthok		Y		Y
Chalamthang			Y	
Pachekani	Y		Y	
Dugalakha		Y		Y

They utilize the money in buying different agricultural tools like *sy*, spade, *kodali*, fork, shovel, etc. For cultivating the land, they hire local people so that the work is done fast, and for that, they have to pay their daily wages. Sometimes, they buy the seeds from the market, which may be costly; otherwise, the department mostly provides them the seeds. As people are practicing organic farming, they have stopped the usage of fertilizers, which is a very good trend (Table 11.6).

11.4 CONCLUSION

Organic farming has been adopted by farmers as a traditional way of farming in Sikkim since ages. Sikkim is rich in biodiversity with abundant plant species because of which the soil is rich in organic matter content and makes the conversion easier. It is therefore advantageous for Sikkim to adopt organic system of farming, keeping in view of protection of soil from degradation, protection of environment and ecology, and healthy living of the people for generations (Bhutia, 2015).

TABLE 11.6 Cost of Production

GPUs	Type of input			
	Seed	Agricultural tools	Fertilizer	Human resources
Karthok		Y		Y
Chalamthang		Y		Y
Pachekani	Y	Y		Y
Dugalakha		Y		Y

From the survey, we have seen that the people have fully adopted the practice of organic farming. More number of trainings and proper knowledge will enable them to be happy in growing their crops organically.

KEYWORDS

- **local germplasm**
- **market**
- **organic farming**

REFERENCES

Bhutia, P. T. (2015). State Policy on Organic Farming Government of Sikkim.

http://smallfarms.ifas.ufl.edu/organic_production/organic_vegetables.html.

http://www.sikkim.gov.in/portal/portal/StatePortal/Government/SikkimOrganic Mission.

Singh, J., (2012). *Basic Horticulture*. Kalyani Publishers, New Delhi, pp. 15.

Treadwell, D. D., (2014). *Introduction to Organic Crop Production*. UF /IFAS Extension, University of Florida.

CHAPTER 12

ORGANIC NUTRIENT MANAGEMENT FOR IMPROVING THE FRUIT QUALITY OF MANGO cv. ALPHONSO

H. LEMBISANA DEVI,[1] Y. T. N. REDDY,[2] THEJANGULIE ANGAMI,[3] PRATIVA SAHU,[4] E. REANG,[5] G. S. YADAV,[1] and B. DAS[1]

[1]*ICAR-RC for NEH Region, Tripura Centre, Lembucherra, West Tripura – 799210, India*

[2]*Division of Fruit crops, Indian Institute of Horticultural Research, Hessarghata Lake Post, Bangalore – 560089, India*

[3]*ICAR-RC for NEH Region, Arunachal Pradesh Centre, Basar, Arunachal Pradesh, India*

[4]*ICAR-Indian Institute of Water Management, Bhubaneswar, Odisha, India*

[5]*College of Agriculture, Lembucherra, West Tripura–799210, India, E-mail: lembihort@gmail.com*

CONTENTS

ABSTRACT

Mango (*Mangifera indica* L.) is a very important and most popular fruit of India due to its delicious taste, attractive fragrance, and health benefits. The present experiment was conducted at the orchard block no. 4, IIHR, Bangalore, India, during the year 2014 on 30-year-old mango trees cv. Alphonso, planted at a distance of 10 m × 10 m in a square system to study the effect of different organic nutrients and biofertilizers on the fruit quality of mango cv. Alphonso. The observation on quality parameters viz. fruit weight, volume, pulp content, peel thickness, pulp/stone ratio, total soluble solids (TSS), acidity, reducing sugar, non-reducing sugar, total sugar, vitamin-C, β-carotene content, and shelf-life were recorded against 14 different organic manures and biofertilizers treatments replicated thrice in a randomized block design having two trees per treatment per replication. Maximum fruit weight (258.60 g), fruit volume (256.67 cc), pulp weight (176.95 g), total sugar content (15.27%), ascorbic acid content (128.00 mg/100g pulp), β-carotene content (22.35 mg/100 g of pulp), and TSS content (17.87 °B) with a minimum fruit acidity (0.030%) were recorded from the fruits of plants supplied with 100% RDF (recommended dose of fertilizer) farm yard manure (FYM) + *Azotobacter* + phosphorous solubilizing bacteria (PSB) + vesicular arbuscular mycorrhiza (VAM). Treatment with 50% RDF FYM + *Azotobacter* + PSB + VAM recorded the maximum shelf life of 18.67 days. It can be concluded that among the different treatments, the application of 100% RDF FYM + *Azotobacter* + PSB + VAM gave the best results in respect of physical and chemical quality of fruit under organic management of mango cv. Alphonso.

12.1 INTRODUCTION

In the last few years, organic farming in India has attracted many farmers across the country, and many farmers have experimented successfully for crops like banana, papaya, pineapple, sapota, cashew nut, coconut, mango, passion fruit, etc. India is one of the leading fruit producers in the world, producing nearly 10% world's fruit production. However, most of the produce is consumed fresh (rather than processed) and also consumed domestically rather than exported. A large variety of tropical and subtropical fruits are grown in India, of which mango, banana, citrus, grape, guava, pineapple, litchi, and papaya are the major ones. India leads the world in the production of mangoes, bananas, sapota, and acid limes and has achieved high productivity in grapes. Given India's production capability in tropical and subtropical fruits, the next step would be to leverage this strength to capitalize on the opportunity created by the high demand for organic fruits across the world.

Among horticultural crops, fruits that have greater scope for export are valued much as organic food. The modern organic agriculture movement evolved in developed countries, mostly in temperate regions. Now, with a growing interest in organic cultivation as a management method for agricultural production in tropical and subtropical countries, greater attention needs to be given to developing standards and guidelines for organic agriculture applicable to tropical products. Export marketing opportunities for developing countries include those organic horticultural products that are not produced domestically in temperate countries such as spices, tropical fruits, and vegetables. A challenge for organic fruit producers is the maintenance of economic feasibility while complying with organic standards (Reddy, 2008).

Mango (*Mangifera indica* L.) is an important fruit crop of India and often refers to as "King of Fruits" grown in an area of 2.52 million ha with the production of 18.43 million metric tons (Indian Horticulture Database, 2014). India ranks first among world's mango producing countries accounting for about 46% of the global area and 40% of the global production. Worldwide production is mostly concentrated in Asia, accounting for 75% of the global production. Among internationally traded tropical fruits, mango ranks only second to pineapple in quantity and value. Major markets for fresh and dried mangoes are Malaysia, Japan, Singapore, Hong Kong and the Netherlands and canned mangoes are the Netherlands, Australia, United Kingdom, Germany, France, and USA. Southeast Asian buyers consume mangoes all year round. Their supplies come mainly from India, Pakistan, Indonesia, Thailand, Malaysia, Philippines, Australia, and most recently from South Africa. The varieties in demand in the international market include Kent, Tommy Atkins, Alphonso, and Kesar. Alphonso is one of the choicest commercial export varieties mainly grown in the states of Karnataka, Maharashtra, Tamil Nadu, Gujarat, and Andhra Pradesh. It is a dual-purpose variety used both for table purposes and processing. Regarding consumer preference, fresh vegetables and fruits are among the most popular organic products. Organically produced mangoes are generally preferred than the traditionally produced ones. With this background, the present experiment was undertaken to study the quality attributes of mango cv. Alphonso under organic management practices.

12.2 MATERIALS AND METHODS

12.2.1 EXPERIMENTAL SITE AND PLANTS

The field experiment was carried out during the year 2014 on 30-year-old mango trees cv. Alphonso, planted at a distance of 10 m × 10 m in a square system at orchard block no. 4 maintained by the Division of Fruit Crops, Indian Institute of Horticultural Research, Bangalore, India, located at 13.58°N and 78°E and elevation of 890 m. The experimental site is located at an elevation of 890 m above MSL at altitude of 13°N and a longitude of 17.37°E. The average maximum temperature in summer is about ~32.8°C and the minimum temperature about ~ 21.7°C, and the average annual relative humidity of this area is about 64% with rainfall of about 760 mm. The experiment was carried out in 14 treatments, viz: T_1 = 100% RDF Farm Yard Manure (FYM), T_2 = 50% RDF FYM, T_3 = 100% RDF FYM + *Azotobacter*, T_4 = 50% RDF FYM + *Azotobacter*, T_5 = 100% RDF FYM + Phosphorous Solubilizing bacteria (PSB), T_6 = 50% RDF FYM + PSB, T_7 = 100% RDF + FYM + *Azotobacter* + PSB, T_8 = 50% RDF + FYM + *Azotobacter* + PSB, T_9 = 100% RDF FYM + VAM, T_{10} = 100% RDF FYM + VAM, T_{11}= 100% RDF FYM + *Azotobacter* + PSB + VAM, T_{12}= 50% RDF FYM + *Azotobacter* + PSB + VAM, T_{13}= 100% RDF Fertilizers, and T_{14}= Control (No manures and fertilizers).

All the treatments were replicated thrice in a randomized block design having two trees per treatment per replication. The trees were maintained under uniform cultural practices. Regular spray of Blitox and Neemyl were done as and when required to control the pests and diseases. Ten uniform mature fruits from each tree were used for recording the various fruit quality parameters. Fruit quality attributes were estimated during the fruiting season (May–June 14) according to standard procedures (Rangana, 1986).

Analysis and interpretation of experimental data were done by employing the completely randomized design (CRD) method for laboratory studies; the data were analyzed statistically and the test of significance was worked out by following the statistical method, as described by Panse and Sukhatme (1985).

12.3 RESULTS AND DISCUSSION

12.3.1 FRUIT PHYSICAL QUALITY

The data recorded for fruit physical quality parameters showed significant variations under different treatments and are presented in Table 12.1.

The average fruit weight varied between 205.58 g and 258.60 g among the different treatments used in this experiment. The data showed the highest fruit weight of 258.60g from the plants supplied with 100% RDF FYM + *Azotobacter* + PSB + VAM (T_{11}) followed by 257.57 g with 50% RDF FYM + *Azotobacter* (T_4) compared with 205.58 g in control. The results with regard to fruit weight were in conformity with Devi et al. (2014) for litchi. The fruit volume was found to be maximum 256.67 cc from the plants supplied with 100% RDF FYM + *Azotobacter* + PSB + VAM (T_{11}) followed by 226.11 cc with 50% RDF FYM + *Azotobacter* (T_4) compared with 175.00

TABLE 12.1 Influence of Organic Nutrient Management on Fruit Physical Quality Parameters of Mango cv. Alphonso

Treatment	Fruit weight (g)	Fruit volume (cc)	Peel weight (g)	Seed weight (g)	Pulp weight (g)	Shelf life (days)
T_1	217.05	176.85	37.84	31.72	147.49	16.33
T_2	210.11	200.83	40.31	28.24	141.56	15.00
T_3	239.98	206.06	46.13	30.47	163.38	17.33
T_4	257.57	226.11	48.44	38.78	170.35	17.00
T_5	226.37	200.56	38.83	31.76	155.78	18.00
T_6	217.74	215.58	44.75	42.38	130.61	17.00
T_7	222.33	198.33	37.67	30.72	153.94	16.67
T_8	236.86	197.50	36.11	31.53	169.23	18.33
T_9	209.88	175.83	34.74	28.25	146.89	15.33
T_{10}	228.63	203.33	40.70	38.36	149.57	14.67
T_{11}	258.60	256.67	43.20	38.46	176.95	17.00
T_{12}	218.22	208.33	41.70	36.10	140.42	18.67
T_{13}	208.45	183.33	32.04	28.27	148.14	16.00
T_{14}	205.58	175.00	37.63	33.60	134.36	14.33
S.Em (±)	20.53	24.31	3.89	4.85	17.56	1.26
CD (5%)	59.15	70.03	11.22	13.98	50.58	3.64

cc in control. The average peel weight of fruit was found to be maximum 48.44 g in plants that received 50% RDF FYM + *Azotobacter* (T_4) and the minimum of 32.04 g with 100% RDF of fertilizers (T_{13}). The average pulp weight was maximum (176.95 g) in fruit from the trees treated with 100% RDF FYM + *Azotobacter* + PSB + VAM (T_{11}) followed by 170.35 g in 50% RDF FYM + *Azotobacter* (T_4). The application of organic manures and biofertilizers significantly influenced the pulp weight of fruit. The maximum seed weight of 42.38 g was observed in T_6 treatment and the minimum was found in T_9 (28.25 g).

The shelf-life of mango fruits was influenced significantly due to integrated nutrient management treatments. The treatments T_5 (100% RDF FYM + PSB, T_8 (50% RDF + FYM + *Azotobacter* + PSB), and T_{12} (50% RDF FYM + *Azotobacter* + PSB + VAM) showed the maximum period of storage or shelf-life (18.00–18.67 days) at room temperature. On the other hand, the control treatment reduced the storage or shelf-life of mango fruits, i.e., only up to 14.33 days.

12.3.2 FRUIT CHEMICAL QUALITY

The different organic treatments showed significant influence in the fruit chemical quality parameters, viz., total soluble solids (TSS), acidity, total sugar, reducing sugar, nonreducing sugar, ascorbic acid, and carotenoid content of fruit (Table 12.2).

The TSS content of fruit ranged between 14.47 and 17.87 °B in 2014 due to different treatments. Improvement in fruit quality was also observed with different treatments, and the maximum TSS content (17.87 °B) of fruit was recorded with the application of 100% RDF FYM + *Azotobacter* + PSB + VAM (T_{11}) followed by 17.70 °B with 100% RDF FYM + PSB (T_5) as compared with 14.47 °B in fruit from the control trees. The fruit acidity content varied between 0.030% and 0.055% among the different treatments. Considerable reduction in fruit acidity (0.030%) was recorded by treatment with 100% RDF FYM + *Azotobacter* + PSB + VAM (T_{11}) compared with 0.043% in fruits from the trees under untreated control. The total sugar content of fruit was maximum (15.27%) in T_{11} (100% RDF FYM + *Azotobacter* + PSB + VAM) treatment, followed by T_3 treatment (14.09%). The reducing sugar content of fruit was maximum (6.00%) in T_{12} treatment (50% RDF FYM + *Azotobacter* + PSB + VAM) and minimum (3.54%) in fruit from the control

TABLE 12.2 Influence of Organic Nutrient Management on Fruit Chemical Quality Parameters of Mango cv. Alphonso

Treatment	TSS (°B)	Total sugars (%)	Reducing sugars (%)	Non reducing sugar (%)	Acidity (%)	Ascorbic acid (mg/100 g)	Carotenoids (mg/100g)
T_1	17.57	11.98	3.98	7.60	0.055	107.52	20.52
T_2	16.17	13.90	5.56	7.93	0.051	94.72	22.21
T_3	16.67	14.09	4.21	9.39	0.047	120.32	19.27
T_4	16.07	11.44	5.70	5.45	0.051	110.08	16.13
T_5	17.70	11.27	5.28	5.69	0.051	58.88	18.85
T_6	17.33	14.05	5.38	9.40	0.051	97.28	15.64
T_7	16.97	11.05	3.60	7.07	0.047	110.08	21.82
T_8	16.67	10.47	4.93	5.27	0.043	79.36	19.90
T_9	15.00	10.13	4.06	5.76	0.038	99.84	20.11
T_{10}	15.80	12.16	4.68	7.10	0.060	81.92	16.06
T_{11}	17.87	15.27	5.19	8.42	0.030	128.00	22.35
T_{12}	17.07	9.95	6.00	3.75	0.047	94.72	15.44
T_{13}	16.60	13.21	4.20	8.56	0.034	97.28	15.05
T_{14}	14.47	13.81	3.54	8.75	0.043	84.48	14.47
S.Em(±)	1.12	1.50	0.83	1.23	0.006	12.50	3.35
CD (5%)	3.23	4.33	2.38	3.53	0.017	35.99	9.64

trees. The maximum non-reducing sugar content (9.40%) was found in T_6 treatment (50% RDF FYM + *Azotobacter* + PSB) followed by 9.39% percent in T_3 treatment (100% RDF FYM + *Azotobacter* + PSB). Application of organic manures and biofertilizers significantly influenced the ascorbic acid content of the fruit. Ascorbic acid content of fruit varied between 58.88 mg/100 g pulp and 128.00 mg/100 g pulp. The analysis showed trees provided with 100% RDF FYM+ *Azotobacter* + PSB + VAM (T_{11}) had the maximum ascorbic acid content of 128.00 mg/100 g pulp followed by 110.08 mg/100 g pulp in T_7 treatment (100% RDF FYM + *Azotobacter* + PSB). At the mature stage, the highest carotenoid content of pulp (22.35 mg/100 g of pulp) was recorded in fruits from the plants treated with 100% RDF FYM+ *Azotobacter* + PSB + VAM (T_{11}) followed by 22.21 mg/100g of pulp in T_2 treatment (50% RDF FYM). The lowest carotenoid content (14.47 mg/100 g of pulp) was observed in control.

Organic manures and biofertilizers have a direct role in nitrogen fixation, production of phytohormone-like substances, and increased uptake of nutrients and hence quality improvement of fruit characteristics. The quality improvement in fruits may be due to proper supply of nutrients and induction of growth hormones, which stimulated cell division, cell elongation, increase in number and weight of fruits, better root development, better translocation of water uptake and deposition of nutrients, which may be attributed to the better fertilizer use efficiency with the application of organic sources of nutrients, biofertilizers (Ranjan and Ghosh, 2006) The results showed that the effect of various organic nutrient application enhanced the conversion of complex polysaccharides into simple sugars, which is in conformity with the findings of Athani et al. (2009) in guava. The decrease in acidity of fruits may be attributed to the conversion into sugars and their derivatives by the reactions involving reversal of glycolytic pathway or might be used in respiration or both. Similar results have also been reported by Dutta et al. (2009) in guava. The consistency of the organic manures effect can be attributed to the fact that these sources were rich in nitrogen, phosphorus, potassium, resulting in greater absorption and translocation of said nutrients to the leaves of the trees, and consequently a larger pool of photoassimilates (Reddy et al., 2001). The function of biofertilizers in combination with different organic sources of manures like FYM and vermicompost in increasing yield and improving fruit quality was observed in litchi (Leu, 2010; Lal et al., 2011), pineapple (Devdas and Kuriakose, 2005), mango (Patel et al., 2005), and guava (Devi et al., 2014). Similar relevance of organic nutrient management was observed by Srivastava et al. (2005) in citrus who reported that the uncertainties in production, arising out of nutritional constraints, could be resolved to a greater extent by organic cultivation of citrus through organic carbon balancing and improved plant metabolism (Kohli et al., 1998; Ghosh, 2000). The mobilization of unavailable nutrients could also be affected by speeding up the rate of mineralization of various organic substrates. Use of microbial biofertilizers, on the one hand, and the utilization of VAM fungi as bioprotectors, bioregulators, and biofertilizers in citrus (Ishii and Kadoya, 1996), on the other hand, are likely to bring a desirable change in the quality of production, apart from the beneficial impact on soil health.

12.4 CONCLUSION

On the basis of the experimental findings, among the different treatments, application with 100% RDF FYM + *Azotobacter* + PSB + VAM gave the best results in respect of physical and chemical quality of fruit under organic management of mango cv. Alphonso. It can be concluded that with the availability of more technical know-how on the efficient use of bulky organic manure, integration with microbial bio-fertilizers, and better understanding of the mycorrhizal symbiosis, organic fruit production would be a pronounced favor among the commercial fruit growers in general and mango in particular, more than ever before, in future. Thus, organic mango would no longer be considered a backyard practice, but rather a real innovative production technique based on environment friendly means with abundant security for soil, environment, and sustainable horticulture.

ACKNOWLEDGMENT

The authors are extremely thankful to the Director of ICAR Research Complex for NEH Region, Umium, India, and Director of Indian Institute of Horticultural Research, Hessaraghatta Lake Post, Bangalore, India, for providing all the facilities required for completion of the work.

KEYWORDS

- **Alphonso**
- **azotobacter**
- **fruit quality**
- **mango**
- **organic**
- **shelf life**

REFERENCES

Devdas, V. S., & Kuriakose, K. P., (2005). Evaluation of different organic manures on yield and quality of pineapple var. Mauritius. *Acta. Hort., 666*, 185–189.

Devi, H. L., Poi, S. C., & Mitra, S. K., (2012). Effect of different organic and biofertilizer sources on Guava (*Psidium guajava* L.) 'Sardar'. *Acta. Hort., 959*, 201–208.

Devi, H. L., Poi, S. C., & Mitra, S. K., (2014). Organic nutrient management protocol for cultivation of 'Bombai' Litchi, *Acta. Hort., 1029*, 215–224.

Dutta, P., Moji, S. B., & Das, B. S., (2009). Studies on the response of biofertilizer on growth and productivity of guava, *Indian J. Hort., 66*, 99–42.

Ghosh, S. P., (2000). Nutrient management in fruit crops. *Fert. News, 45*, 71–76.

Indian Horticulture Database, (2014). National Horticulture Board, Ministry of Agriculture, Government of India 85, Institutional area, Sector-18, Gurgaon. 2015, pp. 91–99.(http// www.nhb.gov.in.).

Ishii, T., & Kadoya, K., (1996). Utilization of *vesicular-arbuscular mycorrhizal fungi* in citrus orchards. *Proc. Int. Soc. Citriculture, 2*, 777–780.

Kohli, R. R., Srivastava, A. K., Huchche, A. D., Paliwal, M. K., & Bhattacharya, P., (1997). Relationship of leaf nutrient status, soil available nutrients and microbial composition with fruit yield of Nagpur mandarin. *Abst. National Symposium on Citriculture*, Nagpur, India, pp. 53.

Lal, R. L., Mishra, D. S., & Rathore, N., (2011). Effect of organic manure on yield and quality of litchi cv. Rose Scented. *J. Eco-friendly Agric.*, 6(1), 16–18.

Leu, A., (2010). Organic lychee, rambutan, star apple, mangosteen and durian production. *Acta. Hort. 873*, 207–210.

Panse, V. G., & Sukhatme, P. V., (1985). *Statistical Methods for Agricultural Workers*, ICAR Publication, pp. 145–148.

Patel, V. B., Singh, S. K., Ram, A., & Sharma, Y. K., (2002). Response of organic manures and bio-fertilizer on growth, fruit and quality of mango cv. Amrapali under high density orcharding. *Karnataka J. Hort. 1*(3), 51–56.

Ram, R. A., & Pathak, R. K., (2007). Integration of organic farming practices for sustainable production of guava: A case study. *Acta. Hort., 735*, 357–363.

Ranganna, S., (1986). *Handbook of Analysis and Quality Control for Fruit and Vegetable Products. 2nd ed.*, Tata McGraw-Hill Publication Co. Ltd., New Delhi.

Ranjan, T., & Ghosh, S. N., (2006). Integrated nutrient management in sweet orange cv. Mosambi (*Citrus sinensis* Osbeck). *Orissa J. Hort.*, *34*, 72–75.

Reddy, P. P., (2008). Fruit Crops. In: *Organic farming for Sustainable Horticulture*. Scientific Publishers, India. pp. 141–186.

Reddy, Y. T. N., Kurian, R. M., Sujatha, N. T., & Srinivas, M., (2001). Leaf and soil nutrients status of mango (*Mangifera indica* L.) grown in peninsular India and their relationship with yield. *J. Applied Hort.*, *3*, 78–81.

Srivastava, A. K., Singh, S., & Marathe, R. A., (2002). Organic citrus: soil fertility and plant nutrition. *J. Sustainable Agric.*, *19*(3), 5–29.

CHAPTER 13

ORGANIC PRODUCTION OF TURMERIC IN NORTHEASTERN INDIA: CONSTRAINTS AND OPPORTUNITIES

MARY CHINNEITHIEM HAOKIP,[1] AKOIJAM RANJITA DEVI,[1] and KHUMBAR DEBBARMA[2]

[1]*Department of Spices and Plantation Crops, Faculty of Horticulture, Bidhan Chandra Krishi Viswavidyalaya, Mohanpur–741252, Nadia, West Bengal, India*

[2]*Department of Agricultural Entomology, Faculty of Agriculture, Bidhan Chandra Krishi Viswavidyalaya, Mohanpur–741252, Nadia, West Bengal, India, E-mail: chinmaryhaokip@gmail.com*

CONTENTS

ABSTRACT

Organic agriculture offers the most sustainable solution for developing the agricultural sector with least negative impacts on the environment. Eight northeastern states of India (18 lakhs ha of land) can be classified as "Organic by Default" as the crops in these states are grown virtually organic. Scope for organic farming in the hills is high as they are one of the mega biodiversities receiving very high rainfall (2000 to 11000 mm/annum), leading to profuse production of biomass. The increasing demand for organic food products in the developed countries coupled with its focus on agri-exports are the drivers for the Indian organic food industries. Indigenous turmeric of this region has vast potential for organic production and export quality. Since time immemorial turmeric has been used to cure liver problems, digestive disorders, skin diseases, and wound, with anti-inflammatory and anticancer properties. A very popular local variety of turmeric, *Lakadong* which is grown in Meghalaya, has high curcumin content of 5–5.05% and meets the export standard. A number of local cultivars exist in this region. Organic turmeric is superior to conventional as it contains high oleoresin and curcumin. Farm yard manure (FYM), neem cake, fish meal, rock phosphate, vermicompost, etc., resulted in the control of rhizome rot up to 53% with an increase in yield. Green manuring, mulching, crop rotation, use of biopesticides, and indigenous technical knowledge (ITK) should be used together in a proper balance for maintaining productivity and farmers' income. However, the main constraints for organic farmers is that they cannot afford to pay the fees required to gain official certification. Identifying certification agency and reduction of certification cost could solve the problem. Thus, this region of the country has potential for organic turmeric production that can meet domestic and international demand for improving livelihood through higher farm income.

13.1 INTRODUCTION

Northeastern India is bestowed with lot of potential to produce all varieties of organic products due to its various agro-climatic regions as this part of the country is "organic by default." This holds promise for the organic producers to tap the market, which is growing steadily in the domestic market related to the export market. The area has great potential of producing organic turmeric *(Curcuma longa L)*. It also has a very long history of medicinal use, dating back nearly 4000 years. In southeast Asia, turmeric is used not only as a principal spice but also as a component in religious ceremonies. Because of its brilliant yellow color, turmeric is also known as "Indian saffron." It is used in diversified forms as a condiment, flavoring, and coloring agent and as a principal ingredient in Indian culinary as curry powder. Turmeric is also used in clinical treatments of the digestive organs: the intestine, for the treatment of diseases such as familial adenomatous polyposis (Cruz-Correa et al., 2006); in the bowels, for the treatment of inflammatory bowel disease (Hanai and Sugimoto, 2009); and in the colon, for the treatment of colon cancer (Naganuma et al., 2006). It has many uses in cosmetic industry. It finds a place in offerings on religious and ceremonial occasions. A type of starch is also being extracted from a particular type of turmeric. The increasing demand for natural products as food additives makes turmeric as ideal produce as a food colorant. Organic farming is gaining gradual movement across the world. The turmeric produced in this region contains high oleoresin and curcumin content. The product is mostly marketed in the fresh form. The local demand being very limited, roughly 70–80% of the total production is reportedly available as marketable surplus from the region. Growing awareness of health and environmental issues in agriculture demanded for more production of organic food, which is emerging as an attractive source of rural income generation. The role of organic farming in Indian rural economy can be leveraged to mitigate the ever-increasing problem of food security in India.

13.2 PRESENT SCENARIO

In terms of area, turmeric is the third largest crop in the region. However, its productivity in the region is only 1.5 tons against 3.9 tons/ha in the country. The area under turmeric in the region is 17.27 thousand ha with a total production of 32.36 thousand tones. The productivity of the crop is much lower (1.87 t/ha) compared to the national productivity of 3.47 t/ha (Spices Statistics, Spice Board, 2004). The production of turmeric is highest in Meghalaya, followed by Assam, Tripura, and Nagaland. However, the productivity

TABLE 13.1 State-Wise Area, Production, and Productivity of Ginger and Turmeric in the North Eastern Region (2004–2005)

State	Area ('000 ha)	Production ('000 t)	Productivity (t/ha)
Arunachal Pradesh	0.40	1.50	3.75
Assam	12.00	8.00	0.67
Manipur	0.37	2.09	5.69
Meghalaya	1.60	8.70	5.44
Mizoram	0.30	2.97	9.9
Nagaland	0.60	3.10	5.17
Sikkim	0.50	1.70	3.40
Tripura	1.50	4.30	2.87
N.E. Region	17.27	32.3	1.87
India	150.5	521.9	3.47
2005–2006			
Assam	11,700	8,400	0.72
Tripura	1,108	3,750	3.38
Meghalaya	850	9,000	10.59
Nagaland	850	9,000	10.59
Sikkim	670	3,600	5.37
Arunachal Pradesh	427	1,631	3.82
Manipur	200	140	0.70
Mizoram	200	1,650	8.25

Source: www.kiran.nic.in/pdf/publications/Organic_Farming.pdf.

is highest in Mizoram (Table 13.1). The most popular cultivated variety in the region is Lakadong (7.5%) and Megha Turmeric-1 (6.8%) that possesses higher curcumin content and has maximum demand.

13.3 EXPORT POTENTIAL AND VALUE-ADDED PRODUCTS

India is the largest producer, consumer, and exporter of turmeric in the world, and it dominates the world production scenario, contributing 78% of the production. India exported 5150 tons of turmeric valued at Rs. 164.80 crores during 2006–2007. The country not only produce fresh rhizome but is also the global leader in value-added products of turmeric and exports. With its inherent qualities and high content of the important bioactive compound curcumin (5.4–7.6%) and high content of oleoresin, Indian turmeric is considered to be the best in the world market. Around 300 MT of curcumin cost Rs. 35 crores in the world market. The areas like the North Eastern Region (NER) of India and other hilly areas, where a lot of biomass is available from forest, weeds, crops, etc., organic farming would be more economical. Moreover, organic produce is expected to fetch premium price (at least 5%t) and therefore should be economical to the poor farmers. Apart from improved varieties like Lakadong and Megha Turmeric-1, a number of local cultivars exist in the NER. The product is mostly marketed in the fresh form. Lakadong Turmeric is available in whole dry form and powder. Because the local demand is very limited, roughly 70–80% of the total production is reportedly available as marketable surplus from the region. As it is abundantly available in the region, different products like turmerones (turmeric oil), oleoresin, and powder can be prepared for export, which are very common in developed countries. The farm gate sale price of cured turmeric has been considered at Rs. 25/kg.

Price difference between organic and nonorganic turmeric

Year	Nonorganic (Rs/kg)	Organic (Rs/kg)
2002	24.2	90.7

Sources: Spice board, 2002.

13.4 CONSTRAINTS

13.4.1 BIOTIC

One of the important biotic factors attributing to low productivity in turmeric is the non-availability of quality planting material. The serious diseases of turmeric are seed borne, viz., rhizome rot (*Pythium* sp., *Rhizoctonia* sp., *and Sclerotium rolfsii*), leaf blotch (*Taphrina maculans*), and leaf spot (*Colletotricum capsici*). Some of these, once introduced into cultivated fields, are very difficult to eradicate. The supply of quality planting material free from diseases can contribute enormously to enhance the productivity. Turmeric shoot borer (*Conogethes punctiferalis*) and rhizome scale (*Aspidiella hartii*) also causes large crop damage.

13.4.2 ABIOTIC

Turmeric is mostly grown in sub-tropical hill zones where soil is acidic in nature. Cultivation is being practiced on steep slopes under *jhum/bun* (raised beds) system in rainfed conditions without adoption of soil and water conservation. Deep virgin soils of forest brought under the *jhum* system are giving higher yields in the first and second years of cultivation even under zero nutrient management conditions. But heavy rains and earthing works associated with the cultural operations and harvesting accelerate the erosion, reducing the fertile soils into abandoned wasteland. In the second cycle of cultivation on such fields after a gap of 3–5 years, very low yields (5–8 t/ha) are obtained. Farmers apply only FYM at planting and no other nutrient application strategies are followed. These factors lead to low productivity.

According to Dr. S.V. Ngachan, Director, ICAR Research Complex for North Eastern Hill Region, "after much research and demonstration, they identified two varieties of turmeric, Lakadong and Megha varieties for their higher yield and quality." The biggest challenge faced by the turmeric growers in the region is lack of premium price for the produce. Postharvest losses of almost all the farm produce in the region are very high due to near zero facility for their handling, processing, value addition, packaging, and even organized marketing. Although the region produces best quality turmeric and also some other horticultural crops, there are very few processing units for any of these crops.

13.5 CERTIFICATION

To make a farming system organic, it does not only mean nonuse of chemicals, but it also includes managing the entire food production chain and observing strict farming regime. Besides, there are other aspects involved in the entire organic farming systems and trade, and one of the most important of all is the certification of the organic zones, production processes and the organic farm products. The northeastern states managed to achieve the distinction of emerging as a hub for the production and export of organic spices in India. The global certifying agency Indocert has certified 300 hectares of land in Meghalaya for turmeric and ginger cultivation, Manipur for ginger and turmeric cultivation, and Arunachal Pradesh for black pepper. According to the Spice Board, currently, the total area under conversion or certification process is 2220.58 hectare (ha). Demand for organically produced foods is growing rapidly in developed countries and the products command a premium. The board estimates that the region can create exportable surpluses at competitive prices so that the top slot occupied by the country in the international spice market would be maintained. By default, the land and agriculture practice is organic, and with some efforts from the state, the region could export organic spices, board source said. Organic spice exports from this region are likely to get a big boost after receiving this Organic Certification. Spices Board officials said in the North East, spices are traditionally grown with organic inputs. Already, the board has entrusted Indocert to process organic certification for the region's spices, apart from providing a range of subsidy for its cultivation. The board will label all spices produced in the region as organic to enhance export. The Spice Board has identified Arunachal Pradesh, Mizoram, and Meghalaya as having the maximum potential for organic ginger cultivation, while Assam and Manipur have potential for chilies. Nagaland has potential for both chilies and ginger cultivation, and Tripura can produce organic turmeric.

13.6 INDIGENOUS PLANT PROTECTION PRACTICES

Shoot borer (*Conogethes punctiferalis* Guen.) is the most important pest of turmeric larvae bore of the pseudostem and feed on the growing shoot resulting in yellowing and drying of the infested shoots. The farmers in this state have been followed indigenous pest and disease management practices. Some farmers plant rhizomes just after burning the field to avoid soil-borne disease and insect damage. The adult of shoot borer after emergence from the soil settle on the tree, and farmers collect and destroy them. Farmers reported that spraying neem oil @ 0.5%during July–October (at 21-day intervals) is effective against the shoot borer. Leaf blotch, a fungal disease caused by *Taphrina maculans*, appears as small oval rectangular brown spots on either side of the leaves. They soon become dirty yellow or dark brown. Tribal farmers removed mud from the bottom of the diseased plant to expose to the roots to the sun. This practice was found to reduce disease (rhizome rot) infestation. Progressive farmers are also deep plowed their field during summer to reduce the disease. Rotten plant roots scratched by farmers in Kandhamal and Keonjhar and applied wood ash as well as vermicompost @ 2 t/ha in field to manage the incidence. Farmers applied *Trichoderma viride*, *Beauveria bassian*, and pseudomonas to control rhizome rot. Farmers in Kandhamal district planted turmeric in red soil and under the shade of tree-like sal, mango, and jackfruit to reduce rhizomes diseases. It has been observed that progressive farmers used own seeds for planting change seed source every 2–3 years to reduce the spread of seed-borne diseases. Turmeric planted in the red soil was found to have less incidence of insect pest and diseases during the storage period (Naresh et al., 2015).

13.7 MAINTENANCE OF FERTILITY

Maintenance of soil fertility may be achieved through organic matter recycling, enrichment of compost, vermicompost, animal manures, urine, FYM, litter composting, use of botanicals, green manuring, etc. Organic inputs like FYM, neem cake, stera meal, rock phosphate resulted in the control of rhizome rot to 53% with an increase in yield. Use of biofertilizers like Azolla, *Azospirillium*, *Azotobacter*, *Rhizobium* culture, PSB, etc. can also be adopted. Blood meal, bone meal, and human excrement may be applied with the approval of the Certification Agency (CA). Saw dust from untreated wood, calcified seaweed, limestone, gypsum, chalk, magnesium rock and rock phosphate and various sprays like vermi wash and liquid manures etc. can be used in crops for nourishing the soil and plant.

13.8 PLANT PROTECTION MEASURES

If shoot borer incidence is noticed, such shoots may be cut open and larvae picked out and destroyed. If necessary neem oil 0.5% may be sprayed at fortnightly intervals for controlling shoot borer, and the application of *Trichoderma* at the time of planting can check the incidence of rhizome rot. No major disease is noticed in turmeric. Leaf spot and leaf blotch can be controlled by restricted use of Bordeaux mixture 1%.

13.9 CONCLUSION

Organic cultivation although not very new in the NER is becoming important in the agriculture sector in the northeast India, largely through the efforts of small groups of farmers. Considering the traditional way of growing spices, steps also have been taken to promote organic cultivation of turmeric. The board, under the Commerce and Industry Ministry, has decided to promote in the overseas market spices such as large cardamom and Lakadong turmeric grown organically in the seven sister states, Northeast can take advantage of this opportunity to transform its underused farmlands into a highly remunerative enterprise, thereby creating rural jobs and environmental sustainability. In addition to this, it will help in export earnings by applying the indigenous methods of crop production as well as crop protection measures thereby increasing the economic return of the country. With the sizable acreage under naturally organic/default organic cultivation, he NER has tremendous potential to grow crops organically and emerge as a major producer of organic products.

KEYWORDS

- **export**
- **ITK**
- **northeastern India**
- **organic farming**
- **turmeric**

REFERENCES

Cruz-Correa, M., Shoskes, D. A, Sanchez, P., (eds.). et al., (2006). Combination treatment with curcumin and quercetin of adenomas in familial adenomatous polyposis. *Clin Gastroenterol Hepatol., 4*, 1035–1038. [PubMed].

For sale organic spice from Northeast India, Jan 6, (2012). http://www.agricultureinformation.com/forums/sale/78337-sale-organic-spices-north-east-india.html.

Hanai, H., & Sugimoto, K., (2009). Curcumin has bright prospects for the treatment of inflammatory bowel disease. *Curr. Pharm. Des., 15*, 2087–2094. [PubMed].

Herbal Medicine: Biomolecular and Clinical Aspects. 2nd edition – Turmeric, the Golden Spice. http://www.ncbi.nlm.nih.gov/books/NBK92752/.

Jha, A. K., & Deka, B. C. *Present Status and Prospects of Ginger and Turmeric in NE States, Kiran* [online] www.kiran.nic.in/pdf/publications.

Munda, G. C., Das, A., & Patel, D. P. *Organic Farming in Hill Ecosystems–Prospects and Practices*. *Kiran* [online] www.kiran.nic.in/pdf/publications/Organic_Farming.pdf. Spice Board of India-Spices statistics, Cochin.

Naganuma, M., Saruwatari, A., Okamura, S., & Tamura, H., (2006). Turmeric and curcumin modulate the conjugation of 1-naphthol in Caco-2 cells. *Biol. Pharm. Bull, 29*, 1476–1479. [PubMed].

Naresh Babu, A. K., Shukla, Pali Marwar, & Manoranjan Prusty, (2015). *Journal of Engineering Computers & Applied Sciences* (JECAS), vol. *4*(2). http//www.borjournals.com.

North East spices to be promoted overseas by Spices Board-PTI (online).

Organic tokri, Northeast India- Roots of organic farming in Northeast.

Prospects of Organic Farming in Northeast, (2015) Rajesh [online] www.nelive.in/north-east/business/prospects-organic-farming-north-east.

Shukla, U. N., Mishra, M. L., & Bairwa, K. C., (2013). Organic Farming: Current Status in India. *Popular Kheti*, *1*(4). (October-December), [online] www.popularkheti.info.

Spices board to tap potential of N-E states. Sajeev [online] www.thehindubusinessline.com.

Spices Board, (2004). *Spices Statistics* . ginger_and_turmeric.pdf [online] http://mdoner.gov.in/sites/default/files/10_1.pdf.

The Economic Times. http://articles.economictimes.indiatimes.com/2015–07–26/news/64880536_1_spices-board-lakadong-turmeric-organic-spices (accessed Jun 23, 2018).

www.organictokri.in/aboutne.php.

CHAPTER 14

ORGANIC HORTICULTURE FOR SUSTAINABLE DEVELOPMENT: STRATEGIC OPPORTUNITIES AND RELEVANCE IN THE INDIAN PERSPECTIVE

T. K. HAZARIKA

Department of Horticulture, Aromatic and Medicinal Plants, School of Earth Sciences and Natural Resources Management, Mizoram University, Tanhril, Aizawl, Mizoram, India, E-mail: tridip28@gmail.com

CONTENTS

ABSTRACT

Adverse effects of modern horticultural practices on the health of all living beings and on the environment have been well documented all over the world. Their negative effects on the environment are manifested through soil erosion, water shortages, soil contamination, etc. Due to the negative impacts of modern horticulture, the concept of sustainable horticulture has developed. The successful management of resources for horticulture to satisfy changing human needs while maintaining or enhancing the quality of environment and conserving natural resources is sustainable horticulture. Organic horticulture is one of the several approaches found to meet the objectives of sustainable horticulture. It is considered the most coherent and stringent system that is designed and maintained to produce horticultural products by using methods and substances that maintain the integrity of organic horticultural products until they reach the consumer. This is accomplished by using substances to fulfill any specific fluctuation within the system so as to maintain long-term wastes to return nutrients to the land and handle the horticultural products without the use of extraneous synthetic additives or processing in accordance with the act and the regulations in this part. But, till now, India is lagging far behind in the adoption of organic horticulture. So far, the implementation of organic horticulture seems to be limited to a few horticultural crops. There is an urgent need that the government should pay attention at the policy-making levels for the spread of organic horticulture in the country. Substantial financial support by governments is absolutely necessary to promote organic farming. Similarly, market development for the organic products is a crucial factor to promote domestic sales. An important role of the government in this direction is giving various supports to the producer and consumer associations to market the products. The producer organizations must be encouraged to get accredited for inspection and certification in accordance with the National Standards for Organic Production.

14.1 INTRODUCTION

The importance of modern agriculture has been acknowledged in alleviating hunger from the world, because the world population more than doubled itself during the last half of the 20th century (Lal, 2000). However, even now globally, almost 800 million people still go hungry. Famines and scarcities have been known in India from the earliest times (Randhawa, 1983; Swaminathan, 1996). This was all in the era of organic agriculture. In contrast, there was no scarcity of food after the severe droughts of 1972 and 1987 (FAI, 2004) due to modern agriculture. India's own achievements in agricultural production after the Green Revolution in 1967–1968 has been exemplary and mainly due to increased use of the components of modern agriculture, namely, fertilizers, pesticides, and farm machinery. Nevertheless, overuse of pesticides especially in vegetables and fruits resulted in residues much above the safety levels (Carson, 1963; HAU, 2003), and this brought to the attention the ill-effects of modern agriculture.

Conventional agriculture is highly dependent on nonrenewable resources, namely, chemicals fertilizers, and pesticides. A growing segment of people in the world, including farmers, are questioning the chemicals and management systems used in conventional agriculture and the impact on their environmental, economic, and social well-being. Consequently, many individuals are seeking alternative practices that would make agricultural productivity more sustainable. Organic agriculture aims to reduce reliance on high levels of fertilizer and chemical inputs.

According to Codex Alimentarius (FAO, 2001) organic agriculture is a holistic production management system that promotes and enhances agroecosystem health, including biodiversity, biological cycles, and soil biological activity. The primary goal of organic agriculture is to optimize the health and productivity of interdependent communities of soil life, plants, animals, and people (Scialabba and Hattam, 2002).

It is a method of farming system that primarily aimed at cultivating the land and raising crops in such a way, as to keep the soil alive and in good health by use of organic wastes (crop, animal and farm wastes, aquatic wastes) and other biological materials along with beneficial microbes (biofertilizers) to release nutrients to crops for increased sustainable production in an eco-friendly, pollution-free environment. Organic farming is the form of agriculture that relies on techniques such as crop rotation, green manure, compost, and biological pest control.

Organic agriculture is internationally regulated and legally enforced by many nations, based in large part on the standards set by the International Federation of Organic Agriculture Movements (IFOAM), an international umbrella organization for organic farming organizations established in 1972 defines the overarching goal of organic farming as: "Organic agriculture is a production system that sustains the health of soils, ecosystems and people. It relies on ecological processes, biodiversity and cycles adapted to local conditions, rather than the use of inputs with adverse effects. Organic agriculture combines tradition, innovation and science to benefit the shared environment and promote fair relationships and a good quality of life for all involved."

A large number of terms are used as an alternative to organic farming. These are: biological agriculture, ecological agriculture, bio-dynamic, organic-biological agriculture, and natural agriculture. According to the National Organic Standards Board of the US Department of Agriculture (USDA), the word "Organic" has the following official definition (Lieberhardt, 2003): "An ecological production management system that promotes and enhances biodiversity, biological cycles and soil biological activity. It is based on the minimal use of off-farm inputs and on management practices that restore, maintain and enhance ecological harmony."

According to the definition of the USDA study team on organic farming, "organic farming is a system which avoids or largely excludes the use of synthetic inputs (such as fertilizers, pesticides, hormones, feed additives, etc.) and to the maximum extent feasible rely upon crop rotations, crop residues, animal manures, off-farm organic waste, mineral grade rock additives and biological system of nutrient mobilization and plant protection." FAO suggested that "Organic agriculture is a unique production management system which promotes and enhances agro-ecosystem health, including biodiversity, biological cycles and soil biological activity, and this is accomplished by using on-farm agronomic, biological and mechanical methods in exclusion of all synthetic off-farm inputs."

14.2 PRINCIPLES OF ORGANIC FARMING

The International Federation of Organic Agriculture Movements (IFOAM) in 1972 internationally recognized principles of organic farming. Some of these are:

1. To produce foodstuffs of high nutritional quality and sufficient quantity.
2. To interact in a constructive and life-enhancing way with natural systems and cycles.

3. To consider the wider social and ecological impact of the organic production and processing systems.
4. To encourage and enhance biological cycles within the farming system, involving microorganisms, soil flora and fauna, plants, and animals.
5. To maintain and increase the long-term fertility of soils.
6. To maintain the genetic diversity of the production system and its surroundings, including the protection of wildlife habitats.
7. To promote the healthy use and proper care of water, water resources, and all life therein.
8. To use, as far as possible, renewable resources in locally organized production systems.
9. To give livestock conditions of life that confirm to their physiological need.
10. To avoid all forms of pollution that may result from agricultural techniques.
11. To allow everyone involved in organic production and processing a quality of life that meets their basic needs and allows an adequate return and satisfaction from their work, including a safe working environment.
12. To progress toward an entire production, processing, and distribution chain that is both socially just and ecologically responsible.

14.3 WORLD SCENARIO OF ORGANIC FARMING

The total land holdings under organic farming has increased year after year globally. At present, the organic agriculture is now practiced in more than 141 countries of the world. About 32.2 million hectares of agricultural land is managed organically.

The largest share of organic agricultural land is by Oceania (37%), followed by Europe (24%) and Latin America (20%). The proportion of organically compared to conventionally managed land, however, is highest in Oceania and in Europe. In the European Union, 4% of the land is under organic management. Most producers are in Latin America. The total organic area in Asia is 2.9 million ha. This constitutes 9% of the world's organic agricultural land. Among Asian countries, China (1.6 mha) and India (1 mha) are having the maximum share. Global demand for organic products remains robust, with sales increasing by over five billion US Dollars a year. Organic Monitor estimates international sales to have reached 46.1 billion US Dollars. Consumer demand for organic products is concentrated in North America and Europe; these two regions comprise 97% of global revenues. Asia, Latin America, and Australasia are important producers and exporters of organic foods (Surekha et al., 2010).

14.4 ORGANIC AGRICULTURE PERSPECTIVE UNDER INDIAN CONDITIONS

India is bestowed with considerable potential for organic farming due to prevailing trend of integrated farming systems of crops and live stocks, high bio-diversity on account of diverse agro-climatic conditions and large number of small and marginal farmers. Besides, inherited tradition of low input agriculture in many parts of the country, particularly in hilly and rain-fed areas too, is an added advantage and augurs well for the farmers to shift to organic farming and tap the steadily growing domestic as well as overseas markets.

In India, about 74% farmers are small and marginal farmers. Organic farming is most relevant to them because they are resource poor to provide costly inputs for enhancing yield. Their only resource, viz., land, need to be prevented from degradations. In the organic farming system approach, a piece of land is used optimally and to its fullest potential to produce a range of nutritious and healthy food as well as other required commodities in a manner that can healthily feed a small family, and maintain soil health and productivity by agricultural practices based on principles of nature. Both insects and diseases are also controlled using biological control measures.

Unlike other countries, in India also, organic farming has received considerable attention in the recent past. To promote organic farming, the government of India has constituted a task force on organic farming, and the task force emphasized on the need for consolidating the information on organic farming and its benefits. The steering committee constituted by this task force has suggested taking up organic farming as a challenging national task and to take up this as a thrust area. The steering committee gave boost to organic farming in the rainfed areas and in the northeastern states where there is limited use of agricultural chemicals. Madhya Pradesh took early lead in this regard and Uttaranchal and Sikkim followed the suit, and these states have declared themselves as organic states (Marwaha and Jat, 2004). The Ministry of Commerce launched the National Organic Program in April 2000, and Agricultural and Processed Food Products Exports (APEDA) is implementing the National Programme of Organic Production (NPOP) (Gouri, 2004). Under the NPOP, documents like national standards, accreditation criteria for accrediting inspection and certification agencies, have been prepared and approved by the National Steering Committee.

In India, at present, there is around 76,000 ha of certified organic food production at the farm level and 2.4 million ha of certified forest area for collection of wild herbs in India (Bhattacharya and Chakraborty, 2005), but the actual area under organics is much more. In Maharashtra alone, about 0.5 million ha area is under organic farming since 2003; out of this, only 10,000 ha is the certified area. In Nagaland, 3,000 ha is under organic farming with crops like maize, soybean, ginger, large cardamom, passion fruit, and chili. Rajasthan has 5,631 ha under organic farming with crops like pearl millet, wheat, mungbean, guar, mustard, and cotton.

14.5 ORGANIC MOVEMENT IN INDIA

In India, traditionally, most of the farmers have been using organic inputs without using any chemical fertilizers. Since 1950s, the farmers gradually changed from organic to chemical-based cultivation, and again, chemicals were increasingly applied during the green revolution period. Though the introduction of green revolution agricultural technology in the 1960's reached the main production areas of the country, there were still certain areas (especially mountain areas) and communities (especially certain tribes) that did not adopt the use of agro-chemicals. Therefore, some areas can be classified as *organic by default*, though their significance and extent has been rather overemphasized. However, an increasing number of farmers have consciously abandoned agrochemicals and now produce organically, as a viable alternative to green revolution agriculture.

Organic farming is a combined effect of farmers' efforts, NGOs work, governmental interventions, and market forces. Organic farming in India has reached a stage where it can swiftly move to occupy prominent space in Indian agriculture. Certified organic farming, which was 42,000 ha during 2003–04, has now grown almost by 20 folds. As on March 2009, the total cultivated area under organic certification process has crossed one million ha mark. Besides this, there is another 2.5 million ha under certified wild forest harvest collection area. National Project on Organic Farming (NPOF) and National Horticulture Mission (NHM) scheme of Department of Agriculture and Cooperation have significantly contributed to this growth. For quality assurance, the country has internationally acclaimed certification process in place for export, import, and domestic markets. The National Program on Organic Production (NPOP) provides necessary policy support. Presently, 16 accredited certification agencies are looking after the requirement of certification process and the products certified by them are accepted in many countries including European Union and USA.

Currently, India ranks 33rd in terms of total land under organic cultivation and 88th position for agriculture land under organic crops to total farming area. In India, about 2.8 million hectares area is under certified organic farming with about 1,95,741 farmers engaged in organic farming. The Indian organic farming industry is estimated at US$ 100.4 million and is almost entirely export oriented. According to APEDA (2009), a nodal agency involved in promoting Indian organic agriculture, about 9,76,646 MT of organic products worth Rs. 498 crores are being exported from India.

14.6 ORGANIC STANDARDS

Organic standards prohibit the use of synthetic pesticides and artificial fertilizers and the use of growth hormones and antibiotics in livestock production (a minimum usage of antibiotics is admitted in very specific cases and is strictly regulated). Genetically modified organisms (GMOs) and products derived from GMOs are explicitly excluded from organic production methods.

A revised EU regulation that came into force in 2007 (EC, 2007) added two main new criteria: firstly, food will only be able to carry an organic logo (certified as organic) if at least 95% of the ingredients are organic (nonorganic products will be entitled to indicate organic ingredients on the ingredients list only); secondly, although the use of GMOs will remain prohibited, a limit of 0.9% will be allowed as an accidental presence of authorized GMOs.

In the United States, Congress passed the Organic Foods Production Act (OFPA) in 1990. The OFPA required the USDA to develop national standards for organically produced agricultural products, to assure consumers that agricultural products marketed as organic meet consistent, uniform standards. The OFPA and the National Organic Program (NOP) regulations require that agricultural products labeled as organic originate from farms or handling operations certified by a state or private entity that has been accredited by USDA (Gold, 2007).

14.7 ORGANIC CERTIFICATION AGENCIES

14.7.1 INDIAN CERTIFICATION AGENCIES

Government of India through Director General of Foreign Trade, New Delhi, allowed the export of organic products only if they are produced, processed, and packed under a valid organic certificate issued by a certification agency accredited by one of the accredited agencies designated by the Government of India. The Government of India has already recognized the agencies, viz., Tamil Nadu Organic Certification Department, Agricultural and Processed Food Products Export Development Authority (APEDA), Spice Board, Coffee Board, Tea Board, etc.

14.7.2 INTERNATIONAL CERTIFICATION AGENCIES

Imported organic produce from Latin America is subject to certification standards and guidelines just as stringent as produce produced in the United States. Under the US Organic Foods Production Act of 1990 (OFPA), the USDA is required to review the certifiers of imported organic produce, in order to ensure that they meet the requirements of the US National Organic Program (NOP).

Foreign certification agencies may apply directly to the USDA for recognition and are evaluated on the same criteria as domestic agencies. Alternately, foreign governments may apply to the USDA or the US government for recognition of equivalency in their organic oversight program. Once accreditation or recognition is granted, organic products produced under the supervision of the certifying agent or foreign government will be eligible for import to the US as certified organic. The following are the some of the International agencies involved in certification of organic products.

Argentina's leading certification agency was created in 1992. In 1997, Argencert became the first Argentine agency accredited by IFOAM. Likewise, California Certified Organic Farmers (CCOF) is another organic certifying agency whose purpose is to promote and support organic agriculture in California and elsewhere. International Federation of Organic Agriculture Movements (IFOAM) is another federation whose main function is coordinating the network of the organic movement around the world. IFOAM is a democratic, grass root oriented federation. Organic trade association is a national association representing the organic industry in Canada and the United States.

Community Alliance with Family Farmers, CAFF, is another agency building a movement of rural and urban people who foster family- scale agriculture that cares for the land, sustains local economics, and promotes social justice.

Institute for Market ecology (IMO) is one of the first and most renowned international agencies for inspection, certification, and quality assurance of eco-friendly products. Since more than 20 years, IMO has been active in the field of organic certification. IMO is certifying all types of agricultural products, from traditional produce such as coffee, tea, spices, cocoa, nuts, fruits, vegetables, cereals, pulses, cotton, dairy products, honey, fish, and seafood.

Skal International, Netherlands, is a certification and inspection organization that certifies organic products, processes, and inputs. Skal International operates worldwide in Western and Eastern Europe, South America, and Southern Asia. Through the network of the shareholder, nearly all countries in the world can be covered.

Ecocert is an inspection and certification body accredited to verify the conformity of organic products against the organic regulations of Europe, Japan, and the United States. The Ecocert certification mark is one of the leading international organic certification marks, enjoying a good reputation and trusted by both consumers and the organic industry.

Demeter is another worldwide certification system, used to verify to the consumers in over 50 countries that food or product has been produced by biodynamic methods. The Bio Dynamic Farming and Gardening Association is the certifier in New Zealand.

14.8 CONCLUSIONS

The ill effects of the conventional farming system are felt in India in terms of the unsustainability of agricultural production, environmental degradation, health and sanitation problems, etc. Organic agriculture is gaining momentum as an alternative method to the modern system. Many countries have been able to convert 2–10% of their cultivated areas into organic farming. The demand for organic products is growing fast, i.e., 20% per annum in the major developed countries.

Global demand for organically grown foods is increasing, and organic agriculture is growing fast in recent years. As a result, the area under organic farming and the number of countries practicing it are also increasing every year. India is also not an exception to this trend, with a considerable land area under organic farming, and most of the northeastern states have been declared as "organic by default."

Hence, using easily available local natural resources, organic farming can be practiced with a view to safeguard our own natural resources and environment for a fertile soil, healthy crop and quality food and let our future generations enjoy the benefits of non-chemical horticulture. Given the same profitability, organic farming is more advantageous than conventional farming considering its contribution to health, environment, and sustainability.

KEYWORDS

- **certification**
- **ecological**
- **fertility**
- **horticulture**
- **organic**
- **sustainable**

REFERENCES

Anonymous, (2010). *Technical Brochure on Organic Farming System – An Integrated Approach for Adoption under National Horticulture Mission.* National Horticulture Mission Department of Agriculture and Cooperation Ministry of Agriculture, New Delhi.

Bhattacharya, P., & Chakraborty, G., (2005). Current status of organic farming in India and other countries. *Indian Journal of Fertilizers 1*, 111–123.

Carson, R., (1963). *Silent Springs*, Hamish Hamilton, London.

FAI, (2004). *Fertilizer Statistics 2003–04.* The Fertilizer Association of India, New Delhi.

FAO, (2001). *Codex Alimentarius – Organically Produced Foods*, FAO, Rome. Fertilizer statistics 2003–04, The Fertilizer Association of India, New Delhi. pp. 77.

Gold, M. V., (2007). *Organic Production*. U. S. Department of Agriculture http://www.ers.usda.gov/data/organic/ Accessed on 20 June 2010.

Gouri, P. V. S. M., (2004). National programme for organic production. p. 61–64. (Singh, K. P., Narayansamy, G., Rattan, R. K., & Goswami, N. N., eds.) In. *Bulletin of the Indian Society of Soil Science*, New Delhi, No. 22.

HAU, (2003). Emerging challenge to Haryana's Agricultural Economy vis-à-vis Diversification in Agriculture. CCS Haryana Agricultural University. pp. 52.

Lal, R., (2000). *Controlling Green House Gases and Feeding the Globe Through Soil Management*. University Distinguished Lecture, Ohio State Univ., Columbus.

Lieberhardt, B., (2003). What is organic agriculture? What I learned from my transition. In: *Organic Agriculture, Sustainability, Markets and Policies*, Organization for Economic Cooperation and Development (OECD) and CAB 1, Wallingford, UK. pp. 31–44.

Marwaha, B. C., & Jat, S. L., (2004). Statistics and scope of organic farming in India. *Fertilizer News, 49*, 41–48.

Randhawa, M. S., (1983). *A History of Agriculture in India*. vol. III (1757–1947). Indian Council of Agricultural Research, New Delhi. pp. 422.

Scialabba, E. N., & Hattam, C., (2002). *Organic Agriculture Environment and Food Security*. FAO, Rome pp. 252.

Surekha, K., Jhansi Lakshmi, Somasekhar, V. N., Latha, P. C., Kumar, R. M., Shobha Rani, N., Rao, K. V., & Viraktamath, B. C., (2010). Status of organic farming and research experiences in rice *J. Rice Res. 3*, 23–35.

Swaminathan, M. S., (1996). *Sustainable Agriculture – Towards Evergreen Revolution*. Konark Pub. Pvt. Ltd., New Delhi. pp. 219.

CHAPTER 15

EFFECT OF ZINC AND BORON ON GROWTH, YIELD, AND ECONOMICS OF CULTIVATION OF ONION *(ALLIUM CEPA L.)* cv. AGRIFOUND DARK RED IN NAGALAND, INDIA

KHATEMENLA,[1] V. B. SINGH,[2] TRUDY A. SANGMA,[1] and LALNGAIHWAMI[3]

[1]*Department of Horticulture, Assam Agricultural University, Jorhat-13, Assam, India, E-mail: Khatemenlalongchar@gmail.com*

[2]*Department of Horticulture, SASRD, Nagaland University, Medziphema, Nagaland, India*

[3]*Department of Plant Pathology, Assam Agriculture University, Jorhat-13, Assam, India*

CONTENTS

ABSTRACT

The onion is one of the most important bulbous crop of the world and the most important commercial crop grown all over the country for both spice as well as for vegetable purpose. Micronutrients play an important role in fertilization program to achieve higher and sustainable bulb yields. The northeastern (NE) region is deficient of zinc whose deficiency causes yellowing of leaves and boron whose deficiency causes bolting in onion. But very little information is available on these aspects under subtropical foothill conditions. Keeping in view, the present experiment was undertaken with the objective to study the effect of zinc and boron on growth, yield, and economics of cultivation of onion.

The experiment was performed in a randomized block design with three replications and nine treatment combination, and data recorded were statistically analyzed by the analysis of variance method. The experiment was conducted with soil application of three levels of zinc, viz., 0, 25, and 50 kg/ha and three levels of boron, viz., 0, 2.5, and 5 kg/ha. The results were found to be significant in most of the parameters. The plant height (57.47 cm), number of leaves/plant (12.37), neck thickness (1.68 cm), bulb diameter (6.83 cm), bulb weight (110.80 g), yield/plot (5.73 kg), yield/hectare (254.81 q), dry weight of bulb (14.30 g), net income (Rs. 211,339 ha^{-1}), and benefit:cost ratio (1:4.3) were maximum in treatment combination of Zn 25 + B 5 kg/ha. Therefore, it can be concluded that the treatment combination of Zn 25 kg + B 5 kg/ha was the best for sustainable production and economically the most profitable for onion cultivation under Nagaland condition.

15.1 INTRODUCTION

The onion (*Allium cepa* L.) belonging to the family Alliaceae is one of the most important bulbous crop of the world and the most important commercial crop grown all over the country for both spice as well as for vegetable purpose. Onion has great economic importance due to its medicinal and dietetic values since ancient time. It has diuretic properties, relieves heat sensation, hysterical faintness, insect bites, and heart stimulation. Through many experiments conducted on onion, it has been realized that the better growth and quality and higher yield of onion can be obtained with the adequate and balanced application of both macronutrients and micronutrients under suitable agro-climatic condition. Micronutrients play an active role in the plant metabolic process from cell wall development to respiration, photosynthesis, chlorophyll formation, enzyme activity, nitrogen fixation etc. Micronutrients work as a co-enzyme for a large number of enzymes. They also plays an essential role in improving yield and quality and are highly required for better plant growth and yield of many crops (Alam et al., 2010). Soil application of micronutrients during crop growth (onion) was successfully used for correcting their deficits and improving the mineral status of plants as well as increasing the crop yield and quality (Jawaharlal et al., 1986). The onion, like any other crops, not only needs macronutrients but also needs micronutrients in adequate and balanced amounts (Ballabh et al., 2013; Shahjahan, 2013). Thus, micronutrients play an important role in the fertilization program to achieve higher and sustainable bulb yields. Unfortunately, micronutrients have received less attention in fertilizer management research, development, and extension. Traditionally, emphasis has been given to N, P, and K. It is reported that the northeastern (NE) region is deficient in zinc and boron whose deficiency causes bolting and yellowing of leaves in onion, respectively. In Nagaland condition, some cultivars like Arka Kalayayan, N-53 and Agrifound Dark Red were tested during kharif season and out of which Agrifound Dark Red was found to be most suitable under Medziphema (Nagaland) condition. Improvement in onion growth, yield, quality, and maintenance of soil fertility has been reported through micronutrient by many scientists at different types of soils. But very little information is available on these aspects under subtropical foothill conditions. Keeping in view, the present experiment was undertaken with the following objectives:

1. To study the effect of zinc on growth, yield, and quality of onion.
2. To study the effect of boron on growth, yield, and quality of onion.
3. To study the combined effect of zinc and boron on growth and yield of onion.
4. To study the economics of cultivation of onion under various treatments.

15.2 MATERIALS AND METHODS

The experiment was carried out at the Horticultural Research Farm, School of Agricultural Sciences and Rural Development, Medziphema, Nagaland University, during 2013–2014. The experiment was conducted in a randomized block design with three replications. Medziphema is located at the foothills of Nagaland at an altitude of 310 m above mean sea level, with the geographical location of 25°45'43" N latitude and 93°53'04" E longitude, sub-tropical climate with heavy monsoon rain. The experiment consists of soil application of three levels of zinc, viz., $Zn_1 = 0$ kg ha^{-1}, $Zn_2 = 25$ kg ha^{-1}, and $Zn_3 = 50$ kg ha^{-1} and three levels of boron, viz., $B_1 = 0$ kg ha^{-1}, $B_2 = 2.5$ kg ha^{-1}, and $B_3 = 5$ kg ha^{-1}. Thus, the total number of treatment were 9, viz., T_1 (Zn 0 kg ha^{-1} + B 0 kg ha^{-1}), T_2 (Zn 0 kg ha^{-1} + B 2.5 kg ha^{-1}), T_3 (Zn 0 kg ha^{-1} + B 5kg ha^{-1}), T_4 (Zn 25 kg ha^{-1} + B 0 kg ha^{-1}), T_5 (Zn 25 kg ha^{-1} + B 2.5 kg ha^{-1}), T_6 (Zn 25 kg ha^{-1} + B 5 kg ha^{-1}), T_7 (Zn 50 kg ha^{-1} + B 0 kg ha^{-1}), T_8 (Zn 50 kg ha^{-1} + B 2.5 kg ha^{-1}), and T_9 (Zn 50 kg ha^{-1} + B 5 kg ha^{-1}). Zinc was applied in the form of zinc sulfate and boron in the form of borax. Bulbs were planted on September 16, 2013, at a spacing of 15 cm × 15 cm in a plot size of 1.5 m × 1.5 m. Farm yard manure (FYM) @ 25 t ha^{-1} along with N, P, and K @ 100:60:60 kg ha^{-1}, respectively, and was applied uniformly in all experimental plots in the form of urea, SSP, and MOP. Urea and MOP was applied in two split doses, i.e., at the time of land preparation and second dose 30 days after planting. All intercultural operations were uniformly followed during the whole period to raise a successful crop. During the initial growth stage, randomly five plants were selected to record the observations on plant height and number of leaves per plant. Plant height was measured with the help of meter scale. At the time of harvest (February 2, 2014), randomly five plants were selected to record the observation on neck thickness, weight of bulb and dry matter of bulb per 100 kg fresh weight. Neck thickness and bulb diameter were measured with the help of Vernier caliper after harvesting. Yield per plot and yield ha^{-1} was also recorded with the help of an electronic weighing machine. The data recorded during the course of investigation were statistically analyzed by the analysis of variance method (Panse and Sukhatme, 1978), and the significance of different sources of variation were tested by error mean square using Fisher Schedecor "F" test of probability of 5% level of significance.

15.3 RESULTS AND DISCUSSION

15.3.1 GROWTH CHARACTERS

The data presented in Table 15.1 strongly showed that there was a significant effect of soil application of zinc and boron on vegetative growth

TABLE 15.1 Effect of Zinc and Boron on Vegetative Growth Characters of Onion

Treatment Zinc	Plant height 75 DAP	Plant leaves plant^{-1} 90 DAP	Neck thickness (cm)
Zn_0 – Control	42.11	9.84	1.39
Zn_1- Zn @ 25 kg ha^{-1}	52.19	11.87	1.56
Zn_2- Zn @ 50 kg ha^{-1}	49.22	12.34	1.61
SEm (±)	1.165	0.40	0.04
CD at 5%	3.49	1.21	0.10
Boron			
B_0- Control	45.42	11.46	1.44
B_1- B @ 2.5 kg ha^{-1}	48.30	10.82	1.49
B_2- B @ 5 kg ha^{-1}	49.80	11.78	1.62
SEm (±)	1.17	0.40	0.04
CD at 5%	3.49	NS	0.10
Zinc x Boron			
Zn_0 + B_0- Control	39.57	7.97	1.15
Zn_1 + B_1- Zn @ 25 kg ha^{-1} + B @ 2.5 kg ha^{-1}	50.57	10.50	1.43
Zn_1 + B_2- Zn @ 25 kg ha^{-1} + B @ 5 kg ha^{-1}	57.47	12.37	1.68
Zn_2 + B_1- Zn @ 50 kg ha^{-1} + B @ 2.5 kg ha^{-1}	47.63	11.73	1.62
Zn_2 + B_2- Zn @ 50 kg ha^{-1} + B @ 5 kg ha^{-1}	47.63	11.63	1.60
SEm (±)	2.08	0.69	0.06
CD at 5%	6.05	2.09	0.18

parameters of onion plants. Soil application of onion plants with treatment combination of Zn 25 kg ha^{-1} and B 5 kg ha^{-1} resulted in the highest value of vegetative growth parameters. The maximum plant height (57.47 cm) was recorded in treatment combination (Zn 25 kg ha^{-1} + B 5 kg ha^{-1}) and the minimum (39.57 cm) under control. The maximum number of leaves (12.34 plant^{-1}) was obtained from treatment combination (Zn 25 kg ha^{-1} + B 5kg ha^{-1}) and the minimum from control (7.97 plant^{-1}). The highest neck thickness of bulbs (1.68 cm) at maturity stage was found in treatment combination (Zn 25 kg ha^{-1} + B 5 kg ha^{-1}) and lowest (1.15 cm) under control.

The favorable effect of micronutrients on plant growth might be due to its role in many physiological processes and cellular functions within the plants. In addition, they play an essential role in improving plant growth, through biosynthesis of endogenous hormones which are responsible for promoting of plant growth (Hänsch and Mendel, 2009). Therefore, it was found that the combination of zinc and boron plays an active role in vegetative growth of the onion plant. These results are in accordance with the investigation of Alam et al. (2010), Ballabh et al. (2013), Shahjahan (2013), and Umesh et al. (2014).

15.3.2 YIELD CHARACTERISTICS

A perusal of data presented in Table 15.2 revealed that there was a significant effect of soil application of zinc and boron on yield parameters of onion. The interaction effect between zinc and boron exhibited the best yield results of onion in comparison with individual dose of zinc and boron and control. The maximum bulb diameter (6.83 cm) was obtained by applying Zn 25 kg + B 5 kg ha^{-1} with an increase of 35.5% and 30%, respectively, over control. The highest of bulb weight (110.80 g) was recorded from the treatment combination (Zn 25 kg + B 5 kg ha^{-1}) and the lowest bulb weight was found in control. Yield data presented in Table 15.2 shows that zinc and boron produced significant variations for bulb yield of onion. The maximum bulb yield per plot (5.32 kg) was recorded from treatment combination (Zn 25 kg + B 5 kg ha^{-1}) while the lowest was found in control. Similarly, the highest bulb yield per hectare (236.59 q) was recorded in treatment combination (Zn 25 kg + B 5 kg ha^{-1}) and the lowest (102.66 q) was found in control. The treatment combination (Zn 25 kg + B 5 kg ha^{-1}) performed as the highest yielder by 56% over control. This effect might be due to the fact that micronutrients played a pivotal role in strengthening plant cell walls and translocation of carbohydrates from leaves to other parts of the plant; this means that there is a possibility of increasing dry matter percentage as well as yield. Hansch and Mendel (2009) reported that micronutrient is highly required for better yield of many crops since it could serve as counter ion. These results are in conformity with works done by Alam (2010), Manna et al. (2011), and Shahjahan (2013).

TABLE 15.2 Effect of zinc and boron on yield attributing characters of onion

Treatment	Diameter of bulb (cm)	Bulb weight (g)	Yield per plot (kg)	Projected yield per hectare (q)
Zinc				
Zn_0 – Control	5.26	76.80	3.11	138.42
Zn_1 - Zn @ 25 kg ha^{-1}	6.09	96.96	4.79	212.94
Zn_2 - Zn @ 50 kg ha^{-1}	6.08	89.93	4.56	202.52
SEm (±)	0.08	0.99	0.06	2.77
CD at 5%	0.25	2.96	0.19	8.30
Boron				
B_0 - Control	5.23	77.03	3.58	159.26
B_1 - B @ 2.5 kg ha^{-1}	5.94	90.33	4.22	187.51
B_2 - B @ 5 kg ha^{-1}	6.24	96.32	4.66	207.11
SEm (±)	0.08	0.99	0.06	2.77
CD at 5%	0.25	2.96	0.19	8.30
Zinc x Boron				
Zn_0 + B_0 - Control	4.40	64.20	2.31	102.66
Zn_1 + B_1 - Zn @ 25 kg ha^{-1} + B @ 2.5 kg ha^{-1}	5.93	95.20	4.76	211.41
Zn_1 + B_2 - Zn @ 25 kg ha^{-1} + B @ 5 kg ha^{-1}	6.83	110.80	5.32	236.59
Zn_2 + B_1 - Zn @ 50 kg ha^{-1} + B @ 2.5 kg ha^{-1}	6.13	89.37	4.57	202.96
Zn_2 + B_2 - Zn @ 50 kg ha^{-1} + B @ 5 kg ha^{-1}	6.30	98.40	4.96	220.30
SEm (±)	0.14	1.71	0.11	4.78
CD at 5%	0.42	5.12	0.32	14.38

15.3.3 ECONOMICS OF CULTIVATION

From the data presented in Table 15.3, the economics analysis of the data from the investigation revealed that the integrated nutrient management had an appreciable impact on the total yield as well as gross income and net return. The highest net income (Rs. 211,339 ha^{-1}) with a benefit:cost ratio of 1:4.3 was obtained from the treatment combination $Zn_1 + B_2$ (Zn 25 kg ha^{-1} + B 5 kg ha^{-1}) followed by the treatment combination $Zn_2 + B_2$ (Zn 50 kg ha^{-1} + B 5 kg ha^{-1}) with a net income of Rs. 191,970 and benefit:cost ratio of 1:3.81. The lowest net income of Rs. 66,466 with a benefit:cost ratio of 1:1.43 was found under control. Most of the treatment combinations were found to increase the net income and benefit:cost ratio in comparison to control. These results are in agreement based on experimental evidence by Kumar and Sen (2005), Chhipa et al. (2005) and Nasreen et al. (2009).

15.4 CONCLUSION

Based on the experimental finding, it can be concluded that soil application of micronutrients like zinc and boron in onion significantly improved the growth and yield of the onion crop. Among the different treatments studied, it was found that the treatment combination of Zn @ 25 kg ha^{-1} + B @ 5 kg ha^{-1} was found to be better in terms of enhancing better growth and yield and also in terms of economics of cultivation of onion in comparison to individual doses of zinc and boron, though all the different combinations of treatment were found to be superior over control.

Thus, based on the support warranted from the above data, it can be concluded that growing of onions by judicious use of combined application of micro- and macronutrients is the best practice in sustaining productivity and profitability and hence can be practiced by the growers effectively under the foot hills of Nagaland.

ACKNOWLEDGMENT

The author is thankful to the advisory committee and faculty members of the Department of Horticulture and Department of Agriculture Chemistry & Soil Science (School of Agricultural Sciences & Rural Development,

TABLE 15.3 Economics of the Treatment

Treatments	Cost of cultivation (Rs. ha^{-1})			Yield (q ha^{-1})	Gross income @ Rs.1100 q^{-1}	Net income (Rs. Ha^{-1})	Benefit: cost ratio
	Fixed cost	Treatment cost	Total				
T_1 – Control	46,460	-	46,460	102.66	112,926	66,466	1:1.43
T_2 – NPK @ 100:60:60 kg ha^{-1} + FYM @ 2.5 t^{-1} + B @ 2.5 kg ha^{-1}	46,460	500	46,960	148.15	162,965	116,005	1:2.47
T_3 – NPK @ 100:60:60 kg ha^{-1} + FYM @ 2.5 t^{-1} + B @ 5 kg ha^{-1}	46,460	1000	47,460	164.45	180,895	133,435	1:2.81
T_4 – NPK @ 100:60:60 kg ha^{-1} + FYM @ 2.5 t^{-1} + Zn 25 kg ha^{-1}	46,460	1450	47,910	190.81	209,891	161,981	1:3.38
T_5 – NPK @ 100:60:60 kg ha^{-1} + FYM @ 2.5 t^{-1} + Zn 25 kg ha^{-1} + B 2.5 kg ha^{-1}	46,460	1950	48,410	211.41	232,551	184,141	1:3.80
T_6 – NPK @ 100:60:60 kg ha^{-1} + FYM @ 2.5 t^{-1} + Zn 25 kg ha^{-1} + B 5 kg ha^{-1}	46,460	2450	48,910	236.59	260,249	211,339	1:4.3
T_7 – NPK @ 100:60:60 kg ha^{-1} + FYM @ 2.5 t^{-1} + Zn 50 kg ha^{-1}	46,460	2900	49,360	184.30	202,730	153,370	1:3.1
T_8 – NPK @ 100:60:60 kg ha^{-1} + FYM @ 2.5 t^{-1} + Zn 50 kg ha^{-1} + B 2.5 kg ha^{-1}	46,460	3400	49,860	202.96	223,256	173,396	1:3.48
T_9 – NPK @ 100:60:60 kg ha^{-1} + FYM @ 2.5 t^{-1} + Zn 50 kg ha^{-1} + B 5 kg ha^{-1}	46,460	3900	50,360	220.30	242,330	191,970	1:3.81

Nagaland University, Medziphema Campus) for their supportive encouragement for the smooth completion of the work.

KEYWORDS

- **boron**
- **economics**
- **growth**
- **yield**
- **zinc**

REFERENCES

Alam, M. N., Abedin, M. J., Azad, M. A. K., (2010). Effect of micronutrients on growth and yield of onion under calcareous soil environment. *Int. Res. J. Plant Sci.*, *1*(3), 56–061.

Ballabh, R., & Rawat, S. S., (2013). Effects of foliar application of micronutrients on growth, yield and quality of onion. *Indian J. Hort.*, *70*(2), 260–265.

Chhipa, B. G., (2005). Effect of different levels of sulphur and zinc on growth and yield of cauliflower (*Brassica oleracea* var. *botrytis* L.). *M. Sc. (Ag.) Thesis,* S. K. N. College of Agriculture, Jobner, RAU, Bikaner.

Hansch, R., & Mendel, R. R., (2009). Physiological functions of mineral micronutrients (Cu, Zn, Mn, Fe, Ni, Mo, B, Cl). *Curr. Op*In. *Plant Biol.*, *12*, 259–66.

Jawaharlal, M., Sundararajan, S., & Veeraragavathatham, D., (1986). *South Indian Hort.*, *34*(4), 236–239.

Kumar, M., & Sen, N. L., (2005). Effect of zinc boron and gibberellic acid on yield of okra (*Ablemoschus esculentus* (L.) Moench). *Indian J. of Hort.*, *62*(3), 308–309.

Manna, D., Maityand, T. K., & Ghosal, A., (2011). Influence of foliar application of boron and zinc on growth, yield and bulb quality of onion (*Allium cepa* L.). *J. Crop and Weed.*, *10*(1), 53–55.

Nasreen, S., Yousuf, M. N., Mamun, A. N. M., Brahma, S., & Haquc, M. N., (2009). Response of garlic to zinc, boron and poultry manure application. *Bangladesh J. Agrl. Res.*, *34*(2), 239–245.

Panse, V. G., & Sukhatme, P. U., (1978). Statistical methods for agricultural workers. *Indian Council Agrl. Res.*, New Delhi, 145–152.

Shahjahan, A., (2013). Effects of micronutrients on growth, yield and quality of three varieties of summer onion. *M. Sc (Agri.) Thesis,* submitted to Bangladesh Agriculture University, Mymensingh.

Umesh, A., Venkatesan, K., Saraswathi, T., & Subramanian, K. S., (2014). Effect of Zinc and boron application on growth and yield parameters of multiplier onion (*Allium cepa* L. var *aggregatum* Don.) var. CO (On)5. *Int. J. Res.*, *2*, 757–765.

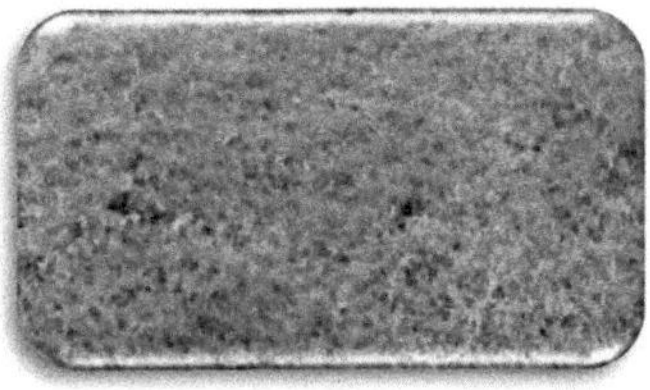

Selection-1

Tifdwarf-419

Bargusto

Panama

Panam

Palma

FIGURE 3.1

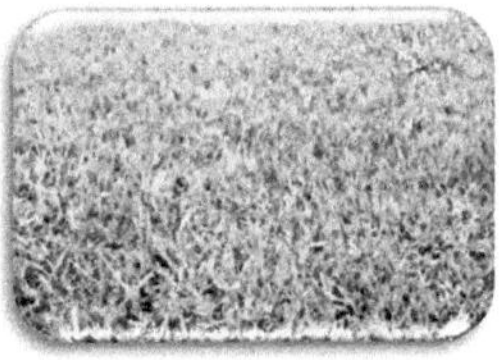

St. Augustine

St. Augustine (Striped)

Bahia grass

FIGURE 3.2

Love grass

Dichondra

FIGURE 3.3

Zoysia japonica

Zoysia matrella

FIGURE 3.4

Spinacia oleracea

Nasturtium officinale

Sechium edule

Uritca dioca

Brassica juncea

Trigonella foenum-graecum

Brassica campestris var. toria

FIGURE 4.1 Some of the local landraces of Sikkim.

FIGURE 4.2 Interview with a seller in Lal Bazaar.

FIGURE 4.3 Interview with a vegetable seller in Tadong.

FIGURE 4.3 Commercial vegetables in Lal Bazaar market.

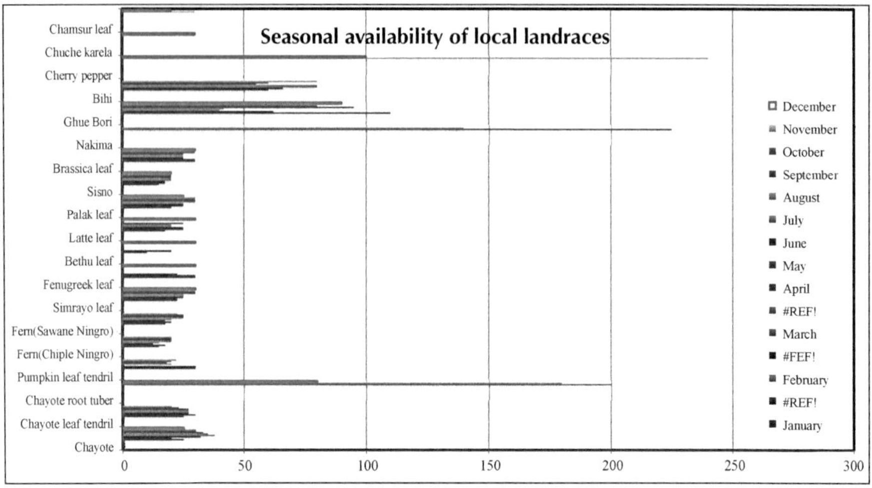

FIGURE 4.5 Seasonal availability of local vegetables in Sikkim.

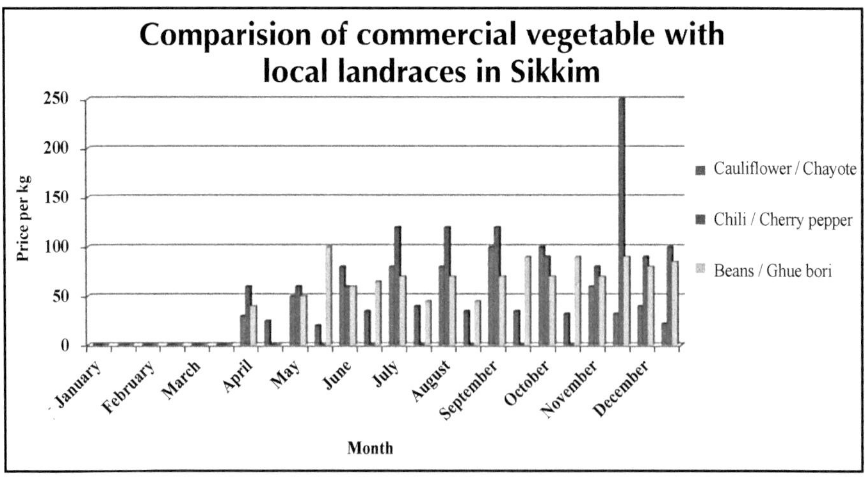

FIGURE 4.6 Comparison of commercial vegetables with local landraces.

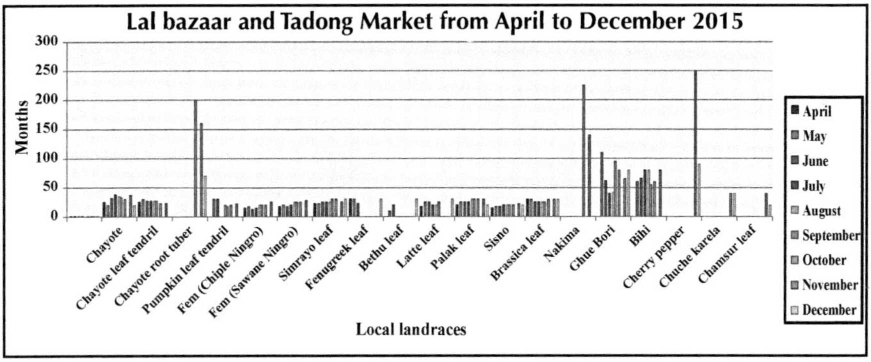

FIGURE 4.7 Lal bazaar and Tadong market price from April to December 2015.

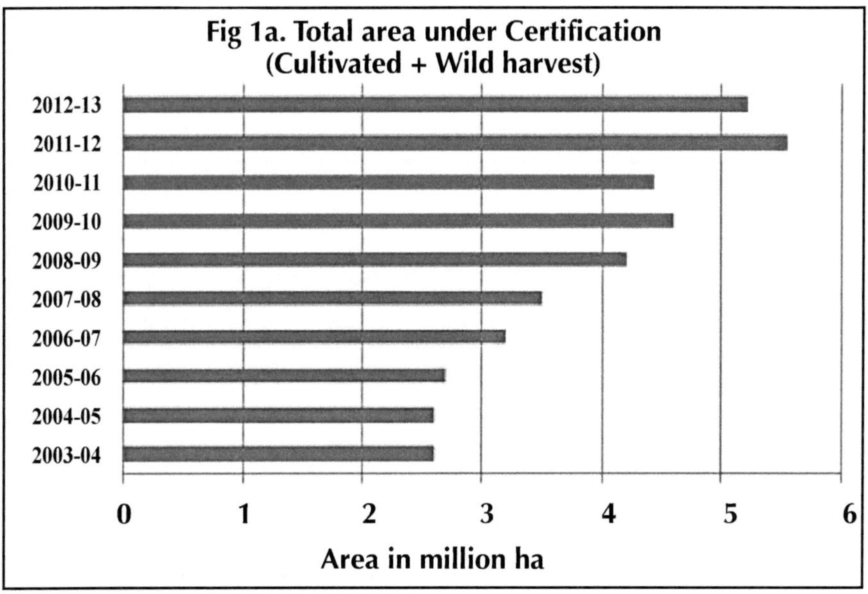

FIGURE 6.1A Total area under certification (cultivated + wild harvest).

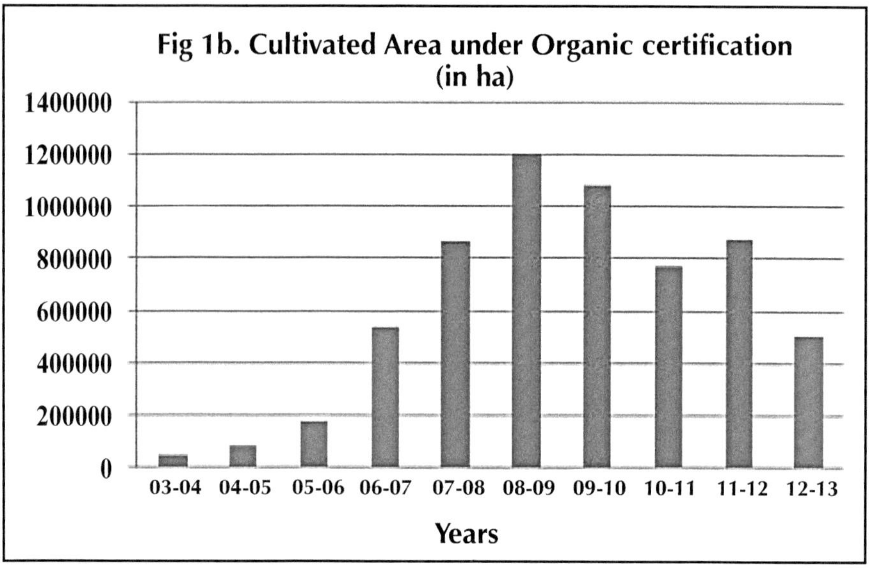

FIGURE 6.1B Cultivated area under organic certification (in ha).

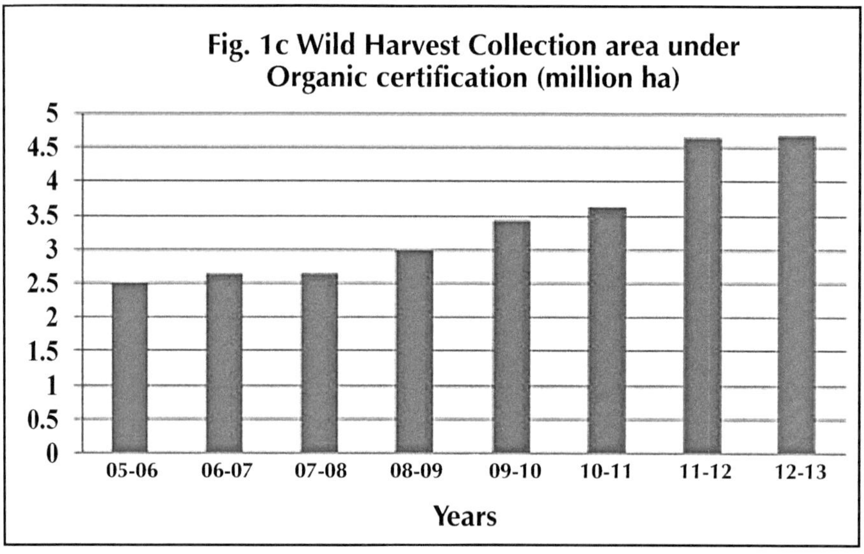

FIGURE 6.1C Wild harvest collection area under organic certification (million ha).

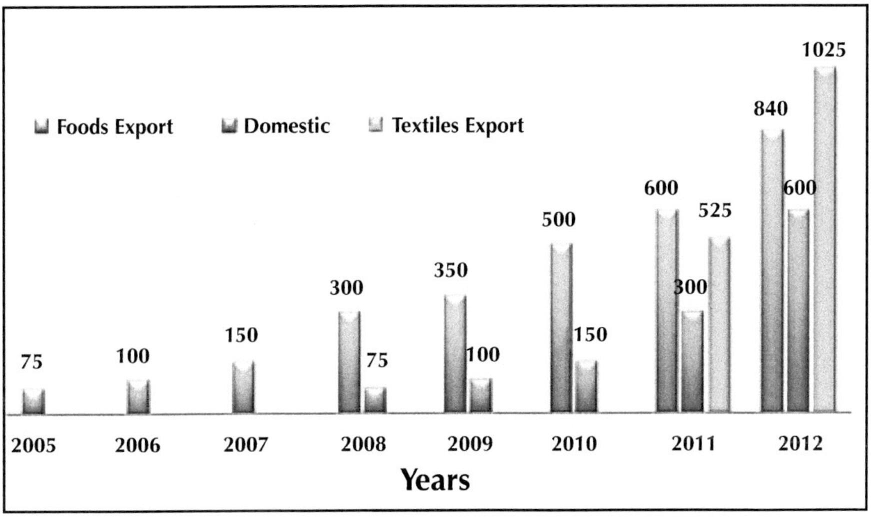

FIGURE 6.2 Export and domestic market for organic products (Rs crores).

Hedgerows are created on steep slopes

A farmer making hedgerows along contour lines

Intercropping of banana with pineapple

Tephrosia candida are planted as hedgerows

Citrus grandis

Citrus grandis

Fruit of Citrus grandis

Fruit of Citrus grandis

Fruit of Citrus grandis cut in half

Fruit pulp of Citrus grandis

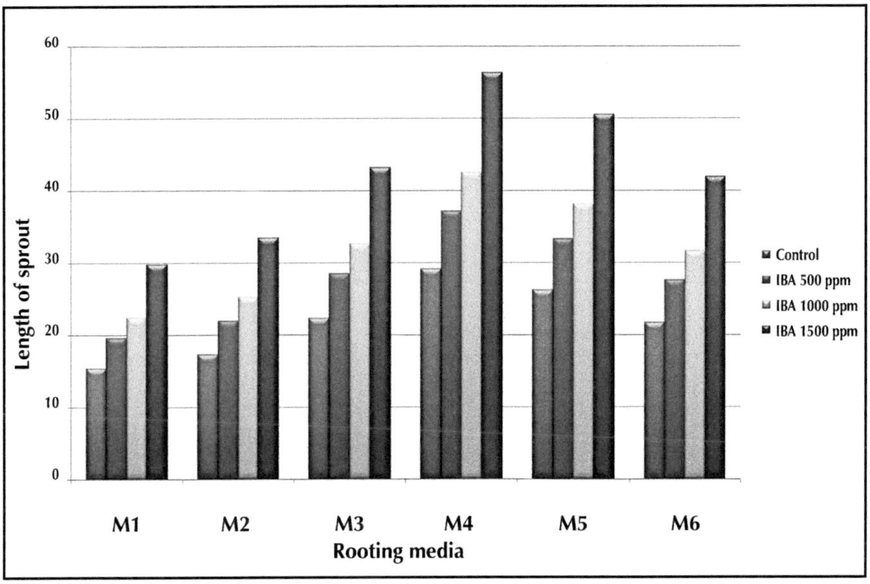

FIGURE 30.1 Effect of IBA and rooting media on length of sprout (cm) in fig.

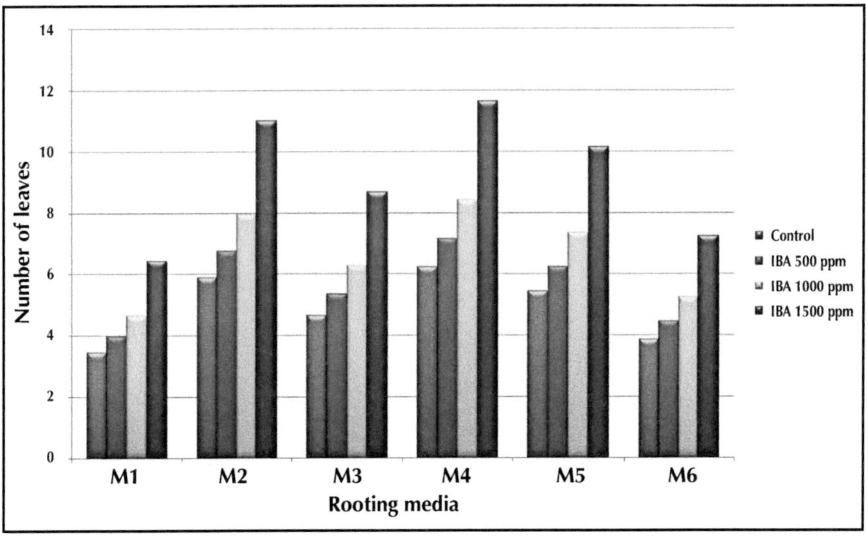

FIGURE 30.2 Effect of IBA and rooting media on number of leaves in fig.

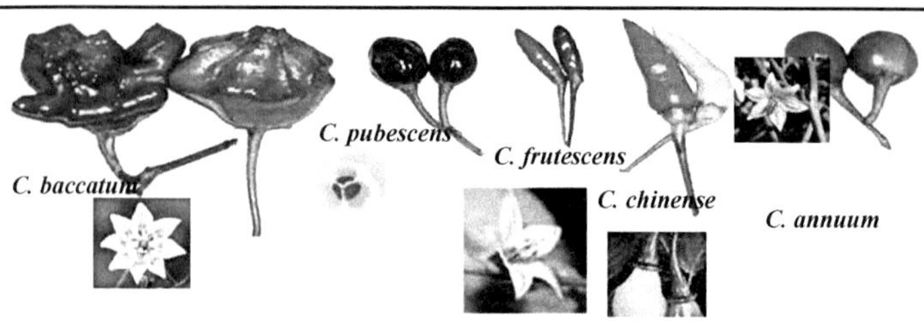

Cultivated species	Distinguishable morphology
C. annuum	White corrolla and white filaments
C. fruitescens	Yellow/greenish corolla and purple filaments
C. chinense	Annular constriction on pedicel attachment and yellow/greenish corolla
C. baccatum	Yellow or greenish yellow spots on corolla
C. pubescens	Hairy stems/leaves and black/brown seeds

FIGURE 32.1 Distinguishable morphology of five domesticated species of *Capsicum*.

FIGURE 32.2 Fruits of wild (deciduous) and cultivated (non-deciduous) *C. annuum*.

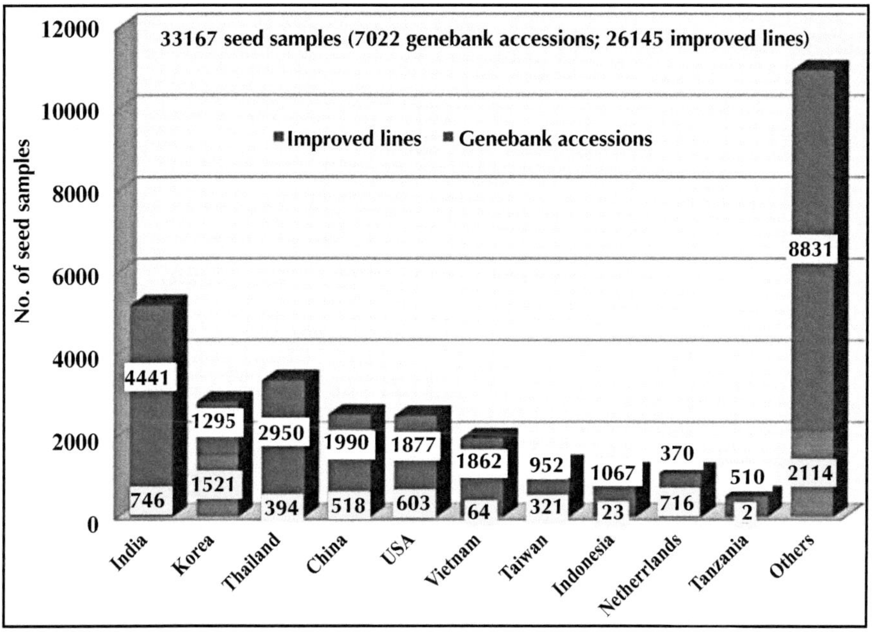

FIGURE 32.3 Seed samples distributed during the period 2001-2013 (top ten recipient countries).

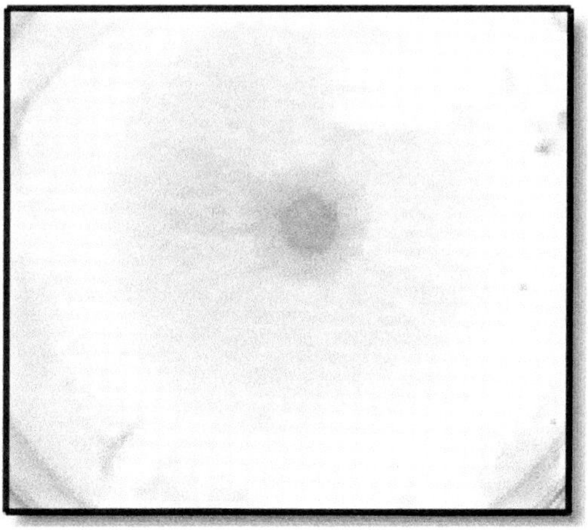

FIGURE 33.1 Culture plate of *Phytophthora* spp.

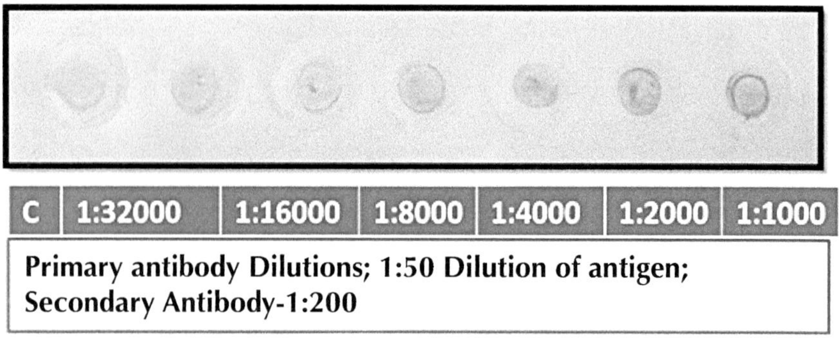

FIGURE 33.4 Standardized DOT-ELISA for *Phytophthora* detection.

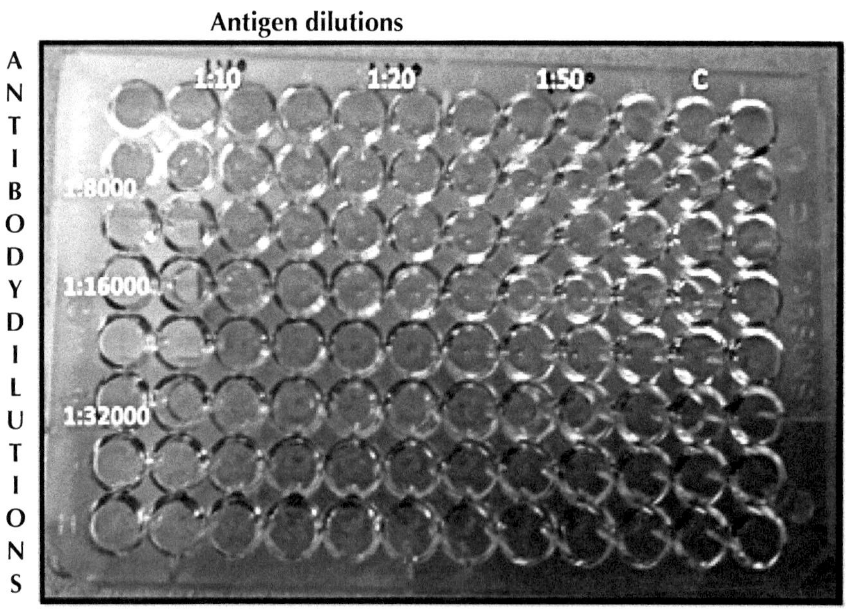

FIGURE 33.5 Standardized DOT-ELISA for *Phytophthora* detection.

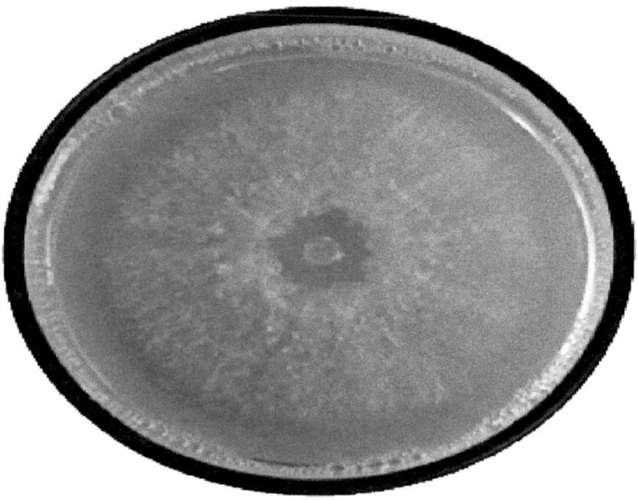

FIGURE 34.1 Pure culture plate of *Phytophthora nicotianae.*

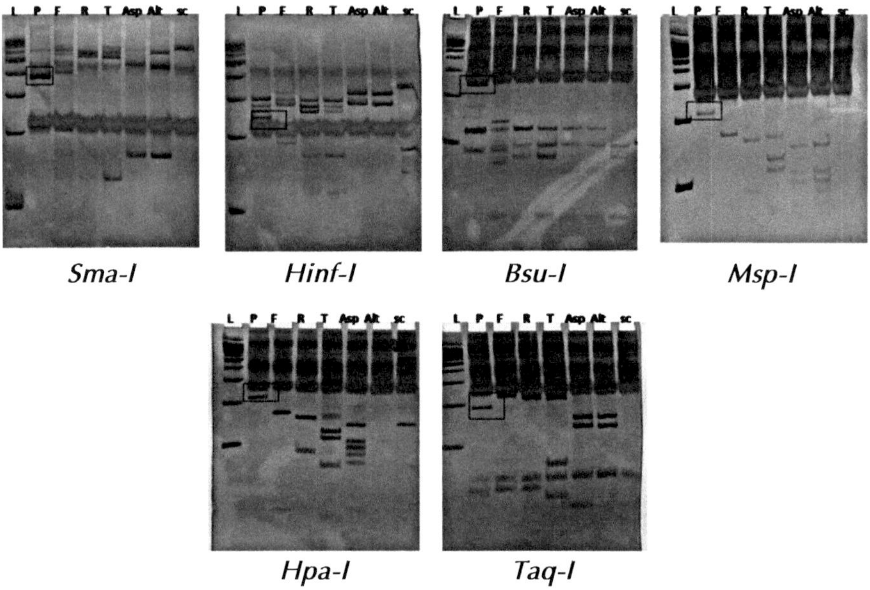

FIGURE 34.3 PAGE showing differentiation of *Phytophthora* from other fungi (L-100-bp ladder. P-*Phytophthora* spp. F-*Fusarium* spp. R-*Rhizoctonia* spp. T-*Trichoderma* spp. Asp-*Aspergillus* spp. Alt-*Alternaria* spp. Sc-*Sclerotium* spp.)

CHAPTER 16

EFFECT OF DIFFERENT NUTRIENT MANAGEMENT ON PRODUCTIVITY OF POTATO (VAR. KUFRI JYOTI) AND SOIL NUTRIENT STATUS IN NEW ALLUVIAL ZONE OF WEST BENGAL, INDIA

PRIYANKA IRUNGBAM and MAHADEV PRAMANICK

Department of Agronomy, Faculty of Agriculture, Bidhan Chandra Krishi Viswavidyalaya, Mohanpur–741252, Nadia, West Bengal, India

CONTENTS

ABSTRACT

A field experiment was conducted at BCKV, Regional Research Station, New Alluvial Zone, Gayeshpur, Nadia, West Bengal during the *rabi* season of 2013–2014 to study "the effect of different nutrient management on productivity of potato (var. KufriJyoti) and soil nutrient status in New Alluvial

Zone of West Bengal." The experimental site area was sandy clay loam soil in texture having good drainage with medium soil fertility. The experiment was carried out in a randomized block design by replicating thrice of eight treatments, i.e., T_1: 50% recommended NPK (inorganic) + 50% N as FYM; T_2: 1/3rd recommended N each from farm yard manure (FYM), vermicompost (VC), neem cake (NC); T_3: T_2 + intercropping (potato + coriander–1:1); T_4: T_2 + straw mulch for weed management; T_5: 50% N as FYM + Rock phosphate + phosphorus solubilizing bacteria (PSB) (+ *Azotobacter*); T_6: T_2 + biofertilizers containing N and P carriers (same as T_5); T_7: 100% recommended NPK; and T_8: Control (without manures and fertilizers). Among the different treatments, T_7 (100% recommended NPK) gave the highest plant height, dry matter accumulation, CGR, number of tubers per hill, and tuber bulking rate but was found statistically at par with integrated nutrient management, i.e., T_1 (50% recommended NPK (inorganic) + 50% N as FYM). The maximum tuber yield of 22.19 t ha^{-1} was produced in T_7, which was statistically at par with T_1 (20.99 t ha^{-1}). Soil chemical properties like organic carbon, N, P, and K content were found appreciably higher under organic treatments over the initial status. Among the different organic-based nutrient managements, T_6 gave the highest organic carbon (0.91%), N (0.076%), P (20.58 kg ha^{-1}), and K (175.57 kg ha^{-1}). Therefore, it was concluded that in order to sustain the productivity of soil and crops for long-term basis, organic nutrient management are the best options followed by integrated nutrient management.

16.1 INTRODUCTION

The green revolution technology involving indiscriminate use of fertilizers and pesticides with the adoption of nutrient-responsive high-yielding crop varieties have boosted the productivity. Of late, concern has been raised over its adverse effect on soil erosion, depletion of organic matter in soil, low availability of water, and contamination of food and water due to agrochemicals and its impact on biodiversity. Organic farming is one of the options to restore the productivity of degraded soils (Ghosh et al., 2007). But organic farming alone may not meet the requirements of agricultural productivity and solve food and other problems. A proportionate combination of organic and inorganic sources of plant nutrients has been found to be the best option for increasing productivity and maintaining sustainability in crop production.

As no single source is capable of supplying the required amount of plant nutrients, integrated use of all sources of plant nutrients is a must to supply balanced nutrition to the crops (Arora, 2008). Potato is an economical food, and it provides a source of low cost energy to the human diet. Potato is the rich source of starch, vitamin C and B, and minerals. It contains about 20.6% carbohydrates, 2.1% protein, 0.3% fat, 1.1% crude fiber, and 0.9% ash. It also contains a good amount of essential amino acids like *leucine*, *tryptophan*, and *isoleucine* (Khurana and Naik, 2003). Potato is a high nutrient requirement crop, and ever since the beginning of the green revolution, the nutrient requirement of this crop is mainly made through synthetic fertilizers. As time progressed, extensive dependence on chemical farming has shown its darker side. The land is losing its fertility and is demanding larger quantities of fertilizers to be used. Pests are becoming immune, requiring the farmers to use stronger and costlier pesticides. Due to increased cost of farming, farmers are falling into the trap of money lenders, who are exploiting them no end, and forcing many to commit suicide. Hence, it is necessary to establish alternative solution to overcome the ill-effects of chemicals in all aspects such as soil, health, and environment, which can be achieved through practicing of organic or integrated nutrient management in potato production.

Being a heavy feeder of nutrients, potato requires high amount of nitrogen, phosphorus, and potassium. Chemical fertilizers are the main source of nutrients used for potato cropping. However, continuous dependence on chemical fertilizers causes nutritional imbalance and adverse effects on physico-chemical and biological properties of the soil. Biofertilizers like phosphorus solubilizing bacteria (PSB)/*Azotobacter* may be useful for improving P and N nutrition in potato. The beneficial effects of organic manures are manifested through increase in soil organic matter and humus over the period. Soil organic matter and humus acts in several ways; it serves as slow release source of plant nutrients to the crops and increases water holding capacity to maintain the water regime of the soil and acts as a buffer against change in soil pH (Upadhayay and Singh, 2003). Fertilizer requirement of potato is very high as compared to cereal crops. It responds well to applied fertilizers and gives good yield per unit area and time. Application of organic manures in conjunction with fertilizers improves physical, chemical, and biological properties of the soil besides improving fertilizer use efficiency and crop yield. This evidence indicates that the use of synthetic fertilizer in combination with organic manures could be a key factor for achieving and maintaining high productivity of potato. Hence, a field experiment was carried out to study the effect of different nutrient managements on productivity of potato (var. Kufri Jyoti) and soil nutrient status in the New Alluvial Zone of West Bengal.

16.2 MATERIALS AND METHODS

The field experiment was conducted at the Regional Research Station, New alluvial Zone, Bidhan Chandra Krishi Vishwavidyalaya, Gayeshpur, Nadia, West Bengal, during the period of December 2013 to March 2014. The experiment site is situated at 23°5' N latitude and 89° E longitude with an altitude of 9.75 m above the mean sea level, and topographically, the land was medium in nature. The place comes under subhumid, subtropical zone and lies in Indo-Gangetic Alluvial agro-ecological zone. The average annual rainfall is around 1450 mm, and the monthly mean temperature ranges from 9°C to 37°C. The soil of the experimental field was sandy clay loam (39.2, 29.0 and 31.8% sand, silt, and clay respectively) with good drainage facilities and medium soil fertility having pH 7.30, organic carbon 0.68%, total nitrogen 0.060% and available Phosphorus and potassium of 16.36 kg/ha and 163.51 kg/ha, respectively.

The experiment was carried out in a randomized block design with three replications in a 4.5 m × 4.5 m plot. The experiment consisted of eight different treatments, namely T_1: 50% recommended NPK (inorganic) + 50% N as FYM; T_2: 1/3rd recommended N each from farm yard manure (FYM), vermicompost, neem cake; T_3: T_2 + Intercropping (potato + coriander- 1:1); T_4: T_2 + rice straw mulch for weed management; T_5: 50% N as FYM + Rock phosphate + PSB + *Azotobacter*; T_6: T_2 + Biofertilizers containing N and P carriers (same as T_5); T_7: 100% recommended NPK; and T_8:Control (without manures and fertilizers). The recommended dose of fertilizer was 100:150:150 kg $N:P_2O_5:K_2O$ ha^{-1} were computed as per treatments. The sources of fertilizers used were urea for nitrogen, single super phosphate for phosphorus, and muriate of potash for potassium. A total of 3/4th of N, full dose of P_2O_5, and K_2O_5 were applied before planting; the remaining 1/4th of N was applied during earthing up. Organic manures were applied at the final land preparation according to treatments. Applied FYM contained 0.63% N, 0.25% P_2O_5, and 0.54% K_2O, whereas neem cake contained 5.20% N, 1.0% P_2O_5, and 1.4% K_2O; vermicompost contained 1.92% N, 0.70% P_2O_5, and 1.4% K_2O.

Inoculation of tuber with *Azotobacter* was also carried out. Rock phosphate was mixed thoroughly with phosphate solubilizing biofertilizer (PSB) before applying to the field. Cut tubers were planted in 45 cm apart in furrows with plant-to-plant spacing of 20 cm in the first week of December. Irrigation for the crop was given when necessary. Earthing up was done twice: first was done at 15–20 DAP and second at 15 days after first earthing up.

Observations on growth attributes at different stages, yield components, and crop productivity were recorded at harvest. Bulk soil samples were collected from five points of each plot with the help of a spiral auger, at the depth of 0–15 cm before planting and after the harvest of the crop for determination of various soil physical and chemical soil properties. The data obtained as described earlier were subjected to statistical analysis of variance method in a completely randomized block design (Panse and Sukhatme, 1967; Gomez and Gomez, 1984) and the significant of different sources of variations were tested by Error Mean Square by Fisher and Snedecor's "F" test at the probability level $P = 0.005$.

16.3 RESULTS AND DISCUSSION

16.3.1 DRY WEIGHT OF TUBER

Tuber dry weight of potato was influenced significantly with different nutrient managements at different growth stages (Table 16.1). Dry weight of potato was found to increase progressively from 50 DAP and reached the maximum value at 90 DAP. Highest dry weight of tuber was obtained under inorganic treatment, T_7 (167.47 gm^{-2}, 364.63 gm^{-2}, and 547.95 gm^{-2}) followed by integrated nutrient management T_1 (163.92 gm^{-2}, 347.28 gm^{-2}, and 520.34 gm^{-2}), which were found to be statistically at par with each other at 50 and 70 DAP but differed significantly at 90 DAP. Among the organic treatments, T_6 recorded higher tuber dry weights of 148.77 gm^{-2}, 331.57 gm^{-2}, and 484.52 gm^{-2}, respectively, followed by T_4 (147.84 gm^{-2}, 322.03 gm^{-2}, and 471.44 gm^{-2}, respectively) at 50, 70, and 90 DAP with no significant difference among them. However, the lowest dry weight of tubers was obtained under T_3 (113.21 gm^{-2}, 244.03 gm^{-2}, and 367.77 gm^{-2}) at 50, 70, and 90 DAP, respectively, after control T_8.

TABLE 16.1 Effect of Different Nutrient Managements on Tuber Dry Weight (gm^{-2}) of Potato and Number of Tubers/Hill at Harvest

Treatments	Dry weight of tubers (gm^{-2})			Number of tubers $hill^{-1}$ at harvest
	50 DAP	70 DAP	90 DAP	
T_1	163.92	347.28	520.34	5.35
T_2	141.26	288.82	431.53	4.55
T_3	113.21	244.03	367.77	4.18
T_4	147.84	322.03	471.44	4.53
T_5	124.33	263.53	386.56	4.05
T_6	148.77	331.57	484.52	4.94
T_7	167.47	364.63	547.95	5.57
T_8	62.15	164.92	252.55	3.07
S.Em ±	4.34	15.06	4.86	0.198
CD (at 5 %)	13.16	45.69	14.73	0.601

16.3.2 NUMBER OF TUBERS/HILL

Highest number of tubers $hill^{-1}$ (5.57) was obtained under inorganic treatment T_7, which was found to be statistically at par with integrated nutrient management T_1 (5.35) followed by organic nutrient management T_6 (4.94). The lowest number of tubers per hill was obtained in control T_8 (3.07), and the percentage increase over the control (T_8) under T_7, T_1, and T_6 were 81.43, 74.27, and 60.91, respectively. (Table 16.1).

T_1: 50% recommended NPK (inorganic) + 50% N as FYM; T_2: 1/3rd recommended N each from FYM, Vermicompost, Neemcake; T_3: T_2 + Intercropping (potato+ coriander- 1:1); T_4: T_2 + rice straw mulch for weed management; T_5: 50% N as FYM + Rock phosphate + P.S.B + *azotobacter*; T_6: T_2 + Biofertilizers containing N and P carriers (same as T_5); T_7: 100% recommended NPK; T_8: Control (without manures and fertilizers).

16.3.3 TUBER BULKING RATE

A significant influence on tuber bulking rate of potato due to different nutrient managements was observed in between 50–70 DAP and 70–90 DAP (Table 16.2). The result revealed that tuber bulking rate was higher at 50–70

DAP than at 70–90 DAP. The maximum tuber bulking rate was observed under T_7 (9.86 gm^{-2}day^{-1} and 9.33 gm^{-2}day^{-1}) which was found statistically at par with T_1 (9.17 gm^{-2}day^{-1} and 8.65 gm^{-2}day^{-1}) and T_6 (9.14 gm^{-2}day^{-1} and 7.65 gm^{-2}day^{-1}) at 50–70 DAP and 70–90 DAP, respectively. The lowest tuber bulking rate was observed in control T_8 (5.14 gm^{-2} day^{-1} and 4.38 gm^{-2} day^{-1}, respectively).

16.3.4 TUBER YIELD:

Tuber yield of potato also differed significantly due to different nutrient managements (Table 16.2). The maximum tuber yield was produced under inorganic treatment T_7 (22.19 t ha^{-1}), which was found to be statistically at par with integrated nutrient management T_1 (21.72 t ha^{-1}). Among the different organic treatments, T_6 produced highest yield (20.04 t ha^{-1}), which was statistically at par with T_4, T_2, and T_5 (19.42 t ha^{-1}, 18.84 t ha^{-1}, and 18.25 t ha^{-1}), respectively. Lowest yield was recorded in T_3 (17.77 t ha^{-1}) after control T_8 (10.87 t ha^{-1}). The results are in conformity with the findings of Baishya et al. (2010). The favorable effect of integrated nutrient management and inorganic nutrient management on tuber yield of potato was also observed by Kumar (2012), Kumar et al. (2008, 2011), and Das et al. (2009).

TABLE 16.2 Effect of Different Nutrient Managements on Tuber Bulking Rate (g/m^2/day) and Tuber Yield (t ha^{-1}) of Potato

Treatments	Tuber bulking rate (g/m^2/day)		Tuber yield (t/ha)
	50-70 DAP	70-90 DAP	
T_1	9.17	8.65	21.72
T_2	7.38	7.14	18.84
T_3	6.54	6.19	17.77 (4.02)*
T_4	8.71	7.47	19.42
T_5	6.96	6.15	18.25
T_6	9.14	7.65	20.04
T_7	9.86	9.33	22.19
T_8	5.14	4.38	10.87
S.Em ±	0.683	0.567	0.651
CD (at 5 %)	2.073	1.722	1.974

*yield of coriander leaf in terms of potato equivalent yield

16.3.5 SOIL NUTRIENT STATUS

Organic carbon, total nitrogen, available phosphorus, and available potassium content were found to be significantly influenced by different nutrient management treatments (Table 16.3). Organic-based nutrient management treatments showed higher organic carbon content, total nitrogen, available phosphorus, and available potassium content in soil in comparison with inorganic nutrient management T_7 and integrated nutrient management T_1. The maximum organic carbon content was found under T_6 (0.91%), which was statistically at par with T_5 (0.89%) and T_4 (0.87%) followed by T_2 (0.85%) and T_3 (0.79%). All the organic-based treatments and integrated treatment showed an increase in organic carbon content over the initial value of 0.68%. The results corroborate the findings of Manna et al. (2006) and Ghosh et al. (2009). Similarly, Vidyavathi et al. (2011) also observed that the application of organic manures resulted in significantly higher organic carbon in the soil.

The maximum increase in total nitrogen content in soil over the initial (0.060%) was found under T_6 (0.076%), which was statistically at par with T_4 (0.074%) followed by T_2 (0.073%), T_5 (0.072%), and T_3 (0.072%) where T_4, T_2, T_5, and T_3 were statistically at par with each other. The percentage

TABLE 16.3 Effect of Different Nutrient Management on Soil Organic Carbon (%), Total N (%), Available P (kg ha^{-1}), and Available K (kg ha^{-1}) after Potato Harvest

Treatments	Organic carbon (%)	Total nitrogen (%)	Available phosphorus (kg/ha)	Available potassium (kg/ha)
T_1	0.76	0.068	18.48	166.60
T_2	0.85	0.073	19.18	170.91
T_3	0.79	0.072	17.01	141.11
T_4	0.87	0.074	19.77	172.94
T_5	0.89	0.072	18.87	168.92
T_6	0.91	0.076	20.58	175.57
T_7	0.65	0.065	17.89	171.15
T_8	0.63	0.054	12.40	139.75
Initial	0.68	0.06	16.36	163.51
S.Em ±	0.018	0.001	0.499	0.961
CD (at 5 %)	0.054	0.003	1.514	2.916

increases of T_6, T_4, T_2, and T_5 over the control T_8 were 26.67, 23.33, 21.67, and 20%, respectively. Inorganic treatment T_7 recorded lower N content (0.065%) followed by control (0.054%). Increase in available nitrogen content in the soil under organic nutrient managements might be due to release of nitrogen after decomposition of organics and direct addition through FYM to the available pool of the soil. It could also be attributed to the greater multiplication of soil microbes, which could convert organically bound nitrogen into inorganic form.

Available phosphorus content in soil was found to be higher under all the organic treatments except T_3, followed by integrated nutrient management (T_1) and synthetic fertilizer-based nutrient management (T_7). The maximum value of available phosphorus was obtained in T_6 (20.58 kgha^{-1}), which was statistically at par with T_4 (19.77 kgha^{-1}) and T_2 (19.18 kgha^{-1}) and their percentages of increase over the initial value (16.36 kgha^{-1}) were 25.79, 20.84 and 17.24%, respectively. Intercrop-based nutrient management treatment T_3 recorded lower phosphorus content in soil (17.01 kgha^{-1}) after inorganic treatment T_7 (17.89 kgha^{-1}). Control treatment T_8 (12.40 kgha^{-1}) recorded lowest phosphorus content in soil, which was 24.21% lower than the initial. Appreciable increase in available phosphorus content of soil in the organic management treatments may be due to the influence of organic manure, which could have enhanced the labile phosphorus in soil by complexing the cations like Ca, Mg, and Al responsible for the fixation of phosphorus.

Available K content in soil was found to be increased significantly over the initial value of 163.51 kg ha^{-1} in all the treatments, except for intercrop-based treatment T_3 and control T_8. The maximum available K in soil was recorded under organic-based nutrient management T_6 (175.57 kgha^{-1}) which was statistically at par with T_4(172.94 kgha^{-1}) followed by T_7(171.15 kgha^{-1}), T_2 (170.91 kgha^{-1}), T_5 (168.92 kgha^{-1}), and integrated nutrient management T_1 (166.60 kgha^{-1}). Babu et al. (2008) observed organic carbon built up and increase in available potassium in the soil with 100% recommended dose of nutrients applied through organic source. Zaman et al. (2011) also reported that FYM @30 t/ha along with biofertilizers recorded the maximum soil fertility build-up after harvest of the crop.

T_1: 50% recommended NPK (inorganic) + 50% N as FYM; T_2: 1/3rd recommended N each from FYM, Vermicompost, Neemcake; T_3: T_2 + Intercropping (potato+ coriander- 1:1); T_4: T_2 + rice straw mulch for weed management; T_5: 50% N as FYM + Rock phosphate + P.S.B + *Azotobacter*;

T_6: T_2 + Biofertilizers containing N and P carriers (same as T_5); T_7: 100% recommended NPK; T_8 : Control (without manures and fertilizers).

16.4 CONCLUSION

The results showed that inorganic nutrient management and integrated use of inorganic and organic sources of nutrients significantly improved the yield of potato. Regarding soil nutrient status, organic-based nutrient managements showed significant improvement over inorganic nutrient management. Application of organic manures either alone or in conjunction with inorganic fertilizer improved fertility status of soil over the years. Thus, organic manures and biofertilizers should be a part of the agronomic practices for potato cultivation. Hence from the above findings, it can be concluded that in order to sustain the productivity of soil and crop for long-term basis, organic packages with one-third recommended nitrogen each from farmyard manure, vermicompost, and neem cake along with *Azotobacter*, rock phosphate, and PSB is the best option followed by integrated nutrient management.

KEYWORDS

- **alluvial zone**
- **biofertilizer**
- **nutrient management**
- **potato**
- **productivity**
- **vermicompost**

REFERENCES

Arora, S., (2008). Balanced nutrition for sustainable crop production. *Krishi World*, 1–5.

Babu, M. V. S., Balaguravaiah, D., Adinarayana, G., Prathap, S., & Reddy, T. Y., (2008). Effect of tillage and nutrient management on rainfed groundnut production and soil properties in alfisols. *Legume Res.*, *31*, 192–195.

Baishya, L. K., Kumar, M., & Ghosh, D. C., (2010). Effect of different proportion of organic and inorganic nutrients on productivity and profitability of potato (*Solanumtuberosum*) varieties in Meghalaya hills. *Indian J. Agron.*, *55*, 230–234.

Das, P. P., Sarkar, A., & Zamen, A., (2009). Response of organic and inorganic sources of nutrients on growthand yield of potato in Gangetic alluvial plains of West Bengal. *Procceeding96th Indian Science Congress,* held on 3–7th January at NEHU, Shillong, Meghalaya.

Ghosh, P. K., Saha, R., Das, A., & Munda, G. C., (2007). Organic farming in India: potential and prospects. In: *Advancing Organic Farming Technology in India,* Munda, G. C., Ghosh, P. K., Das, A., Nagchan, S. V., Bujarbaruah, K. M., eds., 1–17.

Ghosh, P. K., Saha, R., Gupta, J. J., Ramesh, T., Das, A., Lama, T. D., Munda, G. C., Bordoloi, J. S., Verma, M. R., & Ngachan, S. V., (2009). Long term effect of pastures on soil quality in acid soil of North-East India. *Australian J. Soil Res.*, *47*, 372–379. http://dx.doi.org/10.1071/SR08169.

Gomez, K. A., & Gomez, A. A., (1984). *Statistical Procedures for Agricultural Research.* John Wiley and Sons, New York.

Khurana, P. S. M., & Naik, P. S., (2003). The Potato: an overview. In: *The Potato Production and Utilization in Sub- Tropics,* Paul Khurana, S. M., Minas, J. S., Pandey, S. K., eds. Mehta Publication, New Delhi, pp.1–14.

Kumar, M., Baishya, L. K., Ghosh, D. C., & Gupta, V. K., (2011). Yield and quality of potato (*Solanumtuberosum*) tubers as influenced by nutrient sources under rainfed condition of Meghalaya. *Indian J. Agron., 56*, 260–266.

Kumar, M., Baishya, L. K., Ghosh, D. C., Gupta, V. K., Dubey, S. K., Anup, D., & Patel, D. P., (2012). Productivity and soil health of potato (*Solanumtuberosum* L.) field as influenced by organic manures, inorganic fertilizers and biofertilizers under high altitudes of eastern Himalayas. *J. Agril. Sci*., *4*(5). http://dx.doi.org/10.5539/jas.v4n5p223.

Kumar, M., Jadav, M. K., & Trehan, S. P., (2008). Contributing of organic sources to potato nutrition at varying nitrogen levels. *Global Potato Conference,* New Delhi. 9–12.

Kumar, V., Jaiswal, R. C., & Singh, A. P., (2001). Effect of biofertilizers on growth and yield of potato. *J. Indian Potato Assoc., 28*, 6–7.

Manna, M. C., Swarup, A., Wanjari, R. H., Singh, Y. V., Ghosh, P. K., Singh, K. N., Tripathi, A. K., & Saha, M. N., (2006). Soil organic matter in West Bengal inceptisol after 30 years of multiple cropping and fertilization. *Soil Sci. Soc. Am. J., 70*, 121–129. http://dx.doi.org/10.2136/sssaj2005.0180.

Panes, V. G., & Sukhatme, P. V., (1967). *Statistical Methods for Agricultural Workers, Indian Council of Agri. Res.* New Delhi, vol. *8*, 65.

Upadhayay, N. C., & Singh, J. P., (2003). *The Potato (Production and Utilization in Sub-Tropics).* Paul Khurana, S. M., Minhas, J. S., & Pand, S. K., Eds, Mehta Publishers, A-16(East), Naraina II, New Delhi-110028, India, pp.154–160.

Vidya Vathi, Dasog, G. S., Babalad, H. B., Hebsur, N. S., Gali, S. K., Patil, S. G., & Alagawadi, A. R., (2011). Influence of nutrient management practices on crop response and economics in different cropping systems in a vertisol. *Karnataka Journal of Agricultural Sciences*, *24*, 455–460.

Zaman, A., Sarkar, A., Sarkar, S., & Devi, W. P., (2011). Effect of organic and inorganic sources of nutrients on productivity, specific gravity and processing quality of potato (*Solanumtuberosum*). *Indian J. Agril. Sci.*, *81*, 1137–1142.

CHAPTER 17

IMPACT OF ENRICHED COMPOSTS AND BIO-INOCULANTS ON SYMBIOTIC, GROWTH, YIELD, AND QUALITY PARAMETERS IN GARDEN PEA (*PISUM SATIVUM* L.) UNDER ORGANIC FARMING CONDITIONS

SANJAY CHADHA, AKASHDEEP SINGH, and SANGEETA KANWAR

Department of Vegetable Science and Floriculture, CSK Himachal Pradesh Krishi Vishvavidyalaya, Palampur, Himachal Pradesh, India, E-mail: schadha_113@yahoo.co.in

CONTENTS

ABSTRACT

Organic farming is an age-old concept, but during the last two decades, it has attracted worldwide attention. Balanced and efficient use of plant nutrients is essential for realizing the full potential of garden pea. The present investigation was carried out by evaluating 12 treatment combinations comprising four levels of composts, viz., E_3-3% enriched compost, E_2-2% enriched compost, E_1-1% enriched compost, and E_0-control (ordinary vermicompost) and three levels of bio-inoculations, viz., I_2-inoculation with *Rhizobium* + PSB + *Trichoderma*, I_1-inoculation with *Rhizobium* + PSB, and I_0-control (no inoculation) in factorial RBD with three replications in garden pea (variety-Punjab-89) at the Model Organic Farm, Department of Organic Agriculture, CSK HPKV, Palampur during *rabi* 2011–2012 and 2012–2013. Addition of different mineral substrates like oilseed cake for N enrichment, rock phosphate for P enrichment, and Patent Kali especially for K enrichment along with fortification with different types of bio-inoculants like *Rhizobium*, PSB, and *Trichoderma* in composts significantly improved yield, net profit, and B-C ratio under organic farming conditions. In the present investigation, the treatment combination E_2I_2, *i.e.*, 2% enriched compost @ 10t/ha along with bio-inoculation with *Rhizobium* + PSB + *Trichoderma* was found to be the most promising treatment combination which gave the maximum and significantly higher values for nitrogenase enzymes, dry matter, average pod weight, number of grains/pod, and pod yield/ha. E_2I_2 also exhibited the maximum plant height, number of pods per plant, pod length and shelling percentage but was statistically at par with other treatment combinations, viz., E_3I_1, E_1I_2, E_3I_2, E_2I_0, E_3I_0, and E_1I_1 for plant height, E_3I_2, E_3I_1, E_2I_1, and E_1I_2 for number of pods per plant, E_3I_0 for pod length, and E_3I_2 and E_3I_0 for shelling percentage. For nodule weight, total soluble solids and protein percentage, E_2I_2 was also found statistically at par with the highest treatment combinations. For these three traits, E_3I_1, E_0I_2, and E_1I_2 revealed the maximum values, respectively. The treatment combinations were statistically at par with E_1I_2 for nodule weight and total soluble solids, E_2I_1 for nodule weight, E_3I_2 for total soluble solids, and E_3I_1 for protein percentage.

17.1 INTRODUCTION

Garden pea (*Pisum sativum* L.) is one of the most important off-season vegetable crops of Himachal Pradesh and ranks at first position in acreage (23904 ha) with production potential of 271,057 MT (Anonymous, 2013–14). Indiscriminate use of pesticides in vegetables has attracted worldwide attention, and the consumers prefer to pay higher premium for safe food. Consumption of fertilizers and pesticides in the hilly states like Himachal Pradesh is still low as the majority of the farming communities are small and marginal for whom the purchase of fertilizers and pesticides is constrained by their high costs (Gopinath and Meena, 2011). This may be one of the reasons that the consumers have their special preference for hill-grown peas because of their characteristic flavor, sweetness, and freshness. Organic foods are considered safe because the use of synthetic fertilizers and pesticides are strictly prohibited. The essential concept and philosophy of organic farming is to feed the soil rather than the crops to maintain its health and save environment, i.e., giving back to the nature what has been taken from it (Funtilana, 1990). To maintain the fertility of the soil for sustainable yield production is one of the great challenges in organic farming system. The success of organic farming system is very much dependent upon the availability of cheap and good quality organic manures for maintaining the soil health for a long term, but its availability on large scale is always a limiting factor. Hence, in the organic farming system, there is a need to reduce the quantity of compost required for sustainable production by increasing the quality of compost through enrichment and/or fortification with mineral substrates and bio-inoculants.

17.2 MATERIALS AND METHODS

The present investigation was carried out at the Model Organic Farm, Department of Organic Agriculture, CSK Himachal Pradesh Krishi Vishvavidyalaya, Palampur (Himachal Pradesh). Department of Organic Agriculture, CSK HPKV, Palampur, has been identified as a "Niche Area of Excellence on Organic Farming" by Indian Council of Agricultural Research (ICAR), New Delhi, and has been exclusively managed with organic inputs since 2006. The details of the different treatments are given in Table 17.1.

The experiment was conducted in factorial RBD with three replications during *rabi* of 2011–2012 and 2012–2013 in 36 plots of each 12 m^2 size in garden pea (variety-Punjab 89). The data were taken on different symbiotic, growth, yield, and quality parameters. The cultural practices for raising organic crop of pea were followed according to ad-hoc guidelines published (Rameshwar et al., 2011) by the CSK HPKV, Palampur.

TABLE 17.1 Details of the Treatments Comprising of Different Levels of Compost and Bio-inoculations

Treatments		
Compost		**Bio-inoculation**
E_0 Control (Ordinary vermi-compost)		I_0: No inoculation
E_1: 1% Enriched Compost		I_1: Inoculation with *Rhizobium* + PSB
E_2: 2% Enriched Compost		I_2: Inoculation with *Rhizobium* + PSB+*Trichoderma*
E_3: 3% Enriched Compost		
Treatment combinations		
T_1	$E_0 I_0$	Vermicompost+ No inoculation
T_2	$E_0 I_1$	Vermicompost+ Inoculation with *Rhizobium* + PSB
T_3	$E_0 I_2$	Vermicompost+ Inoculation with *Rhizobium* + PSB+ *Trichoderma*
T_4	$E_1 I_0$	1% Enriched Compost + No inoculation
T_5	$E_1 I_1$	1% Enriched Compost + Inoculation with *Rhizobium* + PSB
T_6	$E_1 I_2$	1% Enriched Compost + Inoculation with *Rhizobium* + PSB+ *Trichoderma*
T_7	$E_2 I_0$	2% Enriched Compost + No inoculation
T_8	$E_2 I_1$	2% Enriched Compost + Inoculation with Rhizobium + PSB
T_9	$E_2 I_2$	2% Enriched Compost + Inoculation with *Rhizobium* + PSB+ *Trichoderma*
T_{10}	$E_3 I_0$	3% Enriched Compost + No inoculation
T_{11}	$E_3 I_1$	3% Enriched Compost + Inoculation with *Rhizobium* + PSB
T_{12}	$E_3 I_2$	3% Enriched Compost + Inoculation with *Rhizobium* + PSB+ *Trichoderma*

17.2.1 ENRICHED COMPOST

Enriched composts were prepared by adding the different natural substrates like oil seed cake (for N enrichment), rock phosphate (for P enrichment), and Patent Kali (for K, S, and Mn enrichment) all @ 1% (for 1% enriched compost), 2% (for 2% enriched compost), and 3% (for 3% enriched compost) of the total biomass. *Rhizobium* and PSB were first mixed with different kinds of composts according to treatment combinations @ 10 g/10 kg of the compost. The details of the nutrient analysis of different enriched composts are given in Table 17.2. After mixing biofertilizers with composts, these were moistened with sprinkling water, covered with gunny bag, and kept to incubate for 15 days before applying to the soil along with seed sowing. Fortnightly sprays of vermiwash or compost tea (10%) before pod filling stage were also given. Two hoeing along with hand weeding were followed to keep the field weed free.

17.3 RESULTS AND DISCUSSION

Significant differences for all the symbiotic, growth, yield, and quality parameters were recorded with the application of different composts, bio-inoculants (*Rhizobium*, PSB, and *Trichoderma*), and their interactions in both the years of study (2011–2012 and 2012–2013) and pooled analysis (Tables 17.3 and 17.4)

A critical observation of pooled analysis of different characteristics in Tables 17.3 and 17.4 was that 3% enriched compost (E_3) exhibited the maximum and significantly higher number of nodules/plant (38.4), nitrogenase enzymes (183.5 μC2H4/dry weight of nodule in g/h), and pod yield/plant (62.7) than all other compost levels. E_3 produced the maximum number of pods per plant, average pod weight, and shelling percentage, but it was statistically at par with E_2 for these traits but significantly higher than E_1 and E_0. E_2 revealed the maximum nodule weight (38.0), number of grains per pod, pod length, and total soluble solids (16.8 °Brix), and for all these traits, it was statistically at par with E_3 (37.2 mg and 16.5 °Brix, respectively) but significantly higher than E_1 and E_0. For the traits plant dry matter and plant height, all the enrichment levels viz., E_1, E_2 and E_3 were found statistically at par with each other but significantly higher than the control, i.e., E_0, with

TABLE 17.2 Nutrient Analysis of Enriched Compost Used in Experiment

Compost		Nitrogen (%)	Phosphorus (%)	Potassium (%)
E0	Vermicompost0	98 0	34 0	52
E1	1% Enriched compost 1	01 0	40 0	59
E2	2% Enriched compost 1	23 0	44 0	68
E3	3% Enriched compost 1	35 0	48 0	72

TABLE 17.3 Effect of Composts, Bio-Inoculations, and Their Interactions on Number of Nodules/Plant, Nodule Weight/Plant (mg) Nitrogenase Enzyme, Plant Height (cm), Plant Dry Matter/Plant (g), and Number of Pods/Plant during 2011–2012 and 2012–2013 and Pooled Analysis in Garden Pea

Treat-ments	Number of nodules/ plant			Nodule weight/ plant(mg)			Nitrogenase enzyme (μC2H4/dry weight of nodule in g/h)			Plant height (cm)			Plant dry matter/ plant(g)			Number of pods/plant		
	2011-12	2012-13	Pooled	2011-12	2012-13	Pooled	2011-12	2012-13	Pooled	2011-12	2012-13	Pooled	2011-12	2012-13	Pooled	2011-12	2012-13	Pooled
Enriched compost*																		
E_0	13.7	13.0	13.3	34.0	34.8	34.4	178.8	178.4	178.6	50.3	49.8	50.0	4.0	3.9	3.92	10.0	10.4	10.2
E_1	16.2	16.8	16.5	36.4	36.1	36.3	181.0	180.8	180.9	58.1	57.6	57.9	5.0	4.9	4.95	11.4	10.9	11.1
E_2	17.6	17.0	17.3	37.9	38.1	38.0	181.8	181.5	181.7	59.1	58.3	58.7	5.3	5.0	5.14	11.3	11.7	11.5
E_3	18.8	17.7	18.2	38.4	37.2	37.8	183.7	183.2	183.5	58.0	58.5	58.3	5.3	4.9	5.09	12.0	12.1	11.9
$CD_{0.05}$	0.68	1.13	0.73	1.47	0.98	0.85	1.07	1.12	0.68	1.36	1.03	0.80	0.3	0.2	0.15	0.67	0.58	0.41
Bio-inoculation**																		
I_0	14.5	13.2	13.8	35.3	34.8	35.0	180.2	179.6	179.9	53.7	53.0	53.4	4.7	4.4	4.6	10.1	10.3	10.2
I_1	17.8	17.1	17.5	37.7	37.7	37.7	180.7	180.8	179.9	57.1	56.8	56.9	4.8	4.7	4.8	11.5	11.7	11.6
I_2	17.3	18.1	17.7	37.1	37.5	37.2	183.1	182.5	182.8	58.3	58.3	58.3	5.1	4.9	5.0	11.9	11.9	11.9
$CD_{0.05}$	0.59	0.98	0.63	1.27	0.85	0.74	0.92	0.97	0.59	1.17	0.89	0.70	0.23	0.2	0.1	0.6	0.5	0.4
Interaction (E x I)																		
$E_0 I_0$	11.3	10.3	10.8	33.7	33.0	33.3	177.0	176.5	176.8	46.1	46.0	45.9	3.8	3.8	3.8	9.3	9.3	9.3
$E_0 I_1$	14.7	13.0	13.8	35.3	36.7	36.0	179.5	179.1	179.3	51.1	50.4	50.7	3.9	3.9	3.9	10.0	10.7	10.3
$E_0 I_2$	15.0	15.7	15.3	33.0	34.7	33.8	180.0	179.5	179.7	53.7	53.1	53.4	4.1	4.0	4.1	10.7	11.3	11.0
$E_1 I_0$	13.7	13.3	13.5	33.3	33.3	33.3	180.5	180.0	180.2	56.8	54.7	55.8	4.6	4.6	4.6	10.3	9.7	10.0
$E_1 I_1$	18.3	18.3	18.3	37.3	36.7	37.0	179.4	180.1	179.7	58.6	58.2	58.4	4.9	4.9	4.9	11.7	11.3	11.5

TABLE 17.3 (continued)

Treat-ments	Number of nodules/ plant			Nodule weight/ plant(mg)			Nitrogenase enzyme (µC2H4/dry weight of nodule in g/h)			Plant height (cm)			Plant dry matter/ plant(g)			Number of pods/plant		
	2011-12	**2012-13**	**Pooled**	**2011-12**	**2012-13**	**Pooled**	**2011-12**	**2012-13**	**Pooled**	**2011-12**	**2012-13**	**Pooled**	**2011-12**	**2012-13**	**Pooled**	**2011-12**	**2012-13**	**Pooled**
$E_1 I_2$	16.7	18.7	17.7	38.7	38.3	38.5	183.0	182.5	182.8	58.9	60.1	59.5	5.4	5.3	5.3	12.3	11.7	12.0
$E_2 I_0$	16.3	14.7	15.5	37.3	36.7	37.0	79.4	178.9	179.2	57.0	55.6	56.3	4.8	4.7	4.7	9.7	10.3	10.0
$E_2 I_1$	18.0	16.7	17.3	37.7	39.0	38.3	180.1	179.9	180.0	59.2	58.3	58.7	5.3	4.9	5.1	12.0	12.3	12.2
$E_2 I_2$	18.3	19.7	19.0	38.7	38.7	38.7	186.0	185.7	185.9	61.1	61.0	61.0	5.7	5.5	5.6	12.3	12.3	12.3
$E_3 I_0$	16.7	14.3	15.5	37.0	36.0	36.5	183.9	182.9	183.4	55.0	56.0	55.5	5.5	4.8	5.1	11.3	11.7	11.5
$E_3 I_1$	20.3	20.3	20.3	40.3	38.3	39.3	184.0	184.2	184.1	59.6	60.3	60.0	5.2	5.0	5.1	12.3	12.3	12.3
$E_3 I_2$	19.3	18.3	18.8	38.0	37.3	37.7	183.3	182.4	182.9	59.6	59.2	59.4	5.2	5.0	5.1	12.3	12.3	12.3
$CD_{0.05}$	**1.2**	**2.0**	**1.3**	**2.6**	**1.7**	**1.5**	**1.9**	**1.9**	**1.2**	**2.4**	**1.8**	**1.4**	**0.5**	**0.3**	**0.3**	**1.2**	**1.0**	**0.7**

TABLE 17.4 Effect of Composts, Bio-Inoculations, and Their Interactions on Number of Average Pod Weight (g), Number of Grains/Pod, Pod Length (cm), Shelling Percentage, and Pod Yield/Plant (g) during 2011–2012 and 2012–2013 and Pooled Analysis in Garden Pea

Treatments	Average pod weight (g)			Number of grains/ pod			Pod length(cm)			Shelling percentage			Pod yield per plant(g)		
	2011-12	2012-13	Pooled	2011-12	2012-13	Pooled	2011-12	2012-13	Pooled	2011-12	2012-13	Pooled	2011-12	2012-13	Pooled
Enriched compost*															
E_0	4.47	4.47	4.47	6.44	6.33	6.38	8.11	8.08	8.10	36.82	36.41	36.62	44.84	46.85	45.84
E_1	4.68	4.99	4.83	6.66	6.44	6.55	8.76	8.34	8.55	38.48	38.58	38.53	53.66	54.44	54.05
E_2	5.14	5.16	5.15	7.11	7.00	7.05	9.01	8.92	8.96	41.33	41.01	41.17	58.31	60.30	59.30
E_3	5.19	5.20	5.19	7.22	6.77	7.00	9.26	8.62	8.94	42.65	42.04	42.35	62.37	62.94	62.65
$CD_{0.05}$	0.19	0.21	0.15	0.47	0.47	0.29	0.52	0.15	0.31	1.16	1.23	0.76	1.91	0.77	1.32
Bio-inoculation**															
I_0	4.66	4.73	4.69	6.25	6.16	6.20	8.34	8.45	8.40	39.21	39.09	39.15	47.47	48.65	48.06
I_1	4.83	4.82	4.82	6.91	6.58	6.75	8.82	8.37	8.59	38.50	38.49	38.50	55.69	56.43	56.07
I_2	5.12	5.31	5.22	7.41	7.17	7.29	9.20	8.65	8.92	41.75	40.96	41.36	61.21	63.32	62.27
$CD_{0.05}$	0.16	0.18	0.13	0.41	0.41	0.25	0.45	0.13	0.27	1.00	1.07	0.66	1.65	0.67	1.14
Interaction (E x I)															
$E_0 I_0$	4.24	4.25	4.24	6.00	6.00	6.00	8.00	8.07	8.03	36.68	35.47	36.08	39.49	39.57	39.53
$E_0 I_1$	4.53	4.36	4.44	6.33	6.33	6.33	7.90	8.07	7.98	36.00	36.53	36.27	45.34	46.43	45.89
$E_0 I_2$	4.67	4.82	4.74	7.00	6.67	6.83	8.43	8.13	8.28	37.80	37.25	37.53	49.69	54.57	52.13
$E_1 I_0$	4.63	4.74	4.68	6.00	6.00	6.00	7.77	8.17	7.97	36.77	37.88	37.33	47.77	45.75	46.76

TABLE 17.4 (continued)

Treatments	Average pod weight (g)			Number of grains/ pod			Pod length(cm)			Shelling percentage			Pod yield per plant(g)		
	2011-12	2012-13	Pooled	2011-12	2012-13	Pooled	2011-12	2012-13	Pooled	2011-12	2012-13	Pooled	2011-12	2012-13	Pooled
$E_1 I_1$	4.69	4.79	4.74	6.67	6.33	6.50	8.97	8.47	8.72	36.69	36.71	36.70	54.65	54.18	54.42
$E_1 I_2$	4.75	5.44	5.10	7.33	7.00	7.17	9.57	8.40	8.98	42.00	41.17	41.59	58.57	63.41	60.99
$E_2 I_0$	4.98	4.92	4.95	6.33	6.33	6.33	8.47	8.63	8.55	40.95	40.85	40.90	48.14	50.71	49.43
$E_2 I_1$	4.87	4.77	4.82	7.33	7.00	7.17	8.73	8.80	8.77	39.34	38.93	39.14	58.14	58.72	58.43
$E_2 I_2$	5.57	5.80	5.69	7.67	7.67	7.67	9.83	9.33	9.58	43.72	43.26	43.49	68.65	71.49	70.07
$E_3 I_0$	4.81	5.03	4.92	6.67	6.33	6.50	9.13	8.97	9.05	42.45	42.17	42.31	54.51	58.58	56.55
$E_3 I_1$	5.25	5.39	5.32	7.33	6.67	7.00	9.70	8.13	8.92	42.01	41.80	41.90	64.66	66.42	65.54
$E_3 I_2$	5.52	5.18	5.35	7.67	7.33	7.50	8.97	8.77	8.87	43.51	42.17	42.84	67.95	63.83	65.89
$CD_{0.05}$	0.33	0.37	0.26	0.82	0.82	0.51	0.90	0.27	0.54	2.01	2.14	1.33	3.30	1.34	2.29

the maximum in E_3 for dry matter and E_2 for plant height. For protein, E_1 exhibited the maximum value (14.9%), but it was statistically at par with E_3 (14.2%) and E_2 (14.0%). Riffaldi et al. (1992) and Lopes et al. (1996) had also reported the increase in number of nodules/plant in cowpea with the application of vermicompost.

Over both the years of study, inoculation with *Rhizobium* + PSB + *Trichoderma* (I_2) exhibited the maximum and significantly higher values for all the traits studied than dual inoculation with *Rhizobium* + PSB (I_1) and control (I_0), i.e., no inoculation, except number of pods/plant for which I_2 was statistically at par with I_1 and for nodule weight for which I_1 revealed the maximum value but was statistically at par with I_2 and significantly higher than I_0. I_1 was found significantly higher than I_0 for all the traits studied except total soluble solids and protein percentage for which I_1 was statistically at par with I_0. Various earlier researchers have also noticed significantly desirable effects on symbiotic and quality parameters by inoculation with different bio-inoculants in pea. Singh et al. (2012) also noticed an increase in number of nodules in pea after inoculation with *Rhizobium* and PSB. Ali et al. (2008) recorded higher nodule weight with the application of *Rhizobium* in pea. Khondaker et al. (2003) recorded the maximum nitrogenase activity with the application of vermicompost.

A perusal of contents of Tables 17.3 and 17.4 over both the year of study (pooled) showed that E_2I_2, i.e., 2% enriched compost along with bio-inoculation with *Rhizobium* + PSB + *Trichoderma* was found to be the most promising treatment combination, which gave the maximum and significantly higher values for nitrogenase enzymes (185.9 µC2H4/dry weight of nodule in g/h), dry matter (5.63 g), average pod weight (5.69 g), number of grains per pod (7.67), and pod yield/plant (70.07 g). E_2I_2 also exhibited the maximum plant height (61.01 cm), number of pods per plant (12.33), pod length (9.58 cm), and shelling percentage (43.49%) but was statistically at par with other treatments combinations, viz., E_3I_1, E_3I_2, E_1I_2, E_2I_0, E_3I_0, and E_1I_1 for plant height; E_3I_2, E_3I_1, and E_2I_1 and E_1I_2 for number of pods per plant; E_3I_0 for pod length; and E_3I_2 and E_3I_0 for shelling percentage. For nodule weight, total soluble solids and protein percentage, E_2I_2 was also found statistically at par with the highest treatment combinations. E_3I_1 (39.3mg) for nodule weight, E_0I_2 & E_3I_2 (17.0 °Brix both) for total soluble solids and E_1I_2 (16.3%) for protein percentage were the highest performing treatment combinations, but they were statistically at par with E_2I_2 for respective traits, i.e., 38.7 mg, 16.8 °Brix, and 15.2%, respectively (Table 17.1). The increase in number of nodules/plant might be attributed to the increase in nitrogenase activity (Chattopadhyay et al., 2003). De et al. (2006) and Ahmed et al. (2007) obtained the maximum total soluble solids and protein in pea with combined application of *Rhizobium* and organic manures. Rajput and Pandey (2004) and EL-Etr et al. (2004) recorded the highest pod yield by the application of bio-fertilizers and compost.

ACKNOWLEDGMENT

The authors highly acknowledge ICAR, New Delhi, for providing the necessary funding under the Niche Area of Excellence in Organic farming.

KEYWORDS

- **bio-inoculation**
- **enriched compost**
- **organic farming**
- **pea**
- ***Pisum sativum***

REFERENCES

Ahmed, R., Solaiman, A. R. M., Halder, N. K., Siddiky, M. A., & Islam, M. S., (2007). Effect of inoculation methods of *Rhizobium* on yield attributes, yield and protein content in seed of pea. *J. Soil Nat.*, *1*(3), 30–35.

Ali, M. E., Khanam, D., Bhuiyan, M. A. H., Khatun, M. R., & Talukder, M. R., (2008). Effect of *Rhizobium* inoculation to different varieties of garden pea (*Pisum sativum* L.). *J. Soil Nat.*, *2*(1), 30–33.

Anonymous, (2013–14). *Area and Production of Vegetable Crops*. Department of Agriculture, H. P. Government-Shimla.

Chattopadhyay, A., & Dutta, D., (2003). Response of vegetable cowpea to phosphorus and biofertilizers in old alluvial zone of west Bengal. *Leg. Res.*, *26*(3), 196–199.

De, N., Rai, M., Singh, R. K., & Singh, J., (2006). Effect of nutrient on yield and root parameters in vegetable pea. *Veg. Sci.*, *33*(1), 88–89.

EL-Etr, W. T., Ali, L. K. M., & EL-Khatib, E. L., (2004). Comparative effects of bio-compost and compost on growth, yield and nutrients content of pea and wheat plants grown on sandy soils. *Egyp. J. Agric. Res.*, *82*(2), 73–94.

Funtilana, S., (1990). Safe, inexpensive, profitable and sensible. *Int. Agric. Dev.*, pp. 24.

Gopinath, K. A., & Mina, B. L., (2011). Effect of organic manures on agronomic and economic performance of garden pea (*Pisum sativum*) and on soil properties *Ind. J. Agric. Sci.*, *81*(3), 236–39.

Khondaker, M., Solaiman, A. R. M., Karim, A. J. M. S., & Hossain, M. M., (2003). Response of pea varieties to *Rhizobium* inoculation, nitrogenase activities, dry matter production and nitrogen uptake. *Kor. J. Crop Sci.*, *48*, 355–360.

Lopes, A. J., Stamford, N. P., Figueired, M. V. B., Burity, N. A., & Ferraz, E. B., (1996). Effect of vermicompost, mineral nitrogen and mineralizing agent on N fixation and yield in cowpea. *Revista – Brasileira – De-Ciencia-Do-Solo*, *20*, 55–62.

Rajput, R. L., & Pandey, R. N., (2004). Effect of method of application of bio-fertilizers on yield of pea (*Pisum sativum*). *Leg. Res.*, *25*, 75–76.

Rameshwar, Saini, J. P., Chadha, S., & Sharma, S., (2011). (eds.). *Jaivik Krishi*. Department of Organic Agriculture, CSKHPKV, Palampur., pp 133.

Riffaldi, R., Levi-Minzi, R., Saviozzi, A., & Capurro, M., (1992). Evaluating garbage compost. *Bio Cycle*, *33*, 66–69.

Singh, R. K., Srivastav, D. K., & Sharma, S., (2012). Effect of single and dual inoculation of biofertilizers on root nodule and shoot parameter of pea (*Pisum sativum*). *Ann. Horti.*, *5*(1), 58–62.

CHAPTER 18

INFLUENCE OF ORGANIC WASTES AMENDMENT ON THE SOIL OF A JHUM FALLOW

ANGOM SARJUBALA DEVI and LALBIAKDIKI ROYTE

Department of Environmental Science, Mizoram University, India, E-mail: angom75@yahoo.com

CONTENTS

ABSTRACT

Three different types of organic wastes that are abundant in Mizoram were selected. They were the sheaths of *Dendrocalamus hamiltonii*, *Amomum dealbatum*, and *Zea mays*. They were shredded and dried to constant weight and amended with soil in a 5-year-old jhum fallow. In the study, it was found that the amendment of *Amomum* residues leads to more enrichment of organic carbon in the soil. By comparing between the unamended and amended soil, there was more organic carbon, nitrogen, and phosphorus in the soil of amended plots. This indicates that jhum fallows can be reclaimed and a new cropping pattern can be established by treatment of organic wastes.

18.1 INTRODUCTION

Large amounts of wastes containing different amounts of organic carbon from agro-industrial and municipal origins that are produced daily in the use of byproducts of vegetables and animal and human origins used to restore or to increase soil fertility has been well known for over 2000 years. However, since the end of World War II, the use of organic waste for land fertilization has decreased, and farmers have markedly increased the use of mineral fertilizers in place of organic fertilizers and amendments. Concurrently, the amounts of organic materials from municipal solid wastes, sewage sledges, and wastes of agro-industrial origins have increased exponentially, and millions of tons of organic matter are land filled or incinerated and transformed into methane, carbon dioxide, nitrogen oxides, sulfur oxides, and other greenhouse gases (Clapp et al., 2007).

Applications of organic waste are recommended for soils where the organic matter content is low. In particular, the addition to the soil of stabilized organic matter rich in humic and humic-like components and nutrients contribute to: the natural closure of the nutrient cycles, the increase in fertility of soils, the enhancement of the carbon sink, and the decrease in greenhouse gas emissions (Clapp et al., 2007).

Organic matter can be added to the soil by incorporating plant material, animal residues, manure, sewage sludge, or municipal solid waste. These additions as well as the agricultural management practices can affect soil microbial communities. Changes in microbial activities and the composition of soil microbial communities can in turn influence soil fertility and plant growth by increasing nutrient availability and turnover, disease incidence, or disease suppression (Hanson, 1994).

The present study was undertaken to observe the changes in the physicochemical characteristics of soil due to amendment of organic wastes mainly vegetable wastes and crop residue in an agricultural field that was abandoned for 5 years. This type of abandoned fields is known as jhum fallows in northeast India.

18.2 MATERIALS AND METHODS

18.2.1 STUDY SITE

A 5-year-old jhum fallow was selected in Tanhril village in Aizawl district. The site is located adjacent to Mizoram University campus lying between 23°43'25" N and 23°45'37" N latitudes and 92°38'39"E and 92° 40'23" E longitudes. In the present jhum fallow cultivation of crops is not done; the only outgrowth is tall grasses, mainly *Imperata cylindrica*. The study site is having a slope range of 5–10% (gentle).

18.2.2 TYPES OF ORGANIC WASTES

Three different types of organic wastes were selected. They were: the sheaths of the shoots of the bamboo *Dendrocalamus hamiltonii* and *Amonium dealbatum*. These shoots are edible and plenty of their sheaths are thrown away as wastes mainly during rainy season. The third type of waste was the outer covering of *Zea mays* or corn. These three types of organic wastes were abbreviated as DHW for *D. hamiltonii* wastes; ADW for *A. dealbatum* wastes; and ZMW for *Z. mays* wastes. The amount of the whole crop by moist weight thrown away as waste was found to be 63%, 68%, and 45% of *Dendrocalamus*, *Amomum*, and corn, respectively. These organic wastes were collected from local markets, shredded into small fragments, and air-dried till constant weight.

18.2.3 ORGANIC WASTE AMENDMENT

In the study site, three pits having area of 2.5 × 2.5 m with a depth of 5 cm were dug up. The organic wastes were mixed separately with the removed soil at a ratio of 10 g waste/kg of the soil and put in the pits separately. One unamended plot was maintained as a control plot.

18.2.4 INITIAL CHEMISTRY

Samples of the air-dried and grinded organic wastes were analyzed for initial C and N content by using the CHNS/O Elementary Analyzer Euro 3000 installed at the Central Instrumental Laboratory of Mizoram University.

18.2.5 SOIL ANALYSIS

Samples of soil and organic wastes mixture from the amended plots were collected from a depth of 0–5 cm. Soil sampling was started monthly from August 2013 for 19 months. The soil samples were air dried and passed through a 2-mm sieve in order to remove undecomposed organic wastes, and determination for physico-chemical characteristics of the soil was done. Soil moisture was determined by following the gravimetric method, digital pH was determined using a pH meter, water holding capacity (WHC) was determined by following the filter paper method, organic C and total N were determined by the CHNS/O analyzer, and available P was estimated by following Olson's method outlined in Anderson and Ingram (1993).

Statistical calculations of correlation was done by using Microsoft Excel 2007.

18.3 RESULTS

The initial C, N, and C:N ratio are given in Table 18.1. ADW had the highest initial C and N content and DHW had the lowest.

TABLE 18.1 Initial C, N, and C:N Ratio of the Wastes Residues

Type of waste residue	Initial C (%)	Initial N (%)	C:N ratio
Dendrocalamus waste residue	3.0	0.3	10
Corn waste residue	8.5	0.3	28
Amomum waste residue	10.0	0.5	19

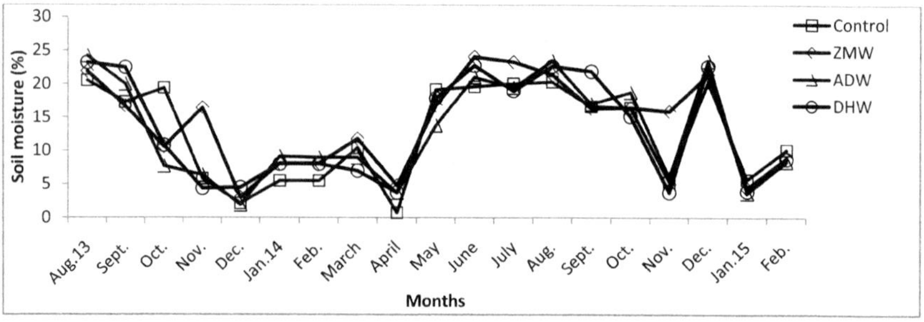

FIGURE 18.1 Soil moisture (%) in the waste amended and unamended plots.

The soil moisture ranged from 5.50% (January 14) to 20.53% (April 13) in the control plot (Figure 18.1); from 3.00% (December 13) to 24.07% (June 14) in the ZMW amended plot; from 1.90% (December 13) to 24.20% (August 13) in the ADW amended plot, and from 8.00% (January 14) to 23.23% (August 13) in the DHW amended plot. On comparing between the amended and unamended plots, the ZMW amended plot had more amount of average soil moisture (14.23%) followed by DHW (13.18%), ADW (12.93%), and least was in the control plot (12.68%). There was an increase in the soil moisture level in all the wastes amended plots. For all the parameters, the change due to waste amendment in soil was calculated by deducting the value of control from the amended plots (Table 18.2). The ZMW amendment leads to more enhancement in the soil moisture, followed by DHW amendment, and least increase was observed in ADW amended plot. The soil pH ranged from 4.81 (October 14) to 6.08 (July 14) in the control plot (Figure 18.2); it varied from 4.70 (October 14) to 5.80 (July 14) in the ZMW amended plot; from 4.70 (June 14) to 6.10 (July 14) in the ADW amended plot, and it ranged from 4.50 (June 14) to 6.10 (July 14) in the DHW amended plot. Within the wastes amended plots and unamended plot,

TABLE 18.2 Change in Average Physico-Chemical Properties of Soil due to Treatment in Different Sites in the Three Slopes

	Soil moisture	Soil pH	Water holding capacity (%)	Organic C (%)	Total N (%)	Available P (%)
Corn	1.59	0.02	4.87	0.71	0.041	25.79
Amom.	0.30	–0.07	8.35	1.35	0.072	31.11
Dendro.	1.50	–0.05	5.52	0.92	0.056	47.79

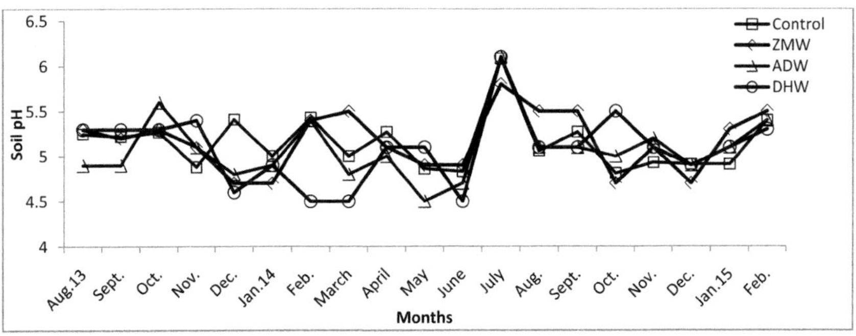

FIGURE 18.2 Soil pH in the waste amended and unamended plots.

the average soil pH was highest in the ZMW amended plot (5.16) followed by control (5.14), ADW (5.07), and DHW amended plot (5.09). There was an increase in the level of soil pH in the ZMW amended plot, i.e., lesser acidity of soil. In the ADW and DHW amended plots, there was a decrease in soil pH from the control plot, i.e., more acidity found in these amended plots.

The WHC of soil varied from 34.33% (March 14) to 68.64% (January 14) in the control plot (Figure 18.3); it varied from 38.89% (March 14) to 72.59% (January 14) in the ZMW amended plot; from 38.53% (December 14) to 88.80% (April 13) in the ADW amended plot and from 32.75% (December 13) to 83.73% (April 13) in the DHW amended plot. The average WHC was least in the control (49.39%) followed by ZMW (54.26%) and DHW (55.91%) and highest in the ADW amended plot (57.74%). Due to treatment of the waste, there was an increase in all the amended plots where highest increase was obtained in ADW amended plot and least in ZMW amended plot.

The organic C varied from 0.99% (December 13) to 6.10% (October 13) in the control plot (Figure 18.4); it varied from 1.51% (January 15) to 6.56% (October 13) in the ZMW amended plot; from 1.62% (March 14) to 7.14% (August 13) in the ADW amended plot; and from 1.26% (January 14) to 7.51% (September 13) in the DHW amended plot. The average organic C was least in the control (2.59%) followed by ZMW (3.30%) and DHW (3.51%) and highest in the ADW amended plot (3.94%). Due to treatment of the wastes, there was an increase in organic C of soil in all the amended plots where highest increase was obtained in ADW amended plot and least in ZMW amended plot.

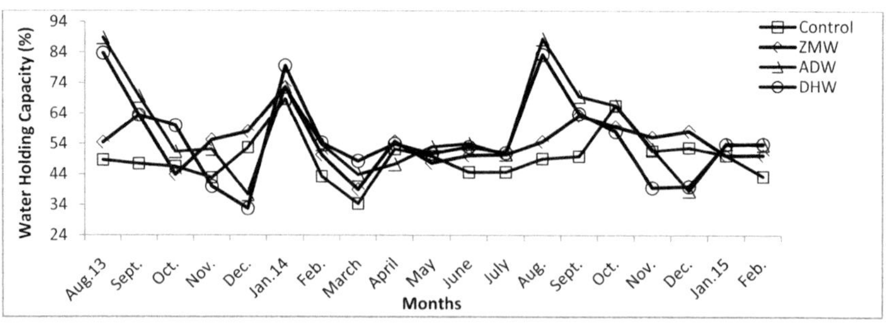

FIGURE 18.3 Soil water holding capacity (%) in the waste amended and unamended plots.

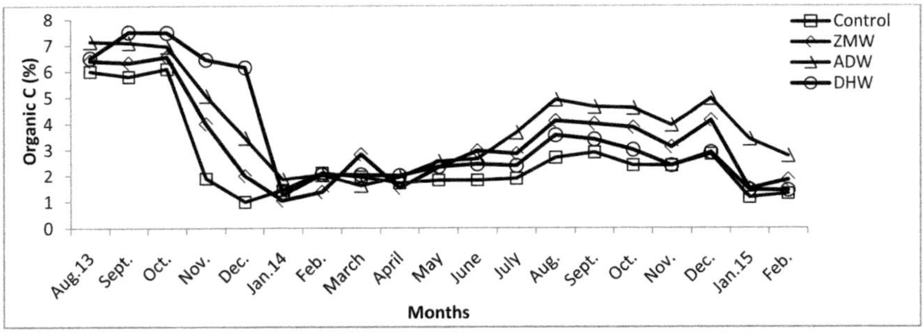

FIGURE 18.4 Soil organic C (%) in the waste amended and unamended plots.

The total N varied from 0.160% (November 13) to 0.410% (August 13) in the control plot (Figure 18.5); it varied from 0.147% (April 14) to 0.414% (August 13) in the ZMW amended plot; from 0.154% (March 14) to 0.401% (August 13) in the ADW amended plot and from 0.157% (January 14) to 0.456%(October 13) in the DHW amended plot. The average total N was least in the control (0.247%) followed by ZMW (0.288%), DHW (0.303%), and highest in the ADW amended plot (0.319%). Due to amendment of the wastes, there was an increase in total N in all the amended plots where highest increase was obtained in the ADW amended plot and least in the ZMW amended plot (Table 18.2).

The available P varied from 63.00 μg/g (October 13) to 143.00 μg/g (May 14) in the control plot (Figure 18.6); it varied from 80.00 (February 15) to 184.00 μg/g (June 14) in the ZMW amended plot; from 75.00 μg/g (October 13) to 173.00 μg/g (June 14) in the ADW amended plot; and from 84.00 μg/g (June 14) to 197.00 μg/g (August 13) in the DHW

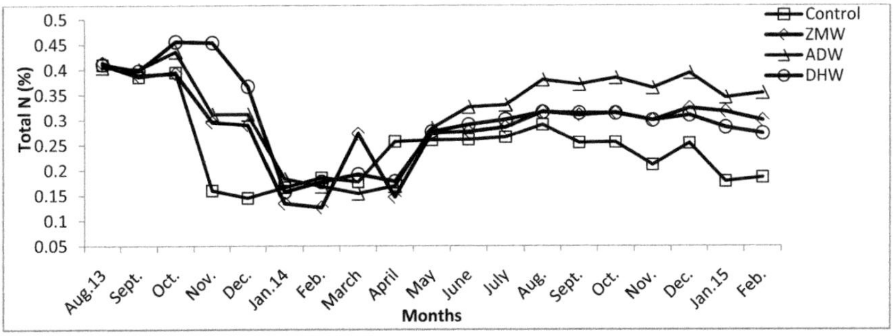

FIGURE 18.5 Total N (%) in the waste amended and unamended plots.

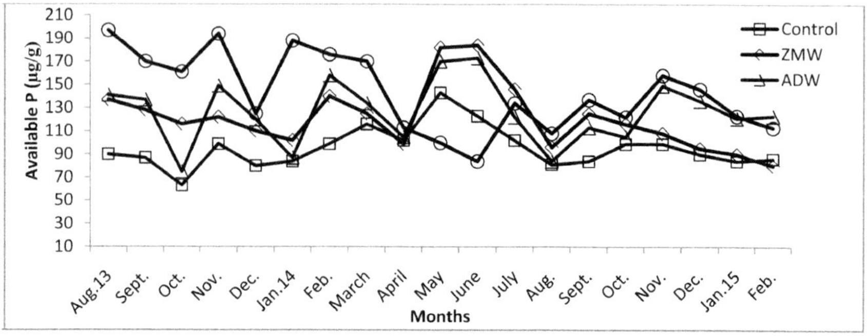

FIGURE 18.6 Available P(µg/g) in the waste amended and unamended plots.

amended plot. The average available P was least in the control and highest in the DHW amended plot. Due to amendment of the wastes, there was an increase in available P in all the amended plots where highest increase was obtained in the DHW amended plot and least in the ZMW amended plot.

While finding correlation results, significant correlation was obtained between organic C and soil moisture only. The organic carbon was found to be significantly and positively correlated with soil moisture in ZMW and ADW wastes amended plots. In the DHW amended plot, there was a positive correlation but not significant (Table 18.3). The change in WHC due to amendment of wastes was found to be positively and significantly correlated with initial C and N (Table 18.4); change in soil organic C due to amendment of wastes with initial C and N; change in total N of soil due to amendment of wastes with initial N and C:N ratio(negative); and change in available P of soil due to amendment of wastes with initial C and C:N ratio both negative and significant.

TABLE 18.3 Correlation between Organic Carbon and Soil Moisture in the Amended and Unamended Plots

Site	r
Control	0.54
ZMW	0.53
ADW	0.50
DHW	0.43

TABLE 18.4 Correlation between Changes in Physico-chemical Properties of Soil due to Amendment of Wastes with Initial C, N, and C:N Ratio of the Organic Wastes

Soil properties	r		
	Initial C	**Initial N**	**Initial C:N**
Waterholding capacity	0.52	0.98	Insignificant
Organic C	0.40	0.94	Insignificant
Total N	Insignificant	0.87	–0.50
Available P	–0.90	Insignificant	–0.95

18.4 DISCUSSION

The amendment of the three types of wastes leads to an increase in soil moisture, organic C, water holding capacity, total N, and available P in the soil. The positive correlation between soil moisture and organic C also indicates the normal pattern of soil quality.

The *Dendrocalamus* waste although having lesser initial C content its amendment has more increase in organic C and WHC content in soil compared to corn waste. The initial N of corn waste and *Dendrocalamus* waste was the same; however, the release of N from *Dendrocalamus* waste was faster than that from corn as there was more increase of total N in the *Dendrocalamus* waste amended plot.

The *Amomum* wastes amended plot had more of soil organic C, N, and WHC. Therefore, there was more enhancement in the WHC, C and N of the soil. The results also indicate that the important role of initial chemical components mainly C and N, because *Amomum* wastes had the highest amount of initial C content (10.0%) and N (0.5%). The decomposition of *Amomum* wastes leads to a release of C and N at a faster rate. The result is also evident by the significant correlation between the change in WHC and organic C and N with initial C and N. The role of other factors, namely initial lignin and cellulose, was not observed in the present study. Therefore, some other factors may be responsible for the slow release of C and N in the corn waste, although it has the ability to retain more soil moisture level.

The change in soil due to amendment of corn waste does not have a definite pattern because its amendment leads to highest increase in soil moisture although contributing to least increase in organic C content in the soil. The amendment of corn waste also leads to an increase in pH level, whereas the other types of amendment leads to a decrease in pH of the soil. Warman et al. (2009) observed a significant increase in soil pH with municipal solid waste compost application that did not have negative impact on plant nutrition or fruit yield in a lowbush blueberry field; because organic C and N release is faster in *Amomum* and *Dendrocalamus* waste amended plots, the WHC was also more in these plots than in the corn waste amended plot. Thus, the release of more C from the wastes leads to more acidity in these two plots.

The amount of available P present in the *Dendrocalamus* waste amended plot was more than that in the *Amomum* waste amended plot, whereas it was least was in the corn waste amended plot. This indicates that the release of available P from the *Amomum* waste amended plot was faster than the release of C and N in *Dendrocalamus* waste residues. The role of the initial C:N ratio is evident here as the release of P was fastest in the organic waste with the least initial C:N ratio, i.e., *Dendrocalamus* waste, and therefore more enhancement was observed in available P. The release of available P was slowest in the organic waste with the highest initial C:N ratio, i.e., corn waste, where there was minimum increase in available P in soil. This relation is proved by the negative and significant correlation between change in available P with the initial C:N ratio.

Overall, the amendment of corn, *Amomum*, and *Dendrocalamus* wastes in abandoned jhum fallows leads to an increase in WHC, organic C, total N, and available P in the soil. Other studies have also reported increase in organic C (Maftown et al., 2005; Pankhrust et al., 2005; Weir and Allen, 2008), and P (Weir and Allen, 2008) apart from ameliorating the poor physical conditions of degraded soil by increasing WHC of soil due to amendment of organic wastes. In order to estimate the rate of release of nutrients from amendment of organic wastes materials in the soil, the role of initial N and C:N ratio of the wastes materials is important. Organic matter releases nutrients in a plant-available form upon decomposition. Therefore, in order to maintain this nutrient cycling system, the rate of organic matter addition from organic wastes, crop residues, manure, and any other source must equal the rate of decomposition and take into account the rate of uptake by plants and losses by leaching and erosion.

KEYWORDS

- **amendment**
- **fertility**
- ***Jhum***
- **organic**
- **soil**
- ***waste***

REFERENCES

Anderson, J., & Ingram, J. S. I., (1993). Tropical soil biology & fertility. *Handbook of Methods.* C. A. B. International, U.K. pp. 30.

Clapp, C. E., Hayes, M. H. B., & Ciavatta, C., (2007). Organic wastes in soils: Biogeochemical and environmental aspects. *Soil Biology and Biochemistry, 39,* 1239–1243.

Hanson, R. G., (1994). *The Soil Management Collaborative Research Support Program.* North Carolina State University. 1–3.

Maufto, M., Moshiri, F., Karimian, N., & Ronaghi, A. M., (2005). Effects of two organic wastes in combination with phosphorus on growth and chemical composition of spinach and soil properties. *Journal of Plant Nutrition, 27,* 1635–1651.

Pankhrust, C. E., Blair, B. L., Magarey, R. C., Sterling, G. B., Bell, M. J., & Gorrie, A. L., (2005). Effect of rotation breaks and organic matter amendments on the capacity of soils to develop biological suppression towards soil organisms associated with yield decline of sugarcane. *Applied Soil Ecology, 28,* 271–282.

Warman, P. R., Burnham, J. C., & Eaton, L. J., (2009). Effects of repeated applications of municipal solid wastes compost and fertilizers to three low bush blueberry fields. *Scientica Horticulturae, 122,* 393–398.

Weir, C. C., & Allen, J. R., (2008). Effects of using organic wastes as soil amendments in urban horticultural practices in the district of Columbia. *Environmental Science and Engineering and Toxicology. 32,* 323–332.

CHAPTER 19

EFFECT OF BIOFERTILIZER CONSORTIUM, FYM, ENRICHED COMPOST, AND CHEMICAL FERTILIZERS ON SUSTAINABLE PRODUCTION AND PROFITABILITY OF GARDEN PEA (*PISUM SATIVUM* L.) IN THE PLAINS OF ASSAM, NORTH-EAST INDIA

TRUDY A. SANGMA,[1] L. SAIKIA,[1] KHATEMENLA,[1] and ROKOZENO[2]

[1]*Department of Horticulture, Assam Agricultural University, Jorhat-13, Assam, India*

[2]*Department of Entomology, Assam Agricultural University, Jorhat-13, Assam, India, E-mail: trudytengse@gmail.com*

CONTENTS

ABSTRACT

The present experiment was carried out to establish the integrated effect of biofertilizer consortium, farm yard manure (FYM), enriched compost, and chemical fertilizers on sustainable production and profitability of garden pea. The experiment was carried out in a randomized block design with seven treatments and replicated thrice. The treatments were T_1: NPK @ 20:46:0 kg ha^{-1} + FYM @ 5 t ha^{-1} (RDF), T_2: NPK @ 20:46:0 kg ha^{-1} + FYM @ 5 t ha^{-1} + biofertilizer consortium, T_3: NPK @ 20:46:0 kg ha^{-1} + FYM @ 10 t ha^{-1}, T_4: NPK @ 10:46:0 kg ha^{-1} + FYM @ 10 t ha^{-1} + biofertilizer consortium, T_5: Enriched compost @ 3 t ha^{-1}, T_6: Enriched compost @ 1.5 t ha^{-1} + biofertilizer consortium, and T_7: Biofertilizer consortium. The data recorded were analyzed through the analysis of variance method. The seed/pod (7.58), seed/plant (52.75), shelling (50.33%), fresh biomass (33.58 g), dry biomass (10.59 g), pod yield/plant (40.88 g), and pod yield/hectare (9.20 t ha^{-1}) were significantly highest in T_3. The plant height (56.25 cm), pod/plant (7.13), nodule number/plant (49.17), nodule fresh weight/plant (379.20 mg), nodule dry weight/plant (96.65 mg), and nodule nitrogen (0.28%) were the highest in T_2. Number of branch, days to 50% flowering, pod length, and pod diameter were found to be nonsignificant. The economic analysis indicated that T_2 gave the maximum benefit-cost ratio of 3.50. Therefore, it is suggested that a judicial combination of biofertilizers along with FYM and inorganic is beneficial for improving yield attributes of pea with maximum profit.

19.1 INTRODUCTION

In order to meet the nutritional demands of the increasing population, efforts are being made at the national and international levels to increase per hectare production. Fertilizers are the vital agriculture input to increase the population, but the main drawback is the use and manufacture of chemicals, viz., energy crisis and unavailability of indigenous effects of chemical fertilizers on our health and environment. All these things have led to the search of alternative renewable source of nutrient for the crop through fertilizers of biological origin (biofertilizers). The biofertilizers are safe, low cost, and easy in application. Biofertilizer application has shown good results for leguminous (pulse) crops, especially exclusive results have been obtained for vegetable pea (garden pea). Garden pea (*Pisum sativum* L.) is a very common nutritive vegetable grown in the cool season throughout the world. Pea is highly nutritive and contains high percentage of digestible protein, along with carbohydrates and vitamins. It is also very rich in minerals, vitamin A, vitamin C, foliate, and dietary fiber. The average productivity of the garden pea in India is 9.5 t ha^{-1} (Anonymous, 2014), which is quite low as compared to some other countries. In India, pea is cultivated over an area of about 421 thousand hectares with a production of 4006 thousand metric tons (Anonymous, 2014). The productivity of pea is higher than that of other pulse crops. Moreover, its high yield potential through balanced fertilization envisages ample scope to increase its yields further (Anonymous, 2014). Because fertilizer nutrients constitute a major costly production inputs, the exploitation of yield potentiality of this crop depend on how effectively and efficiently this input is managed. Moreover, high fertility levels not only put a heavy financial burden to the growers but gradually decrease the partial productivity and thereby, jeopardize the sustenance of the basic system of production. On the other hand, the large-scale use of only chemical fertilizers as a source of nutrients has less use efficiency. Besides low, variable and generally unbalanced nutrient contents, it is difficult to provide the proper nutrient balance to meet crop requirements with bulky organic manure (Kumar et al., 2008). To combat this problem, the use of chemical fertilizers along with organic manures and biofertilizers is probably a nice way to keep up the sustained food production. However, the information on the requirements for an appropriate combination of nutrients through various sources, viz., inorganic fertilizers and organics and biofertilizers, in garden pea is meager in acidic soil of Eastern Himalaya condition of northeast India. In view of the above facts, the present investigation was carried out to ascertain the effect of bio-fertilizers and their interaction on the yield and cost economics of garden pea.

19.2 MATERIALS AND METHODS

The study was conducted using the variety GS-10 under the agro-climatic condition of Jorhat (Assam) during the *rabi* (October–January) season of the year 2014–2015 in the Experimental Farm of the Department of Horticulture, Assam Agricultural University, Jorhat, situated at 26°47' N latitude and 94°12' E longitude at an elevation of 96.6 m above mean sea level. The prevailing climatic condition of Jorhat is humid subtropical with hot summer, cold winter, and high relative humidity and receive a rainfall between 1875 mm and 2146 mm (Anonymous, 2014). A total of seven treatments were replicated thrice in a randomized block design. The treatments were T_1: NPK @ 20:46:0 kg ha^{-1} + FYM @ 5 t ha^{-1} (RDF), T_2: NPK @ 20:46:0 kg ha^{-1} + FYM @ 5 t ha^{-1} (RDF) + biofertilizer consortium, T_3: NPK @ 20:46:0 kg ha^{-1} + FYM @ 10 t ha^{-1}, T_4: ½ RD N & RD PK @ 10:46:0 kg ha^{-1} + FYM @ 10 t ha^{-1} + biofertilizer consortium, T_5: Enriched compost @ 3 t ha^{-1}, T_6: Enriched compost @ 1.5 t ha^{-1} + biofertilizer consortium, and T_7: Biofertilizer consortium (*Rhizobium* + *Azotobacter* + *Azospirillum* + PSB). Garden pea (GS-10) was grown in a plot size of ^{2} maintaining a spacing of . Manures were incorporated as a basal dose at the time of field preparation. Biofertilizer consortium (*Rhizobium* + *Azotobacter* + *Azospirillum* + PSB) was applied as the seed treatment before sowing. In every plot, 10 plants away from the margin were tagged for recording the yield and yield attributes. Observational data recorded during field experimentation at different growth stages of the crop and laboratory determination were subjected to the statistical analysis according to the standard procedures to test the significance by calculating "F" values as described by Panse and Sukhtame (1985).

19.3 RESULTS AND DISCUSSION

19.3.1 CROP YIELD AND YIELD ATTRIBUTES

In the present investigation, the growth and development of pea under different treatments reflected the influence of different integrations upon yield and yield attributes. Growth of garden pea measured in terms of plant height at harvest (Table 19.1) was significantly higher (56.25 cm) with the application of RD NPK + 5 t FYM ha^{-1} + biofertilizer consortium (T_2), while the lowest plant height (52.16 cm) was recorded in the treatment receiving biofertilizer consortium alone (T_7). This might be due to the fact that N-fixer and P-solubilizing microorganism secrete certain organic acids and some biochemicals, which are growth promoting in nature. These results are in close conformity with Patel et al. (1998); Reddy et al. (1998); and Vimala and Natarajan (2000) who also reported great influence of the integrated nutrient management involving *Rhizobium*, chemical fertilizers, and other bulky organic manure (FYM) on the growth of pea and thereby, its physiological parameters. This superior performance of the combined use of FYM and

TABLE 19.1 Effect of Different Integrations on Growth Attributing Characters of Garden Pea

Treatments	Plant height (cm)	Number of branch	Days to 50% flowering
T_1 (NPK @ 20:46:0kg + 5 t FYM)	54.08	9.90	39.33
T_2 (NPK @ 20:46:0kg + 5 t FYM + biofertilizer consortium)	56.25	9.77	36.67
T_3 (NPK @ 20:46:0kg + 10 t FYM)	55.25	10.07	40.67
T_4 (NPK @ 10:46:0kg + 10 t FYM + biofertilizer consortium)	54.78	9.67	39.67
T_5 (3 t Enriched compost)	53.93	9.83	39.00
T_6 (1.5 t Enriched compost + biofertilizer consortium)	53.60	9.73	41.33
T_7 (Biofertilizer consortium)	52.16	10.13	39.33
S.Ed.	0.95	0.22	3.12
$CD_{0.05}$	1.70	NS*	NS*

NS*=Non-significant.

biofertilizer may probably be due to augmented nutrient supply to crop by increasing availability through the exploitation of natural processes like biological nitrogen fixation, solubilization of insoluble P, and decomposition of organic matter. Therefore, artificial seed inoculation is often needed to restore the population of effective strains of *Rhizobium* in rhizosphere (Srivastava and Verma, 1984).

The treatments did not bring about any significant variation in the number of branches. Similarly, the number of days taken for 50% flowering was not influenced significantly by fertility management treatments. Furthermore, pod length and pod diameter were found to be insignificant. This might be attributed to the genetic makeup of the variety used. Branching is an undesirable character. 50% flowering by earliness also indicate the maturity and it is an established fact that pea is a heat unit responsive crop. In the present study, nonsignificance of maturity by 50% flowering is quite normal.

Perusal of data in Table 19.2 revealed that the yield attributes, viz., seed per pod (7.58), seed per plant (52.75), and shelling percentage (50.33%) were significantly higher under RD NPK + 10 t FYM ha^{-1} (T_3), while the maximum pod per plant (7.13) was recorded with the application of NPK @ 20:46:0 kg

TABLE 19.2 Effect of Different Integrations on Pod Formation Characters of Garden Pea

Treatments	Pod length (cm)	Pod diameter (cm)	Pod per plant	Seed per pod	Seed per plant	Shelling%
T_1 (NPK @ 20:46:0kg + 5 t FYM)	8.87	1.81	6.42	6.34	40.70	43.42
T_2 (NPK @ 20:46:0kg + 5 t FYM + biofertilizer consortium)	8.10	1.84	7.13	7.32	52.19	49.27
T_3 (NPK @ 20:46:0kg + 10 t FYM)	9.73	1.85	6.96	7.58	52.75	50.33
T_4 (NPK @ 10:46:0kg + 10 t FYM + biofertilizer consortium)	8.80	1.82	7.07	6.93	48.99	48.62
T_5 (3 t Enriched compost)	7.73	1.83	6.93	6.76	46.84	47.45
T_6 (1.5 t Enriched compost + biofertilizer consortium)	8.70	1.88	6.87	6.58	45.20	46.50
T_7 (Biofertilizer consortium)	8.17	1.80	5.80	6.18	35.84	39.97
S.Ed.	0.62	0.04	0.22	0.14	2.01	0.66
$CD_{0.05}$	NS*	NS*	0.67	0.43	6.05	2.00

NS* = Non-significant.

ha^{-1} + FYM @ 5 t ha^{-1} (RDF) + biofertilizer consortium (T_2). T_7 (biofertilizer consortium) gave the lowest number of pod per plant (5.80), seed per pod (6.18), seed per plant (35.84), and shelling percentage (39.97%). It is clearly evident (Table 19.3) that the maximum pod yield per plant is produced in T_3 (40.88 g), while T_7 (19.95 g) resulted in the lowest pod yield per plant. It was observed (Table 19.4) that the highest pod yield per hectare yield was given by T_3 (9.20 t ha^{-1}). T_7 gave the lowest yield with only 4.49 t ha^{-1}. The yield is the resultant of various growth, development, and yield contributing characteristics. The effect of any factor on these characteristics may result in differences in the final green pod yield. Consequently, upon variation in the yield contributing characters of garden pea, significant differences in the green pod yield were obvious. Datt et al. (2003) obtained significant increase in green pod yield of pea with increasing level of NPK fertilizers up to 150% in the presence of farmyard manure. Enhancement in yield and other attributes with higher dose of FYM may be due to availability of more nutrients to the plants.

It is clearly evident from Table 19.3 that T_3 resulted in highest fresh biomass (33.58 g) and dry biomass (10.59 g), and the lowest fresh biomass (21.77 g) and dry biomass (7.50 g) was observed in T_7. Moderate increase in plant biomass could be ascribed to better availability of nutrients as the FYM dose was doubled to 10 t ha^{-1} (Jaipaul et al., 2011).

TABLE 19.3 Effect of Different Integrations on Yield of Garden Pea

Treatment	Yield per plant (g)	Yield (t ha^{-1})	Percentage increase or decrease over control (T_1) %
T_1 (NPK @ 20:46:0kg + 5 t FYM)	35.77	8.05	-
T_2 (NPK @ 20:46:0kg + 5 t FYM + biofertilizer consortium)	39.57	8.90	10.55
T_3 (NPK @ 20:46:0kg + 10 t FYM)	40.88	9.20	14.38
T_4 (NPK @ 10:46:0kg + 10 t FYM + biofertilizer consortium)	38.64	8.69	7.90
T_5 (3 t Enriched compost)	37.37	8.41	4.47
T_6 (1.5 t Enriched compost + biofertilizer consortium)	36.05	8.11	0.74
T_7 (Biofertilizer consortium)	19.95	4.49	(-) 44.22
S.Ed.	1.52	0.34	-
$CD_{0.05}$	4.55	0.61	-

TABLE 19.4 Effect of Different Integrations on Fresh and Dry Biomass of Garden Pea

Treatments	Total fresh biomass (g)	Total dry biomass (g)
T_1 (NPK @ 20:46:0kg + 5 t FYM)	27.26	9.07
T_2 (NPK @ 20:46:0kg+ 5 t FYM + biofertilizer consortium)	30.75	10.06
T_3 (NPK @ 20:46:0kg + 10 t FYM)	33.58	10.59
T_4 (NPK @ 10:46:0kg + 10 t FYM + biofertilizer consortium)	25.21	8.21
T_5 (3 t Enriched compost)	28.59	9.05
T_6 (1.5 t Enriched compost + biofertilizer consortium)	24.37	8.69
T_7 (Biofertilizer consortium)	21.77	7.50
S.Ed.	1.48	0.53
$CD_{0.05}$	4.45	1.61

19.3.2 ROOT NODULE CHARACTERISTICS

A critical examination of data in Table 19.5 showed that the nodule count was highest (49.17) under RD NPK + FYM 5 t ha^{-1} + biofertilizer consortium (T_2) probably due to the better growth of pea crop, while control (T_1) gave the least nodule count (1.90). The maximum nodule fresh weight was recorded in T_2 (379.20 mg), while the least was in T_1 (5.88 mg). Similarly, the highest nodule dry weight (96.65 mg) and nodule nitrogen (0.28%) was obtained from T_2 and the least nodule dry weight (2.30 mg) and nodule nitrogen (0.13%) in T_1. It was lowest under biofertilizer consortium alone, because the native initial organic matter content of soil was low to maintain the nodulations activity. Hence, lower substrate for the microorganisms could not support nodule formation, because their population might not have increased. In contrast, the treatment that received the highest amount of FYM @ 10 t ha^{-1} (T_3) even could not produce nodules due to lack of potential N-fixing microbial populations. It has been proved that *Rhizobium* inoculation helped in increasing the nitrogen content especially in the form of NH^{2+} in the soil, which is preferred by the plants. The P-solubilizing organisms help in solubilizing the native as well as applied phosphates, thus increasing the phosphate content in soil and subsequently plant uptake. The association

TABLE 19.5 Effect of Different Integrations on Root Nodule Characters of Garden Pea

Treatment	Nodule number/ plant	Nodule fresh weight (mg)	Nodule dry weight (mg)	Nodule N (%)
T_1(NPK @ 20:46:0 kg + 5 t FYM)	1.90	5.88	2.30	0.13
T_2(NPK @ 20:46:0 kg + 5 t FYM + biofertilizer consortium)	49.17	379.20	96.65	0.28
T_3(NPK @ 20:46:0 kg+ 10 t FYM)	2.10	6.33	2.74	0.15
T_4(NPK @ 10:46:0 kg + 10 t FYM + biofertilizer consortium)	23.37	180.71	39.60	0.24
T_5(3 t Enriched compost)	2.60	7.80	2.78	0.16
T_6(1.5 t Enriched compost + bio-fertilizer consortium)	10.40	29.74	9.70	0.22
T_7(Biofertilizer consortium)	4.57	13.70	3.60	0.18
S.Ed.	1.08	0.83	0.86	0.004
$CD_{0.05}$	3.26	2.51	2.59	0.008

of soil microorganisms and organic manures has been known to help in increasing mineralization that promotes the development of crops (Sharma, 1983). The increase in nodulation and nitrogen fixation due to the inoculation leads to lesser days to 50% flowering (Dravid, 1990). The increase in nodulation and nitrogen fixation due to the inoculation leads to significantly more plant height was also reported by Srivastava and Verma (1984) and Dravid (1990). Interaction between organic/bio-fertilizer and chemical fertilizers was significant on the nodule count. These findings are in agreement with Parmar et al. (1998) and Negi et al. (2006). Seed inoculation with *Rhizobium* and PSB resulted in conspicuous nodule number and nodule weight due to increased nitrogenase activity, according to Chattopadhyay and Dutta (2003).

19.3.3 ECONOMIC ANALYSIS

The economic analysis (Table 19.6) of the different treatments produced a unique reflection on the part of cost economics. It was found that T_2 produced the maximum benefit cost ratio of 3.50. T_7 showed the least benefit-cost ratio of only 1.57 against per Re.1.00 invested. Even though the highest

TABLE 19.6 Cost Economics of Various Treatment Combinations

Treatments	Common cost (Rs.)	Treatment cost (Rs.)	Total cost (Rs.)	Yield (Kg)	Gross return (Rs.)	Net return (Rs.)	B:C
T_1 (NPK @ 20:46:0kg + 5 t FYM)	33325.00	4660.00	37985.00	8050	161000	123015	3.23
T_2 (NPK @ 20:46:0kg + 5 t FYM + biofertilizer consortium)	33325.00	6160.00	39485.00	8900	178000	138515	3.50
T_3 (NPK @ 20:46:0kg + 10 t FYM)	33325.00	8660.00	41985.00	9200	184000	142015	3.38
T_4 (NPK @ 10:46:0kg + 10 t FYM + biofertilizer consortium)	33325.00	10060.00	43385.00	8690	173800	130415	3.00
T_5 (3 t Enriched compost)	33325.00	6000.00	39325.00	8410	168200	128875	3.27
T_6 (1.5 t Enriched compost + biofertilizer consortium)	33325.00	4500.00	37825.00	8110	162200	124375	3.28
T_7 (Biofertilizer consortium)	33325.00	1500.00	34825.00	4490	89800	54975	1.57

percentage increase (14.38%) in pod yield over the control was obtained in $T_{3,}$ the pod yield was increased 10.55% over control along with highest benefit-cost ratio (3.50) in the treatment receiving RD NPK + 5 t FYM ha^{-1} + biofertilizer consortium (T_2). This was in agreement with the result of Bairwa et al. (2009). Gopinath et al. (2008) reported that in bell pepper, the highest gross margin and benefit-cost ratio were recorded under integrated nutrient management treatment in comparison to treatments containing only organic sources of nutrients.

19.4 CONCLUSION

This study shows that integrated use of recommended dose of fertilizers along with 5 t ha^{-1} of FYM and biofertilizer consortium (T_2) resulted in an adequate productivity and profitability (3.50 benefit-cost ratio) of garden pea. Though making the FYM dose double resulted in close follow up regarding the benefit-cost ratio, it is not advisable, as both availability and purchasing power of the farmers may not be adequate.

ACKNOWLEDGMENT

The authors would like to express their gratitude to the Department of Horticulture, Assam Agricultural University, Jorhat, for providing the necessary facilities during the entire course of research work.

KEYWORDS

- **biofertilizer**
- **enriched compost**
- **garden pea**
- **profitability**
- **sustainability**

REFERENCES

Anonymous, (2014). National Horticulture Board, Ministry of Agriculture, Government of India 85, Institutional Area, Sector-18, Gurgaon-122 015, INDIA.

Bairwa, H. L., Mahawer, L. N., Shukla, A. K., Kaushik, R. A., & Sudhar Mathur, A. D., (2009). Response of integrated nutrient management on growth, yield and quality of okra (*Abelmoschus esculentus*). *Indian J. Agril. Sci.*, *79*(5), 381–384.

Chattopadhyay, A., & Dutta, D., (2003). Response of vegetable Cowpea to p-enriched compost for improving growth, yield and nodulation of chickpea. *Pakistan J. Bot.*, *40*, 1735–1441.

Datt, N., Sharma, R. P., & Sharma, G. D., (2003). Effect of supplementary use of farmyard manure along with chemical fertilizers on productivity and nutrient uptake by vegetable pea (*Pisum sativum*) and build up of soil fertility in Lahaul valley of Himachal Pradesh. *Indian J. Agril. Sci.*, *73*, 266–268.

Dravid, M. S., (1990). Effect of salinization, *rhizobium* inoculation, genotype variation and P-application on dry matter yield and utilization of P by pea (*Pisum sativum* L.) and lentil (Lens culinarisa Medic.). *J. Nucl. Agril. Biol.*, *19*, 227–231.

Gopinath, K. A., Saha, S., Mina, B. L., Kundu, S., Selvakumar, G., & Gupta, H. S., (2008). Effect of organic manures and integrated nutrient management on yield potential of bell pepper *(Capsicum annuum*) varieties and on soil properties. *Arch. Agron. Soil Sci.*, *54*, 127–137.

Jaipaul Sharma, S., Dixit, A. K., & Sharma, A. K., (2011). Growth and yield of capsicum (*Capsicum annum*) and garden pea (*Pisum sativum*) as influenced by organic manures and biofertilizers. *Indian J. Agril. Sci.*, *81*(7), 37–42.

Kumar, B., Gupta, R. K., & Bhandari, A. L., (2008). Soil fertility changes after long term application of organic manures and crop residues under rice-wheat system. *J. Indian Soc. Soil Sci.*, *56*(1), 80–85.

Negi, S., Singh, R. V., & Dwivedi, O. K., (2006). Effect of biofertilizers, nutrient sources and lime on growth and yield of garden pea. *Legume Res.*, *29*(4), 282–285.

Panse, V. G., & Sukhtame, P. U., (1985). Statistical methods for agricultural workers. *Indian Council of Agril. Res.*, New Delhi, 145–152.

Parmar, D. K., Sharma, P. K., & Sharma, T. R., (1998). Integrated nutrient supply system for 'DPP 68' vegetable pea (*Pisum sativum* var arvense) in dry temperate zone of Himachal Pradesh. *Indian J. Agril. Sci.*, *68*(2), 84–86.

Patel, T. S., Katare, D. S., Khosla, H. K., & Dubey, S., (1998). Effect of biofertilizers and chemical fertilizers on growth and yield of garden pea (*Pisum sativum* L.). *Crop Res. Hisar*, *15*(1), 54–56.

Reddy, R., Reddy, M. A. N., Reddy, Y. T. N., Reddy, N. S., & Anjanappa, M., (1998). Effect of organic and inorganic sources of NPK on growth and yield of pea (*Pisum sativum*). *Legume Res.*, *21*(1), 57–60.

Sharma, H. L., (1983). Studies on the utilization of crop residues, farmyard manure and nitrogen fertilization in rice-wheat cropping system under sub temperate climate. PhD. thesis submitted to HPKV, Palampur (H. P.).

Srivastava, S. N. L., & Verma, S. C., (1984). Effect of nitrogen, Rhizobium and techniques of phosphorus application on yield and quality of field pea (*Pisum sativum* L. var Arvense poir.). *Legume Res.*, *7*(1), 37–42.

Vimala, B., & Natarajan, S., (2000). Effect of nitrogen, phosphorus and biofertilizers on pod characters, yield and quality in pea (*Pisum sativum* L. spp. hortense). *South Indian Hort.*, *48*(1–6), 60–63.

CHAPTER 20

KENDU: AN UNDEREXPLOITED FOREST FRUIT SPECIES FOR POVERTY ALLEVIATION OF TRIBALS*

DEBJIT ROY, SUBHASIS KUNDU, and BIKASH GHOSH

Department of Fruits and Orchard Management, Faculty of Horticulture, Bidhan Chandra Krishi Viswavidyalaya, Mohanpur–741252, Nadia, West Bengal, India, E-mail: debjit.bckv@gmail.com

CONTENTS

*Thematic Area: (G) Horticulture for Food, Health, and Nutrition.

ABSTRACT

Kendu (*Diospyros melanoxylon* Roxb.), an underexploited fruit species in the family of Ebenaceae, is grown as natural wild in the forests and marginal lands of West Bengal, Madhya Pradesh, Orissa, Bihar, Chhattisgarh, Jharkhand, and Andhra Pradesh. In West Bengal, it is commonly found in plateau districts of Paschim Midnapur, Bankura, and Purulia. Being highly economical species for the local inhabitants and tribals, it is naturally being protected by them. All the plant parts like fruits, seeds, leaves, and bark can be used for different commercial purposes. Fruits are good source of carbohydrates, calcium, phosphorus, and carotene and can prevent the malnutrition of tribals. Dried powdered fruit is used as carminative and dried flowers are reported to be useful in urinary, skin and blood diseases. Seeds are prescribed for curing mental disorders, palpitation of heart, and nervous breakdown. Leaves are used as a raw material for the "Bidi" industry, and West Bengal Tribal Development Co-operative Corporation Limited (WBTDCC) has given top priority in this regard. Due to its immense importance, this crop should be given priority so that it can provide the nutritional security and uplift the socio-economic condition of the poor tribal peoples. Hence, a review of this crop on different aspects of nutritional, medicinal, etc., has been done with a view to exploit this underexploited fruit tree to the maximum extent.

20.1 INTRODUCTION

Forestry in India is a significant rural industry and a major environmental resource. According to the Forest Survey of India (2013), the total forest and tree cover of the country is 78.92 million hectares, which is 24.01% of the geographical area of the country. Based on assessment, states with the maximum forest cover are Madhya Pradesh (77,522 km^2), Arunachal Pradesh (67,321 km^2), Chhattisgarh (55,621 km^2), Maharashtra (50,632 km^2), Orissa (50,347 km^2), and West Bengal (3810 km^2) (Anonymous, 2014). India has a thriving non-timber forest products industry that produces latex, gums, resins, essential oils, fragrances and aroma chemicals, handicrafts, thatching materials, and medicinal plants. About 60% of nontimber forest products are consumed locally, which is the major income source to over 400 million people in India, mostly rural tribal. Among the different economically important forest species, *kendu* (*Diospyros melanoxylon* Roxb.) is an important underexploited fruit species under the family of Ebenaceae, which is grown basically as natural wild in the forests and marginal lands of West Bengal, Orissa, Madhya Pradesh, Chhattisgarh, and Andhra Pradesh. It is highly economical species for the local inhabitants and tribal people of this region. All the plant parts like bark, leaves, fruits, and seeds can be used for different commercial purposes. Moreover, it can provide the nutritional security and uplift the socio-economic condition of the poor tribal peoples. Due to lack of awareness of its vast potentialities, this crop has not yet been fully explored. Hence, a review of this crop on different aspects of nutritional, medicinal, etc., has been done with a view to exploit this underexploited fruit tree to the maximum extent.

20.2 SOCIO-ECONOMIC STATUS OF TRIBAL PEOPLE

According to census 2011, the tribal population in India is 10.43 crores, constituting 8.6% of the total population, whereas in West Bengal, it is 5.1% (Anonymous, 2013). Broadly the tribals inhabit two distinct geographical areas, the central India and the -eastern area. The socioeconomic conditions of the tribals are very poor and lagging behind in basic amenities such as housing conditions, availability of drinking water, sanitation facility, type of fuel used, electricity, and communication facilities from the modern society. A large number of tribal population in rural areas of India is still dependent on forests for their livelihood, and therefore, provisions for basic necessities like food, fuel, housing material etc. are made from the forest produce in this forest-based tribal economy. Health issues are of critical concern determining the overall development of tribal population. High prevalence of nutritional deficiency and chronic energy deficiency are observed among the tribal women, men, and children, indicating nutritional problem being more serious for this category.

Among the different forest species, *kendu* (*Diospyros melanoxylon* Roxb.) has very important role in the socio-economic upliftment and poverty alleviation of tribal populations of tropical dry forests of India as well as areas of southwest Bengal, and its fruits and leaves are the main nontimber forest produce.

20.3 DISTRIBUTION

In India, it is the most common species of forests of West Bengal, Madhya Pradesh, Bihar, Jharkhand, Chhattisgarh, Rajasthan, Gujarat, Orissa, Andhra Pradesh, and Tamil Nadu (Malik et al., 2010). In West Bengal, it is commonly found in plateau districts of Bankura, Purulia, and Paschim Midnapur. It generally grows in dry mixed deciduous forests, occurring alongside *Shorea robusta* and *Tectona grandis*.

20.4 BOTANICAL DESCRIPTION

Botanically, *kendu* is known as *Diospyros melanoxylon* Roxb. and belongs to the family Ebenaceae. It is native and endemic tree of India and widely found in the peninsular plains and lower hills, especially in the dry deciduous forests of central, western, and northern India (Stewart and Brandis, 1992).

It has many local names such as *kendu* or *kend* in Bengali, Oriya, and Hindi; *tendu* in Hindi and Marathi; *tamru* in Gujarati; *dirghapatraka* in Sanskrit; *karai* or *thumbi* in Tamil; and *tumi* or *tuki* in Telugu (Orwa, 2009).

Kendu is a middle-sized tree, with height of about 18–24 m. and a girth of 2 m. or above. The bark is pelican in color, exfoliating in rectangular scales. The primary root is long, thick, and fleshy at first, and afterwards woody, grayish, often swollen in the upper part near the ground level. The roots form vertical loops in sucker generated plants. Leaves opposite or alternate and coriaceus, up to 35 cm long, tomentose on both sides when young, becoming glabrous above when fully grown. Male flowers are mauve in color, tetramerous to sextamerous, 1–1.5 cm long, sessile, or nearly sessile in short peduncles, mostly three flowered. Female flowers mauve, mostly extra-axillary or sometimes solitary, larger than the male flowers. Fruits olive green, ovoid or globose 3–4 cm across; 2–8 seeded berries. Pulp yellow, soft and sweet. Seeds compressed, oblong, shiny, often banded (Rathore, 1970).

20.5 IMPORTANCE OF *KENDU*

20.5.1 NUTRITIONAL IMPORTANCE OF FRUITS

The tribal people collect the fruits from forest and gather the ripe fruits, dry them in hot summer, and store them in bamboo containers for use in the rainy season when almost no food grain is available.

The National Institute of Nutrition, Hyderabad, revealed that the *kendu* fruit is very rich in carbohydrates, calcium, phosphorus, and carotene (equivalent of vitamin A) besides having other food value. Moreover, it is a promising organic food (Table 201.).

Bio-chemical analysis of fruits was done under West Bengal condition and revealed that the fruits had moderate amount of TSS (16.3 °Brix), vitamin C (13.86 mg/100 g), acidity (0.16%), total sugar (13.33%), and reducing sugar (9.52%).

20.5.2 MEDICINAL VALUE

All the plant parts like bark, leaves, fruits, and seeds have high medicinal value and can be used for different commercial purposes.

20.5.2.1 Fruits

Dried powdered fruit is used as a carminative and astringent. Its tannin content is 15% and that of half ripe fruit is 23%.

TABLE 20.1 Nutritive Values per 100 Grams of Edible Portion (Gopalan et al., 1989)

Item	Unit	Quantity
Moisture	(g)	70.6
Carbohydrates	(g)	26.8
Protein	(g)	0.8
Fat	(g)	0.2
Fiber	(g)	0.8
Minerals	(g)	0.8
Energy	(K. Cal)	112
Calcium	(mg.)	60.0
Phosphorus	(mg.)	20.0
Iron	(mg.)	0.5
Carotene	(mg.)	361.0
Thiamine	(mg.)	0.01
Riboflavin	(mg.)	0.04
Niacin	(mg.)	0.30
Vitamin C	(mg.)	0.00

Source: National Institute of Nutrition, Hyderabad.

20.5.2.2 Seeds

The seeds are prescribed as cure for mental disorders, palpitation of heart, and nervous breakdown.

20.5.2.3 Flowers

Dried flowers are reported to be useful in urinary, skin, and blood diseases (Hocking, 1993).

20.5.2.4 Barks

Good source of tannin (19%). The bark is burnt by tribals to cure small pox and has significant effects against malaria.

20.5.2.5 Leaves

The leaf of the tree contains valuable flavones (pentacyclic triterpenes) that possess antimicrobial properties.

20.5.3 ECONOMIC IMPORTANCE OF LEAVES

Kendu leaf is a precious and commercially viable natural resource that is used as a raw material for the "Bidi" industry. The valuable leaves are used for wrapping bidis, a popular smoking material especially among poor natives. In the processed form, the *kendu* leaves are graded into different quality that are Grade I to Grade IV according to the specification of color, texture, size, and body condition of the leaf, and packets are made by taking 5 kg of leaves as a bundle. Twelve such bundles are packed in a gunny bag, and 100 such bags equivalent to 60 quintals make one truckload. The *kendu* leaf is collected and processed by the *kendu* leaf wing of the forest department. It is a seasonal operation involving huge number of laborers (both skilled and non-skilled), specially tribals and other forest dwellers during the season of operation (Mishra, 1997). After processing of the kendu leaf, the lots are formed and delivered to forest department for marketing. Forest department call open tender in prescribed formats mentioning the list of lots and the intended tenderers offer their prices against the interested lots subject to specified terms and conditions. After execution of an agreement, the purchaser lifts the *kendu* leaves on a valid price issued by the forest department.

Nationalization of *kendu* leaf was done in 1973 to promote the tribal welfare, which ensures maximum revenue for the governments and better returns to the tribal collectors. *Kendu* leaf provides employment opportunities for millions of tribal people and other landless people during summer when there is no agricultural work or opportunities of wage earnings. West Bengal Tribal Development Co-operative Corporation Limited (WBTDCC) gives top priority in the collection of *kendu* leaves. The government will purchase *kendu* leaf from the poor tribal people in order to pay (Rs. 30 per *Chatta*) them proper price to help them earn their livelihood. A bidi worker earns up to Rs. 80 per day depending upon the hour of binding bidis in West Bengal.

20.6 OTHER USES

20.6.1 FOOD

The fruits and powdered seeds are sold in local markets and eaten.

20.6.2 FODDER

A tolerance to pruning makes *Diospyros melanoxylon* a good fodder species. The leaves are reported to contain 7.12% crude protein and 25.28% crude fiber.

20.6.3 FUEL

Kendu is reported to be a good fuel wood; the calorific value of sapwood is 4957 kcal/kg and of heartwood, 5030 kcal/kg.

20.6.4 TIMBER

Wood is hard, whitish-pink, tough, fairly durable, and used for building, shoulder poles, mine props, and shafts of carriages. The ebony is very heavy and valued for carving and other ornamental works.

20.7 VALUE-ADDED PRODUCTS

20.7.1 FOOD FROM FRESH RIPE KENDU

The fresh ripe fruits do not have long-keeping quality, and as such, they could be used to prepare food during ripening time only. The foods that could be made out of fresh fruits are as follows.

20.7.1.1 Cake

One part of fresh *kendu* meat is added to four or five parts of rice powder or *suji* and a liquid paste is prepared by adding drinking water. It is cooked in oven or in a similar pot with low heat. Sugar, cashewnut, and spices like cardamom could be added to make the cake flavorful and tasteful.

20.7.1.2 Pudding

One part of fresh *kendu* pulp is added to three parts of boiled milk and thoroughly mixed and cooked for about 10–15 minutes. As the content gets thickened, cashew nut, raisin, and cardamom may be added and cooked for 5–7 minutes to have a good pudding.

20.7.1.3 Sherbet

One part of fresh *kendu* pulp with 8–10 parts of drinking water is added and thoroughly mixed. Sugar, rock salt, and black pepper power may be added as per taste to the mixture to have a tasteful drink.

20.7.1.4 Kheer

Pulp of fresh ripe fruits is added with equal quantity of drinking water. The pulp is added to kheer to be made out of rice and *suji* at a ratio of 1:4 or 1:5.

20.7.1.5 Wine

Wine was prepared from fruits and evaluated. Must from *kendu* fruits were made by raising total soluble solids (TSS) to 20 °B, 0.1% $(NH_4)SO_4$ as nitrogen source and fermenting it using 2% *Saccharomyces cerevisiae* var. *ellipsoideus* into wine. The wine had the proximate compositions of TSS, 2 °B; total sugar of 3.78 g/100 ml; titratable acidity, 1.32 g tartaric acid/100 ml; pH, 3.12; total phenolics, 0.95 g/100 ml; β-carotene, 8 μg/100 ml; ascorbic acid, 1.52 mg/100 ml; lactic acid, 0.39 mg/100 mL, methanol, 3.5% (v/v), and ethanol,6.8% (v/v). The wine had a 2,2-diphenyl-1-picrylhydrazyl (DPPH) scavenging activity of 52% at a dose of 250μg/mL (Sahu et al., 2012).

20.7.2 FOOD FROM DRIED KENDU

Various foods that could be made out of dried *kendu* meat in rural areas are as follows.

20.7.2.1 Pickle

Instant pickling could be done for ready use. One part of dry *kendu* meat is soaked in hot water for 10–15 minutes and then boiled with two parts of jaggery (*gur*) by weight about 20 minutes to bring the consistency, and then, while the mixture is hot, 0.1 part of mustard oil and tamarind pulp and 1–2 spoonful of coriander and cumin powder mixture may be added as preservative and for taste. Now, the pickle so prepared is ready for use, and it could be stored for 4–5 years as such in airtight container.

20.7.2.2 Papad

One part of *kendu* meat powder is added to 6–8 parts of rice powder. One or two spoonful of rock salt and black cumin powder may be added to the mixture depending upon the taste and thoroughly mixed. The dough is made by adding water, and with this dough, *papad* is made by sun drying.

20.7.2.3 Upama

One part of *kendu* meat powder is added to 4 to 8 parts of *suji* depending upon the freshness and sweetness of the meat, and *upama* is prepared. For fresh and very sweet meat powder, the proportion of *suji* should be 6–8 parts depending upon the taste.

20.7.2.4 Paratha

One part of *kendu* meat powder could be added to 4–8 parts of wheat flour to prepare the dough for paratha. For roti, the same mixture could be followed. The paratha and roti so prepared are very tasteful and flavorful.

20.8 CONSTRAINS IN *KENDU* CULTIVATION

Commercial exploitation of the species is limited due to lack of knowledge about the fruit, its value addition, and its medicinal uses. There is not much characterization data available for *kendu* as fruits have not been considered of much horticultural importance. Germplasm collected by the National Bureau of Plant Genetic Resources, New Delhi, have been characterized for various fruit and seed characteristics (Malik et al., 2010). The tree takes about 10 years to bear fruit; so, peoples are less interested to cultivate the crop. It is commonly propagated by seeds, but the seed loose viability very quickly (Hocking, 1993). There are no identified cultivars known in this species. Genetic resources of this species have not been given much emphasis and only naturally occurring wild plants are used by local people and tribals inhabited in the forest area. Systematic research on various aspects of the cultivation of *kendu* plant is still lacking.

20.9 ROLE OF GOVERNMENT

1. Forest department should upscale and intensify *kendu* leaf activities in migration prone districts to check labor migration.
2. Government provides scholarship for children of pluckers.
3. Government has planned and launched many schemes, viz., plucker card, plucker passbook, and free health checkup and medicine distribution.
4. *Kendu* Leaf Grant is utilized to give fair price to the pluckers and for the development of *kendu* leaf growing subdivision.
5. Supply of chappal to pluckers for protection from scorching heat.

6. Government involves co-operative societies in the trade of *kendu* leaves in order to give more benefits to forest dwellers.
7. The main objective of NTFP (non-timber forest products) operations is to safeguard the poor tribal people from the exploitation of the unscrupulous private traders.
8. Collection of NTFPs and disposal of the produces to its neighboring societies at the fixed rates abiding by the terms and conditions.
9. The government of West Bengal started a social security scheme 2015 for the tribals who survive on collecting *kendu* leaves in different districts, viz., Paschim Midnapur, Bankura, and Purulia. The beneficiaries would be entitled to compensation for accidental and natural death, partial or permanent disability, and medical assistance.

20.10 FUTURE STRATEGIES

1. Systematic research on various aspects of the cultivation of *kendu* plant should be carried out with a view to increase the yield of leaves per unit area.
2. Research should be done to improve the fruit quality and yield.
3. Collection of leaves in proper time should be done.
4. There is a need for proper storage of leaves to protect them against damp and insect attack.
5. Need attention to improve the marketing system of fruits and leaves.
6. If the trade of *kendu* leaves is organized properly, it will improve considerably the economic status of the village people specially the tribals.
7. As a part of long-term program, the *kendu* species should be planted in a compact block so that they could be easily collected.

20.11 CONCLUSION

Considering the economic potentiality, nutritive, and medicinal values, *kendu* has significant importance for both conservation and sustainable utilization. It has a very important role in overcoming malnutrition and alleviating the poverty in rural and tribal dominant areas of our country. Sustainable utilization and conservation of this plant has gained considerably importance, because it is also the key element of biodiversity. The government should take initiative in research and development of the product, including measures to ensure sustainability of the resource.

KEYWORDS

- **fruits and leaves**
- **nutritional security**
- **socio-economic upliftment**

REFERENCES

Anonymous, (2013). Demographic Status of Scheduled Tribe population and its distribution-Section-I, In: *Statistical profile of scheduled tribes in India*. Ministry of tribal affairs statistics division, Government of India. Chaar Dishayen Printers, Noida, pp. 1–10.

Anonymous, (2014). India State of Forest Report 2013. Press Information Bureau, Government of India, Ministry of Environment and Forests.

Gopalan, C., Rama Sastri, B. V., & Balasubramanian, S. C., (1989). *Nutritive Value of Indian Foods*. National Institute of Nutrition, Hyderabad–500007.

Hocking, D., (1993). *Trees for Drylands*. Oxford & IBH Publishing Co. Pvt. Ltd., New Delhi, pp. 370.

Malik, S. K., Chaudhury, R., Dhariwal, O. P., & Bhandari, D. C., (2010). *Genetic Resources of Tropical Underutilized Fruits in India*. NBPGR, New Delhi, pp. 178.

Mishra, M. R., (1997). *NTFP Prices and State Policy*: Revenue vs. Livelihood concern Report, Vasundhara, Bhubaneswar.

Orwa, C., Mutua, A., Kindt, R., Jamnadass, R., & Anthony, S., (2009). *Agroforestree Database*: A tree reference and selection guide version 4.0. World Agro forestry Centre, Kenya. http://www.worldagroforestry.org/treedb2/speciesprofile.php?Spid=690 (accessed Nov 30, 2015).

Rathore, J. S., (1970). Diospyros melanoxylon, a bread-winner tree of India. *Economic Botany*. *26*(4), pp. 333–339.

Sahu, U. C., Panda, S. K., Mohapatra, U. B., & Ray, R. C., (2012). Preparation and evaluation of wine from tendu (*Diospyros melanoxylon* L.) fruits with antioxidants. *International Journal of Food and Fermentation Technology*, vol. *2*, pp.167–178.

Stewart, J. L., & Brandis, D., (1992). *The forest flora of North-West and Central India*. Reprinted by Bisen Singh, & Mahendra Pal Singh, New Connaught Place, Dehradun, pp. 602.

CHAPTER 21

PROLINE ACCUMULATION UNDER DROUGHT STRESS IN ONION (*ALLIUM CEPA* L.) CULTIVARS

K. RIAZUNNISA and G. SAI SUDHA

Department of Biotechnology and Bioinformatics, Yogi Vemana University, Kadapa, YSR District, Andhra Pradesh, India, E-mail: khateefriaz@gmail.com, krbtbi@yogivemanauniversity.ac.in

CONTENTS

ABSTRACT

Drought is the major abiotic stress, which often adversely affects plant growth and productivity. Plant responses depend on the type of stress, species, and genotype. Proline is one of the most common compatible osmolytes in water-stressed plants. It is a common physiological response in many plants in response to a wide range of biotic and abiotic stresses. This study was undertaken to identify physiological changes in seedlings of four onion cultivars (Prema-178, Arka kirthaman, Bellary, and Arka Lalima). The experimental design was randomized entirely, with different concentrations of polyethylene glycol – 6000, i.e., 0, 25 g/L, 50 g/L, 75 g/L, and 100 g/L for different time intervals (0, 8, 24, 48, and 72 h). Our results in cultivars of onion gave a positive correlation between proline accumulation verses induced drought treatments. We observed a significant increase in proline accumulation up to 48 h and then decrease at 72 h in all cultivars at different concentrations. The increase in proline accumulation was three-fold at 48 h. Further studies are required to identify the tolerant variety suitable for local climatic conditions. The germplasm of the tolerant variety will be used in creating more tolerant variety with high productivity and yield.

21.1 INTRODUCTION

Abiotic stress is the primary cause of crop yield decrease worldwide; among all, water deficit is the major culprit in lowering the crop yield. Drought stress induces various physiological, biochemical, and molecular responses in crop plants, which would help plant to adapt to such limiting environmental barriers (Ashraf, 2004). The change in response include the overproduction of enzymes responsible for the biosynthesis of osmolytes, late embryogenesis, abundant proteins, and detoxification enzymes (Bajaji et al., 1999). The antioxidant defense system of the plant comprises a variety of antioxidant molecules and enzymes (Arora et al., 2002). Drought stress inhibits the photosynthesis of plants (as closure of stomata occurs leading to the nonavailability of CO_2 for photosynthesis); this causes changes in chlorophyll contents and components and damage to the photosynthetic apparatus (Ormaetxe et al., 1998). Before responding to the drought stress, plants first need to identify the stress caused by the conditions like decrease or loss of turgor, change in cell volume or membrane area, loss of membrane stretch, change in solute content and alteration in cell wall plasma membrane connections or protein–ligand interactions. Further, a signal transduction event occurs that triggers either physiological responses or molecular responses. It includes accumulation of certain metabolites to maintain turgor in the cells. Majorly, there are two categories of metabolites, i.e., osmolytes and antioxidants (Bartels et al., 2007). The osmolytes are sugars, sugar alcohols, amino acids, and amines. Proline play a multifunctional role in the defense mechanisms (Sai Sudha et al., 2015). It acts as a mediator of osmotic adjustment, a stabilizer of subcellular structure, a scavenger of free radicles, an energy sink, and a stress-related signal (Nanjo et al., 1999). Non-enzymatic scavenging mechanisms include accumulation of proline and glycine betaine (Ashraf et al., 2007). Generally, the amino acid proline occurs widely in higher plants and is deposited in large quantities in response to environmental stresses (Ozturk et al., 2002); it acts as an osmolyte for osmotic adjustment. Further, proline stabilizes subcellular structures like membrane and proteins scavenging free radicles and buffering cellular redox potential under stress conditions.

Drought stress can be induced with the use of polyethylene glycols (PEGs). PEGs are inert polymers available in different molecular weights. PEGs are non-toxic to mammals and are readily soluble in water (Lawlor, 1970). PEG has been used extremely for the experimental control of water stress in plants growing in vitro (Kaufmann et al., 1971). Larger PEG molecules such as PEG-6000 are more useful for simulating drought stress. PEG modifies osmotic potential of the surroundings and thus induces water stress. PEG has been used to induce drought stress for different plants like colt cherry (Ochatt et al., 1989), rice (Wani et al., 2010), maize (Matheka et al., 2008), and banana (Bidabadi et al., 2012).

Onion (*Allium cepa* L.) belongs to the family Amaryllidaceae; it is the second most important horticultural crop, which is cultivated throughout the world, and is extensively grown and widely consumed in India (Sai Sudha and Riazunnisa, 2015). Therefore, the present investigation is to exploit the genetic variability of the germplasm of onion to identify the tolerant cultivars that may give reasonable yield on drought-affected soils.

21.2 MATERIALS AND METHODS

21.2.1 COLLECTION OF SEEDS

Four cultivars of onion seeds were collected, namely, Agrifound rose, Prema-178, Nasik red (National Horticultural Research and Development Foundation), Arka kirthaman, Arka Lalima (IISC, Hyderabad), and Bellary (local farmers).

21.2.2 SURFACE STERILIZATION

Seeds were selected and transferred into labeled beakers. The seeds were then washed with distilled water and treated with 70% ethanol for 30 seconds, followed by 2% sodium hypochlorite. Finally, they were washed with distilled water.

21.2.3 SEED GERMINATION

The surface sterilized onion seeds were placed in petri plates with wetted filter papers and allowed to germinate for 10 days.

21.2.4 DROUGHT STRESS WITH DIFFERENT TIME INTERVALS

Drought was induced with different concentrations of polyethylene glycol (PEG-6000) at 0, 25 g/L, 50 g/L, 75 g/L, and 100 g/L at different time intervals, i.e., 0, 8, 24, 48 and 72 h.

21.2.5 DETERMINATION OF PROLINE

Proline was determined according to the method of Bates et al. (1973) with modifications. Fresh seedlings of 500 mg were homogenized with 5 mL of 3% (w/v) sulfosalicylic acid 2 mL of extract, 2 mL of acid ninhydrin and 2 mL of glacial acetic acid were added and incubated for 1 h in a boiling water bath at 90°C followed by an ice bath and 4 mL of toluene was then added. The chromophore containing toluene was separated, and the absorbance of red color developed was read at 520 nm against toluene blank.

21.3 RESULTS AND DISCUSSION

The *Allium cepa* L. seedlings exposed to drought stress showed a clear increase in free proline content from zero hours control or uninduced plants to stressed plants. Drought is one of the most important yield limiting factors of crop plants (Choudhary et al., 2015). There was no change in the amount of free proline content at 8 h of stress. Increase in the proline level was observed at 24 h and 48 h stressed seedlings. A slight decrease in proline accumulation was observed after 72 h stress (Figure 21.1C–E).

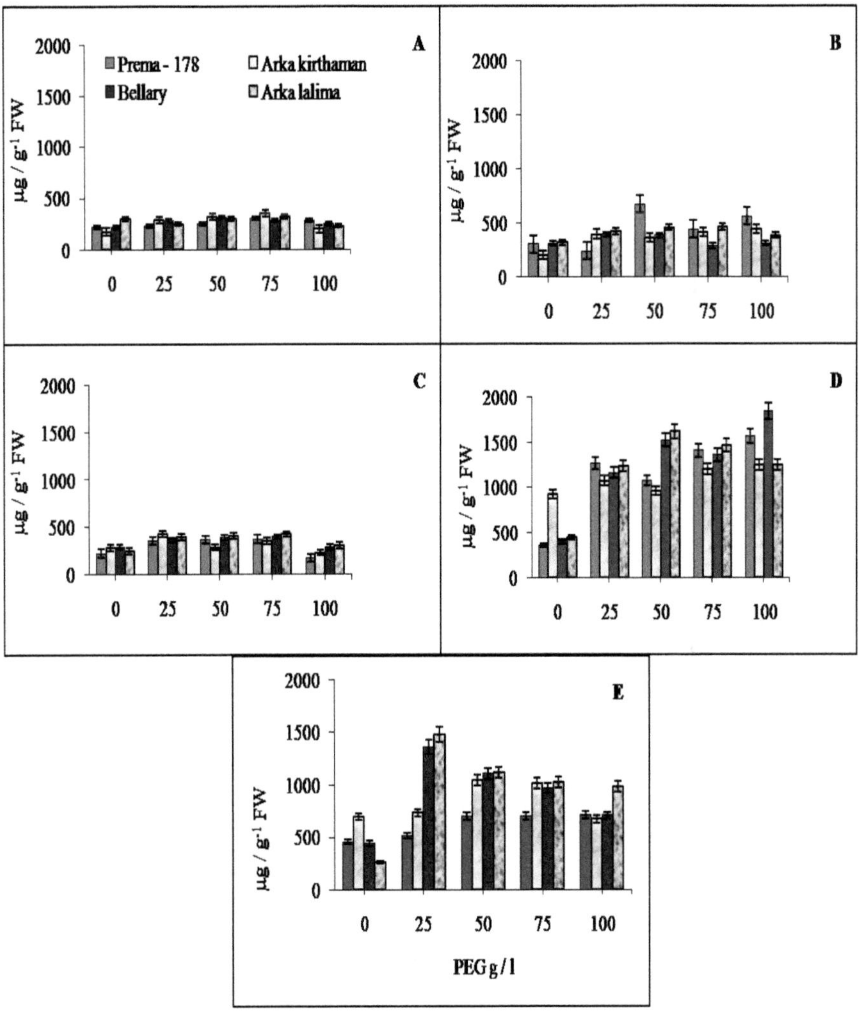

FIGURE 21.1 Effect of drought stress on the accumulation of proline in onion seedlings of Prema-178, Arka kirthaman, Bellary, and Arka Lalima cultivars, in response to varying concentration of PEG-6000 (0, 25, 50, 75, and 100 g/l) at different time intervals (A) 0 h, (B) 8 h, (C) 24 h, (D) 48 h, and (E) 72 h.

Onion seedlings at zero hours of PEG-induced stress showed that the proline content is low (Figure 21.1A). Accumulation of free proline is a typical response to drought stress, which is a first manifestation of abiotic stress in plants (Ahmad et al., 2008). The accumulation of osmolyte, i.e., proline under stress was gradually increased with increasing concentration of PEG. The maximum amount of proline was observed at 48 hours compared to that in 0, 8, and 24 hours drought-stressed seedlings of onion. At 48 h, a 2- to 3-fold increase in proline accumulation was observed compared to 0 h in the seedlings of Prema-178, Bellary, and Arka Lalima cultivars at 100 g/L PEG (Figure 21.1B–D).

Apart from protection of macromolecules, proline has several other functions during stress, e.g., osmotic adjustment, osmoprotectant, free radical scavenger, and antioxidant activity (Sai Sudha and Riazunnisa, 2015). Variation in proline accumulation was observed among the cultivars, and it was affected by PEG concentration and by treatment duration. Proline content was also affected by the interaction between these two factors. In Arka Lalima cultivar, more proline levels were observed than in Prema-178. There was a gradual increase in proline concentration with the increase in treatment time. At 24 h, proline content in PEG-treated seedlings increased significantly when compared to control seedlings with the increase of PEG concentration in all the cultivars. The highest accumulation of proline was observed with Arka Lalima at 100 g/L PEG at 72 h (Figure 21.1E). Our results gave a positive correlation between proline accumulation and treatment duration in different onion cultivars. The above study demonstrates that the accumulation of proline was more in Arka Lalima than in other cultivars.

21.4 CONCLUSION

The current study gives information for the introduction of drought-tolerant plants in water-defected areas. According to the above results, Arka Lalima germplasm can be used as a donor line for developing drought-tolerant cultivars for improving onion crop yield.

ACKNOWLEDGMENTS

The authors are thankful for financial support from the Agri-Science Park Project (State Government) and DST–SERB project (No.SR/FT/LS-352/2012) to Dr. K.R.

KEYWORDS

- ***Allium cepa* L.**
- **drought stress**
- **onion**
- **PEG-6000**
- **proline**

REFERENCES

Ahmad, P., John, R., Sarwt, M., & Umar, S., (2008). Responses of proline, lipid peroxidation and antioxidant responses in two varieties of *Pisum sativum* under salt stress. *Int. J. Plant Prod., 8,* 353–365.

Alscher, R., Erturk, G. N., & Heath, L. S., (2002). Role of superoxide dismutases (SODs) in controlling oxidative stress in plants. *J. Exp. Bot., 53,* 1331–1341.

Arora, A., Sairam, R. K., & Srivastava, G. C., (2002). Oxidative stress and antioxidative system in plants. *Curr. Sci., 82,* 1227–1238.

Ashraf, M., & Foolad, M. R., (2007). Roles of glycine betaine and proline in improving plant abiotic stress resistance. *Environ. Exp. Bot., 59,* 206–216.

Bajaji, S., Targolli, J., Li Fei, L., Tuan Hua, D. H., & Ray, W., (1999). Transgenic approaches TP increase dehydration-stress tolerance in plants. *Mol Breed., 5,* 493–503.

Bartels, D., & Sunkar, R., (2005). Drought and stress tolerance in plants. *Crc. Cr. Rev. Plant Sci., 24,* 23–58.

Bidabadi, B. S., Sariah, M., Zakaria, W., Sreeramanan, S., & Maziah, M., (2012). In vitro selection and characterization of water stress tolerant lines among ethyl methanesulphonate (EMS) induced variants of banana (*Musa* spp, with AAA genome). *Aust J Crop Sci., 6,* 567–575.

Choudhary, M., Manjhi, J., & Sinha, A., (2015). Effect of drought stress in various enzymes of *pennisetum glaucum. Int. J. Appl. Sci. Biotechnol., 3*(1), 134–138.

Iturbe Ormaetxe, I., Escuredo, P. R., Arrese Igor, C., & Becana, M., (1998). Oxidative Damage in Pea plants exposed to water deficit or paraquat. *Plant Physiol., 116,* 17–181.

Lawlor, D. E., (1970). Absorption of polyethylene glycols by plants and their effects on plant growth. *New Phytol., 69,* 501–513.

Matheka, J, M., Magiri, E., Rasha, A. O., & Machuka, J., (2008). In vitro selection and characterization of drought tolerant somaclones of tropical Maize (*Zea mays* L.). *Biotechnol., 7,* 641–650.

Nanjo, T., Kobayashi, M., Yoshiba, Y., Sanada, Y., & Wada, K., (1999). Biological functions of proline morphogenesis and osmotolerance revealed in antisense transgenic Arabidopsis thaliana. *Plant J., 18,* 185–193.

Ochatt, S. J., & Power, J. B., (1989). Selection for salt and drought tolerance in protoplast and explant derived tissue cultures of Colt cherry (*Prunusavium* × *pseudocerasus*). *Tree Physiol., 5,* 259–266.

Ozturk, L., & Demir, Y., (2002). In vivo and in vitro protective role of proline. *Plant Growth Regul., 38,* 259–264.

Sai Sudha, G., & Riazunnisa, K., (2015). Effect of salt stress (NaCl) on morphological parameters of onion (*Allium cepa* L.) seedlings. *Int. J. Pl. An and Env. Sci., 4*(5), 125–128.

Sai Sudha, G., & Riazunnisa, K., (2015). Germination and antioxidant defense system in onion (*Allium cepa* L.) Seedlings under salt stress. *Ann. Biol. Res., 6*(11), 39–46.

Sai Sudha, G., Habeeb Khadri, C., & Riazunnisa, K., (2015). *Antioxidants and Antioxidative Enzymes in Crop Plants Under Salt Str*ess: A review. (Eds. Viswanath, B., & Indravathi, G.), Paramount Publishing House, India, 253–256.

Wani, S. H., Sofi, P. A., Gosal, S. S., & Singh, N. B., (2010). In vitro screening of rice (*Oryza sativ*a L) callus for drought tolerance. Communic. *Biomet. Crop Sci., 5,*108–115.

CHAPTER 22

A STATISTICAL STUDY ON PINEAPPLE IN NORTH-EASTERN STATES OF INDIA FOR SUSTAINABLE POLICY DEVELOPMENT

KRAJAIRI MOG, SH. HEROJIT SINGH, and ANURUP MAJUMDER

Bidhan Chandra Krishi Viswavidyalaya, Krishi Viswavidyalaya P.O., Nadia, West Bengal–741 252, India, E-mail: krajainrimgc@gmail.com

CONTENTS

ABSTRACT

Pineapple (*Ananas comosus* (L.) Merr.) is considered to be one of the most important fruit crop of the world. Total annual world production is estimated at 14.6 million tons of fruits. India is the fifth largest producer of pineapple with an annual output of about 1.2 million tons (2013–2014).

In India, pineapple is a very popular fruit crop. The important pineapple producing states are Assam, Kerala, Tripura, Andhra Pradesh, Bihar, Uttar Pradesh, West Bengal, Manipur, Meghalaya, and Mizoram. It is well established that pineapple is a major agroproduct in almost all the states in northeastern India. India produces overall 1736.74 thousand MT (2013–2014) of pineapple, while the northeastern states (Arunachal Pradesh, Assam, Manipur, Meghalaya, Nagaland, Mizoram, and Tripura) produce 947.19 thousand MT (2013–14), i.e., more than 50% of the total Indian production; even the area under production counts to 70.01 thousand ha, which is nearly 64% (approx.) of total Indian coverage of 109.88 thousand ha (2013–2014). Thus, research and analysis on pineapple statistics in the context of sustainable economic development of northeastern states are very much relevant. The present paper includes a study on statistical information on pineapple for each of the northeastern states of India. However, the area and production of the states are not much smooth over the time periods due to variations in climatic parameters, especially rainfall. Segmented or piecewise regression analysis has been done, and the models are fitted to overcome the problem of interrupted time series data of pineapple. Such segmented regression analysis can be used to develop sustainable policies for improvement and overall development of entire northeastern parts of the country.

22.1 INTRODUCTION

Pineapple (*Ananas comosus* (L.) Merr.) probably originated from Latin American countries like Brazil and Paraguay. Pineapple is a well-accepted fruit crop of the world. According to the National Horticultural Board, Ministry of Agriculture, Government of India, pineapple is a major fruit crop in India, and India contributes nearly 10% of the world's pineapple production. Around 19 to 20 Indian states produce pineapple. Almost all states in northeastern part of India produce pineapple. Agro-climatic conditions of northeastern part of India favors the most for pineapple production. As a result, this part contributes nearly 54% of the country's total production. The northeastern part of India thereby holds a significant position in pineapple production scenario in India. In fact, pineapple production in this part of the country surpasses the production of other states of India, except West Bengal, the state with highest pineapple production in the country. Excepting a few pockets lying in higher altitudes, pineapple is cultivated all most entirely in the northeastern states. Predominant pineapple growing states of this part of the country are Assam, Tripura, Manipur, Arunachal Pradesh, Meghalaya, and Nagaland. As pineapple is intensively and traditionally cultivated in all states (except Sikkim) of northeastern India, it is very much relevant to study and examine the statistical dynamics of pineapple production of this area of the country. The statistics of horticultural crops in India is published every year by the National Horticultural Board, Ministry of Agriculture, Government. of India. The useful data about any horticultural crop in India is available on these databases. The scientific report through statistical analysis on all major horticultural crops based on such published data will be more relevant for socio-economic development of the country. Despite tremendous potentialities of pineapple in northeastern states of India, statistical analysis of pineapple cultivation in the area is still lacking. Such types of study are available in literature for many other agricultural crops all over the world, which helps the planners and stakeholders for future planning and development. Majumder et al. (2004) reported the scenario of jute cultivation in West Bengal. Wasim (2011) presented an in-depth study on trends, growth, and variability of major fruit crops in two different periods of Baluchistan, Period I (1989–1990 to 1998–1999) and Period II (1999–2000 to 2008–2009). Major fruit crops were apple, grapes, pomegranate, dates, apricot, peach, plums, and almonds. Verma et al. reported wheat yield modeling using remote sensing and agro-meteorological data in Haryana state.

Keeping in mind the importance of statistical study on pineapple in northeastern states of India, the present study exposes an in-depth statistical analysis on the database of pineapple cultivation (Area, Production, and Productivity/Yield) with respect to different pineapple-producing northeastern states of India.

22.2 MATERIALS AND METHODOLOGIES

The database of 23 years of area, production, and productivity of pineapple crop in northeastern states except Mizoram and Tripura was used. The initial year is 1991–1992 and the latest year is 2013–2014. The data for Mizoram are not available, and for Tripura, the available data starts from 2001–2002 to 2013–2014 (Source: Indian Horticultural database, published by the National Horticultural Board, Ministry of Agriculture, Government. of India, 2009 to 2014). The data of each state are subjected to analysis of descriptive statistics. The instability of trend over years of pineapple area and production of the study states was also examined through instability index (IX) (Wasim, 2011). IX is given as follows:

IX = CV $(1-R^2)^{1/2}$, where CV is percentage value of coefficient of variation and R^2 is the adjusted correlation coefficient over the time-period under study. Larger the value of such index signifies that the parameter under study is not stable over the years, and it may indicate a steady growth of the parameter under study over the years.

In many cases, crop yield variations over time can be modeled as a linear trend, quadratic trend, or polynomial trend that have been explored by researchers (Just and Weninger, 1999). Piecewise regression model was found to be useful (Skees et al., 1997) when critical breakpoint is present in yield pattern.

When analyzing a relationship of crop yield over time t, it may be apparent that for different ranges of t, different linear relationships occur for the yield. In addition to technological changes, these could also be due to government policy change to improve agricultural productivity. In these cases, a single linear function may not provide an adequate specification of the function. Piecewise linear regression may be a better representative function that allows multiple linear (or nonlinear) models to be fit to the data for different ranges of time. Breakpoints are the values of t (time) where the slope of the linear function changes. The value of the breakpoint may or may not be known in advance. In this study, breakpoint t (any year) is assumed to be known, although they are not same for all parameters (Area or Production or Productivity).

In other words, relationships that have different direction or magnitude of slopes at different time segments in the response variable with time can be modeled using piecewise linear segments of models combined together that has different slopes for different time segments.

The parameters under study, i.e., area, production, and productivity of pineapple over all northeastern states (except Tripura and Mizoram) have been studied by using the methodology of segmented regression. The analysis was done using software SAS, ver. 9.0 (sponsored by project under NAIP, ICAR).

22.2.1 DATABASE

A set of area, production, and productivity data of pineapple for different states of northeastern India. The initial year of data is 1991–1992 and it ends in 2013–2014.

22.3 RESULTS AND DISCUSSION

The analysis is divided into three parts. The first part explains the statistical characteristics of different states of northeast India for area, production, and productivity separately. The second part explains the growth over time for area, production, and productivity through IX values of the same states mentioned above. Piecewise linear segments of models combined together that has different slopes for different time segments for area, production, and productivity of entire northeastern states together is presented in the third part.

The production (in thousand MT) statistics of different northeastern states of India has been presented in Table 22.1. Among the states, Assam has the maximum mean production. The maximum dispersion was also observed for Assam, as revealed by Range and Standard Deviation. Coefficient of variation in percentage is maximum for Nagaland and is minimum for Assam and Meghalaya.

Table 22.2 presents the area statistics of the Northeastern States over the study period. Area under pineapple cultivation mean value over the years is the maximum for Assam, followed by Manipur and Meghalaya. The reason may be the state with larger area and population. However, it is interesting to observe that despite lower population density, Arunachal Pradesh has more area under pineapple cultivation than Nagaland or Tripura or Mizoram.

TABLE 22.1 Production (in '000MT) Statistics of Pineapple for Northeastern States of India

States	Arunachal Pradesh	Assam	Manipur	Meghalaya	Nagaland	Tripura
Mean	31.57	205.49	78.85	84.41	47.42	108.89
Standard Deviation	15.07	48.57	29.88	20.05	32.17	39.78
Kurtosis	1.96	11.22	0.84	8.32	1.92	1.93
Skewness	0.89	–2.60	–0.02	–2.04	1.15	–0.78
Range	64.01	273.00	127.01	107.17	140.09	151.70
Minimum	5.60	15.60	9.30	10.60	2.42	13.30
Maximum	69.61	288.60	136.31	117.77	142.50	165.00
CV(%)	47.73	23.64	37.89	23.75	67.84	36.53

TABLE 22.2 Area (in '000Ha) Statistics of Pineapple for Northeastern States of India

States	Arunachal Pradesh	Assam	Manipur	Meghalaya	Mizoram	Nagaland	Tripura
Mean	7.43	13.36	9.45	9.40	1.23	2.86	5.79
Standard Deviation	2.80	2.04	2.46	0.78	1.02	2.26	2.60
Kurtosis	0.12	14.48	1.33	–0.02	–0.05	4.27	1.88
Skewness	–0.37	–3.39	–0.94	0.39	1.29	2.15	1.56
Range	11.20	11.30	10.50	2.90	2.60	8.00	8.10
Minimum	1.10	4.90	2.60	7.90	0.40	1.00	3.70
Maximum	12.30	16.20	13.10	10.80	3.00	9.00	11.80
CV(%)	37.63	15.27	26.03	8.31	82.90	79.11	44.93

The results of Table 22.3 show that maximum productivity is observed for Nagaland followed by Tripura and Assam. But the coefficient of variation is maximum for Nagaland and minimum for Assam and Meghalaya. This indicates that productivity is static for Assam and Meghalaya over the 23 years of the study period.

Instability index values of the production of pineapple for different Northeastern states are displayed in Table 22.4. The results show that for the period-I (1991–2000), the production of pineapple for all the Northeastern states is more or less static for Assam and Meghalaya. But Nagaland, Manipur and Arunachal Pradesh have much higher index values. The indices

TABLE 22.3 Productivity (in MT/Ha) Statistics of Pineapple for Northeastern States of India

States	Arunachal Pradesh	Assam	Manipur	Meghalaya	Mizoram	Nagaland	Tripura
Mean	4.23	15.15	9.99	9.16	6.44	18.51	16.52
Standard Deviation	0.69	1.39	9.31	0.70	1.00	9.19	5.49
Kurtosis	–0.12	8.56	21.51	–0.13	–1.89	–1.38	–1.40
Skewness	0.33	–2.30	4.57	0.39	–0.83	–0.02	–0.43
Range	2.50	7.50	48.10	3.00	2.13	30.93	14.17
Minimum	3.10	10.00	4.00	7.70	5.17	2.37	8.63
Maximum	5.60	17.50	52.10	10.70	7.30	33.30	22.80
CV(%)	16.24	9.19	93.24	7.62	15.47	49.64	33.26

TABLE 22.4 Instability Index for Production of Different Northeastern States

States	Period I (1991–2000)		Period II (2001–2013)		Total Period	IX
	Cor. Coeff., (r)	IX	Cor. Coeff., (r)	IX	Cor. Coeff., (r)	
Arunachal Pradesh	0.94	13.53	0.35	40.67	0.65	36.97
Assam	0.93	2.70	0.02	32.12	0.12	24.02
Manipur	0.70	15.88	0.16	37.08	0.55	32.50
Meghalaya	0.54	7.02	0.06	31.02	0.22	23.72
Nagaland	0.96	19.05	0.47	69.82	0.40	63.58
Tripura			0.39	35.17		

increase rapidly in period-II (2001–2013) for all the states. In comparison to the results of Table 22.5, the area under pineapple is more or less similar to Table 22.4.

Table 22.6 shows that productivity of pineapple increases abruptly for Manipur from period-I to Period-II. Changes are also observed for Arunachal Pradesh and Nagaland. But the index values are approximately the same for Assam and Meghalaya.

Table 22.7 shows the piecewise regression analysis of all Northeastern states for pineapple production. The model significantly fits with the observed data. Figure 22.1 shows that the production of pineapple for entire Northeastern states jumps after 2010.

Table 22.8 shows the piecewise regression analysis of all northeastern states for area under pineapple cultivation. The model significantly fits with

TABLE 22.5 Instability Index for Area of Different Northeastern States

States	Period I (1991-2000)		Period II (2001- 2013)		Total Period	
	R^2	IX	R^2	IX	R^2	IX
Arunachal Pradesh	0.88	15.88	0.96	5.60	0.95	36.97
Assam	0.65	3.59	0.54	7.88	0.50	24.02
Manipur	0.88	8.96	0.46	25.25	0.59	32.50
Meghalaya	0.92	1.76	0.85	3.92	0.91	23.72
Nagaland	0.88	15.88	0.92	17.67	0.92	35.17
Tripura			0.96	5.60		

TABLE 22.6 Instability Index for Productivity of Different Northeastern States

States	Period I (1991-2000)		Period II (2001- 2013)		Total Period	
	R^2	IX	R^2	IX	R^2	IX
Arunachal Pradesh	0.69	8.07	0.22	17.08	0.53	14.05
Assam	0.75	3.04	0.54	4.17	0.80	3.66
Manipur	0.04	22.26	0.10	105.09	0.18	94.73
Meghalaya	0.07	6.56	0.65	5.62	0.56	6.09
Nagaland	0.77	26.52	0.54	41.96	0.42	45.56
Tripura			0.81	11.48		

TABLE 22.7 Piecewise Regression Analysis of Northeastern States for Pineapple Production

Source	DF	Sum of Squares	Mean Square	F Value	Approx Pr > F
Model	3	259245	86414.9	72.62	<.0001
Error	19	22610.5	1190.0		
Corrected Total	22	281855			

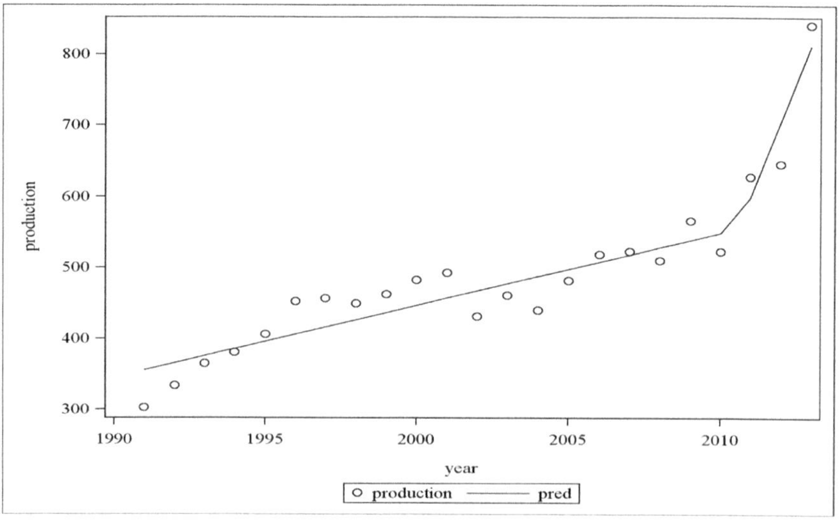

FIGURE 22.1 Piecewise regression lines for pineapple production of northeastern states.

the observed data. Figure 22.2 shows that area under pineapple for the entire northeastern states jumps after 2010 similar to that shown in Figure 22.1.

Table 22.9 shows the piecewise regression analysis of all northeastern states for area under pineapple cultivation. The model significantly fits with the observed data. Figure 22.3 shows that the productivity of pineapple for the entire northeastern states jumps after 2002.

TABLE 22.8 Piecewise Regression Analysis of Northeastern States for Area under Pineapple Cultivation

Source	DF	Sum of Squares	Mean Square	F Value	Approx. Pr > F
Model	3	828.1	276.0	20.58	<.0001
Error	19	254.8	13.4116		
Corrected Total	22	1083.0			

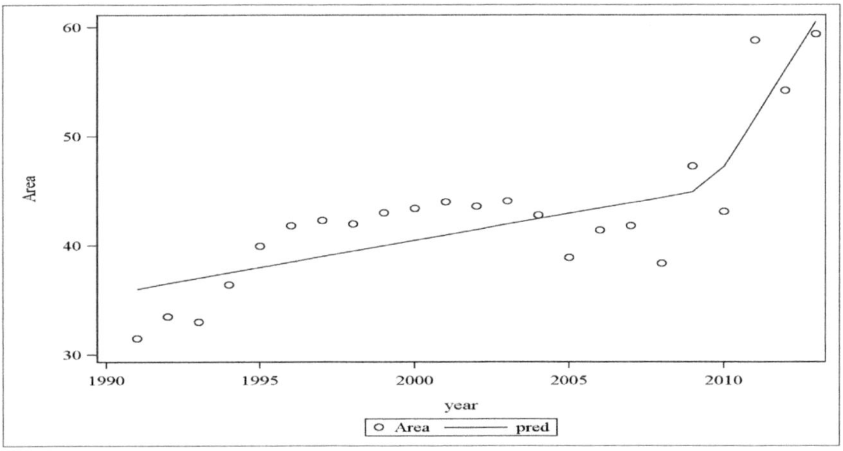

FIGURE 22.2 Piecewise regression lines for area under pineapple cultivation of northeastern states.

TABLE 22.9 Piecewise Regression Analysis of Northeastern States for Productivity of Pineapple

Source	DF	Sum of Squares	Mean Square	F Value	Approx Pr > F
Model	3	16.4273	5.4758	7.79	0.0014
Error	19	13.3635	0.7033		
Corrected Total	22	29.7908			

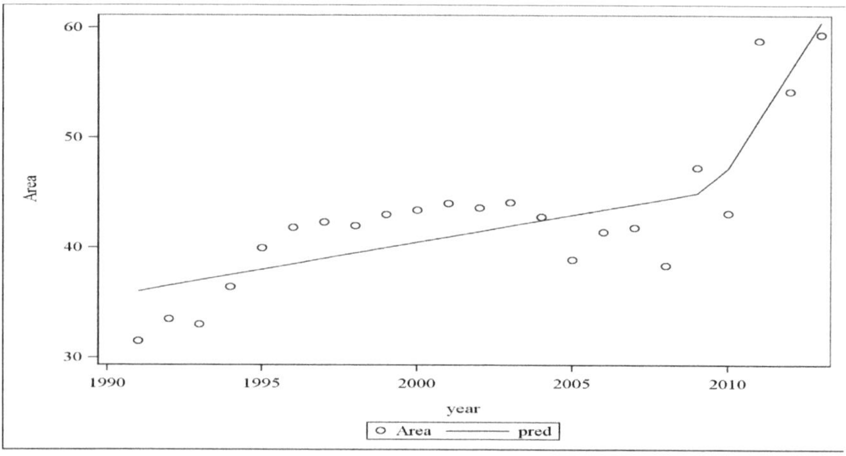

FIGURE 22.3 Piecewise regression lines for productivity of pineapple of Northeastern states.

22.4 CONCLUSION

Pineapple is a well-accepted crop of the entire northeastern part of India. Natural climatic conditions favor the cultivation of pineapple. However, the productivity of pineapple is far lower than the other pineapple growing parts of the country. Improved varieties and improved cultivation practices will improve the total production of pineapple of this region, which in turn will help the economic development of the entire area.

KEYWORDS

- **cultivation**
- **north eastern**
- **pineapple**
- **productivity**
- **regression**
- **statistics**

REFERENCES

Indian horticultural Database (2000–2014). Published by National Horticultural Board, Ministry of Agriculture, Govt. of India.

Just, R. E., & Weninger, Q., (1999). Are crop yields normally distributed?. *American Journal of Agricultural Economics*, *81,* 287–304.

Majumder, A., Pal, S., Chattapadhya, R., & Ghosh, A., (2004). Change in land use pattern and dynamics of area under jute crop in West Bengal, *Textile trends*, *XLVI*(6), 41–43.

Skees, J. R., Black, J. R., & Barnett, B. J., (1997). Designing and rating an area yield crop insurance contract. *American Journal of Agricultural Economics, 79*, 430–438.

Verma, U., Piepho, H. P., Hartung, K., Ogutu, J. O., & Goyal, A., (2015). Linear mixed modeling for mustard yield prediction in Haryana state (India), *Journal of Mathematics and Statistical Science, 1,* 96–105.

Wasim, M. P., (2011): Trends, growth and variability of major fruit crops in Balochistan, Pakistan (1989–2009), *ARPN Journal of Agricultural and Biological Science*, *6*(12), 27–36.

CHAPTER 23

PERFORMANCE OF HORTICULTURE CROPS UNDER SLOPING AGRICULTURE LAND TECHNOLOGY IN HILLY TERRAINS OF MIZORAM, INDIA

F. LALNUNMAWIA,[1] B. GOPICHAND,[2] TAWNENGA,[3] R. LALFAKZUALA,[1] and F. LALRILIANA[2]

[1]*Department of Botany, Mizoram University, Aizawl, India*

[2]*Department of Forestry, Mizoram University, Aizawl, India*

[3]*Department of Botany, Pachhunga University College, Aizawl, India*

CONTENTS

ABSTRACT

The physical features of the hill slopes of Mizoram make agricultural practices extremely difficult and diverse; as such, the region is characterized by heavy soil erosion, loss of soil fertility, and deforestation causing severe ecological imbalance. Moreover, due to shortening of the *jhum* cycle, the secondary forests also do not get adequate time to regenerate. The repeated use of land with short *jhum* cycle finally converts a large area of *jhum* fallows into degraded wastelands. Considering the landscape and climate of Mizoram, a number of experts suggested that horticulture would have better prospects than agriculture, which has a limited growth in the northeast region.

The study site is located at Tachhip, under the Aibawk Rural Development Block (Aizawl district). The project aimed at introducing low-cost integrated horticultural farming systems in the target villages with community participation. Fruit tree-based intercropping system and sloping agricultural land technology (SALT) have been introduced in the farmers field. *Tephrosia candida* is used as hedgerows as well as green manure and mulch, which are useful in dry periods. Permanent horticultural crops, viz., mango and citrus are cultivated at standard spacings. Short-term and medium-term income-producing crops are planted between strips of permanent crops as source of food and regular income, while waiting for the permanent crops to bear fruits. The medium crops introduced are banana, pineapple, and papaya. The annual cash crops such as maize and mustard are also grown between the permanent and medium crops. The choice of crops was made taking into account the edaphic and microclimatic factors and scope of markets. To enrich the soil and effectively control erosion, straws, stalks, twigs, branches, leaves, dry logs, and woods are piled at the base of hedgerows.

The organic manure, viz., cow-dung, chicken manures, and biofertilizers, are distributed to the farmers to enrich the soil nutrient availability. The project also aims at cultivating the land and raising horticultural crops to keep the soil in good health by using organic wastes (crop, animal, and farm wastes) and biofertilizers for an increased and sustainable production in an eco-friendly and pollution-free environment. The horticultural crops have been found to perform well under the SALT, and the farmers started harvesting the medium crops from the third year. It has been observed that the farm income of the beneficiaries has substantially increased with adoption of the improved farming system.

23.1 INTRODUCTION

The land of Mizoram is characterized by heavy soil erosion, loss of soil fertility, and deforestation causing severe ecological imbalance. Shifting cultivation or slash and burn agriculture, locally known as *jhum* cultivation, is the main form of agriculture in this region. Shifting cultivation becomes very devastating in nature causing drastic decline in crop yield, loss of forest wealth, soil fertility, biodiversity, and environmental degradation (Mishra and Saha, 2007). Moreover, due to shortening of the *jhum* cycle, the secondary forests also do not get adequate time to regenerate. The repeated use of land with short *jhum* cycle finally converts a large area of *jhum* fallows into degraded wastelands.

Because there are limited land areas with gentle slopes for agricultural practices in Mizoram, ways need to be explored to intensify production and raise farmers' income on the existing land without threatening the ever-decreasing forests and its biodiversity.

However, with changing requirements of high population pressure on land, *jhum* cultivation becomes very devastating in nature causing drastic decline in crop yield, loss of forest wealth, soil fertility, biodiversity, and environmental degradation.

Jhumming has been the way of life and integral part of rural livelihood in Mizoram. As such, it is not an easy task to develop a viable and widely acceptable land use model that can replace shifting cultivation. However, there is an urgent need to develop alternative land use model for shifting cultivation and analyze the soil characteristics to overcome the problems of ecological imbalance for sustainable crop production.

23.2 SCOPE AND OBJECTIVE OF THE STUDY

The present study was carried out to test the efficiency of horticulture-based intercropping system in hilly terrains by introducing low-cost integrated farming systems in the target villages with community participation. The study aims at comparing the growth and yield of horticultural crops in combination with agricultural crops under sloping agriculture land technology.

23.3 MATERIALS AND METHODS

The study site is located at Tachhip village under the Aibawk Rural Development Block. Permanent fruit crops, viz., mango, citrus, and medium crops like pineapple, papaya, and banana are cultivated at standard spacing. Agricultural crops, viz., maize and mustard are planted between strips of permanent crops while waiting for the permanent crops to bear fruit. The choice of crops was made taking into account the edaphic and microclimatic factors and scope of markets. *Tephrosia candida*, a nitrogen-fixing leguminous shrub is planted along the contour line as hedgerows to control soil erosion and to enrich nitrogen content of soil. Green manuring and mulching are applied to improve the soil, and the straws, stalks, twigs, branches, and leaves are piled at the base of hedgerows. Plant height and basal diameter of tree crops and the yield of annual agricultural crops are recorded for further analysis of efficiency of the cropping systems. A comparative study has been made on the efficiency of hedgerows, mulching and green manuring, and intercropping horticulture crops and annual agricultural crops in various combinations. The report presented in this paper is based on the second year's data of the project.

23.4 RESULTS AND DISCUSSION

23.4.1 MANGO-BASED CROPPING SYSTEM

The present paper is based on the data on growth performance of 3-year-old horticulture tree crops and the yield of annual agriculture crops. Analysis of data collected from a mango-based SALT farm indicates significant growth and yield of crops under different treatments. It was observed that the growth performances of mango seedlings as well as medium crops (banana and papaya) are best under *Tephrosia* hedgerows + green manuring and mulching, followed by *Tephrosia* hedgerow without green manuring, and least in control plots (without application of any treatments). The use of leguminous hedgerows in combination with the application of green manuring and mulching had significant impacts on the growth of the test crops.

Similarly, the yield of agricultural crops and the yield of medium horticultural crops as well as income obtained from them were maximum under *Tephrosia* hedgerows + green manuring and mulching followed by

Tephrosia hedgerow without green manuring, and the same was the lowest in control plots (without application of any treatments). Application of leguminous hedgerows in combination with the application of green manuring and mulching significantly increased the yield of crops and income from the farms (Table 23.1).

23.4.2 CITRUS-BASED CROPPING SYSTEM

The results obtained from a citrus-based SALT farm shows that the growth performance of citrus (Assam lemon) seedlings as well as medium crops (banana & papaya) is the best under *Tephrosia* hedgerows + green manuring and mulching followed by *Tephrosia* hedgerow without green manuring, and least in the control plots (without application of any treatments). The use of leguminous hedgerows in combination with application of green manuring and mulching had significant impacts on the growth of the test crops.

Similarly, the yield of agricultural crops and the yield of medium horticultural crops as well as income obtained under citrus-based intercropping were maximum under *Tephrosia* hedgerows + green manuring and mulching followed by *Tephrosia* hedgerow without green manuring, and the same was the lowest in control plots (without application of any treatments). The application of leguminous hedgerows in combination with the application of green manuring and mulching significantly increased the yield of crops and income from the farms (Table 23.2).

The present study aims at cultivating the land and raising crops in such a way as to keep the soil alive and in good health by using green manuring and mulching for an increased and sustainable production in an eco-friendly and pollution-free environment. In addition to these steps, organic manures, viz., cow dung and chicken manures, may be applied to enrich the soil nutrient content.

The use of fertilizer has become an indispensable input of intensive agriculture in hilly areas like Mizoram. The most common chemical fertilizers used by the farmers in Mizoram are urea, DAP, and MoP. Although the intensive use of agrochemicals in large amounts has resulted in increase in the productivity of farm commodities, the adverse effects of these chemicals are clearly visible on the soil structure. This has led to a decrease in the availability of soil nutrient, a vital element for plant growth. Thus, judicious

TABLE 23.1 Mango + Banana + Papaya + Maize + Mustard Intercropping

Treatments	Mango		Banana			Papaya			Mustard		Maize	
	Height (ft)	Basal diameter (cm)	Height (ft)	Basal diameter (cm)	Yield (Kg)/ha	Height (ft)	Basal diameter (cm)	Yield (Kg/ha)	Quantity/ Year (Kg/ha)	Income (Rs)	Quantity/ Year (cob)	Income (Rs)
Tephrosia hedge row	7.4	2.8	6.37	12.34	5,000	10.3	11.06	3,500	14,000	3,500	22,500	4,200
Tephrosia hedge row + Green manuring	7.6	2.86	6.75	13.14	6,000	10.7	12.66	4,500	16,000	4,000	25,000	5,000
Control	7.2	2.66	5.75	11.14	4,500	9.87	10.98	3,000	10,000	2,500	17,500	3,500

TABLE 23.2 Assam Lemon + Banana + Papaya + Maize + Mustard Intercropping

Treatments	Assam lemon		Banana			Papaya			Mustard		Maize	
	Height (ft)	Basal diameter (cm)	Height (ft)	Basal diameter (cm)	Yield	Height (ft)	Basal diameter (cm)	Yield	Quantity/ Year (Kg)	Income (Rs)	Quantity/ Year/ha (cob)	Income (Rs)
Tephrosia hedge row	5.5	4.14	6.37	12.34	5,000	10.3	11.06	3,500	14,000	35,000	2,500	14,500
Tephrosia hedgerows +	5.5	4.14	6.75	13.14	6,000	10.7	12.66	4,500	16,000	40,000	35,00	15,000
Mulching, Green manuring												
Control	4.5	3.5	5.75	11.14	4,500	9.87	10.98	3,000	10,000	25,000	1,500	10,500

use and application of locally available organic materials should be emphasized in promoting a sustainable farming system.

Hedgerows are created on steep slopes

A farmer making hedgerows along contour lines

Intercropping of banana with pineapple

Tephrosia candida are planted as hedgerows

(See color insert.)

23.5 CONCLUSION

The present study is carried out with an objective to help the farmers to take up sustainable and permanent horticulture-based farming system in hill slopes of Mizoram as an alternative to practice of shifting cultivation.

Appropriate land use and integrated farming system needs to be developed to replace the existing land utilization systems in Mizoram; this is the need of the hour to improve the economic condition of farming community in the state. In this regard, emphasis needs to be given on the potential and prospects of horticulture development in Mizoram. However, more investments are needed to improve irrigation systems and enhance the level of land productivity. Finally, integration of various land-based activities, viz., agriculture, horticulture, and animal husbandry, with proper irrigation at appropriate areas would help in a sustainable farming system in hilly areas of northeast India, and Mizoram, in particular.

KEYWORDS

- **green manuring**
- **horticulture**
- **Mizoram**
- **sloping agriculture land technology**

REFERENCES

Anonymous, (1992). Agro-Climatic Planning for Agricultural Development in Meghalaya. Working group, Zonal Planning Team, Eastern Himalayan Region, AAU, Jorhat.

Grogan, P., Lalnunmawia, F., & Tripathi, S. K., (2012). Shifting cultivation in steeply sloped regions: A review of management options and research priorities for Mizoram State, North–East India. *Agroforestry System, 84*, 163–177.

Jha, L. K., (1995). *Advances in Agroforestry*, APH publishing corporation, New Delhi, pp. 1–665.

Lalnunmawia, F., & Lalzarliana, C., (2012). *Land Use System in Mizoram*. Akansha Publishing House, New Delhi, pp. 188.

Mishra, V. K., & Saha, R., (2007). Characterization of soil health under different land use pattern in hilly eco-system of Meghalaya. *Envis. Bulletin, 15*(1&2), 6–13.

Singh, K. B., & Savant, P. V., (2000). Social forestry for rural development in Mizoram. *Proc. Int. Workshop on Agroforestry and Forest Products*, Aizawl, Jha, L. K., et al., (eds.); Linkmen Publication, pp. 57–59.

CHAPTER 24

CHANGES IN VITAMIN C CONTENT DURING THE VARIOUS STAGES OF RIPENING OF *CITRUS GRANDIS*: A MAJOR FRUIT CROP OF SIKKIM, INDIA

YENGKOKPAM RANJANA DEVI,[1] PUTHEM MINESHWOR,[1] and ROCKY THOKCHOM[2]

[1]*Biochemistry Laboratory, College of Agricultural Engineering and Post Harvest Technology, Central Agricultural University, Ranipool, Gangtok, Sikkim–737135, India, E-mail: y_ranjana@yahoo.co.in, puthem02ae@gmail.com*

[2]*Uttar Banga Krishi Viswavidyalaya, Pundibari, Cooch Behar, West Bengal–736165, India, E-mail: rockythokchomaau@gmail.com*

CONTENTS

24.1 INTRODUCTION

The genus *Citrus* belongs to the family Rutaceae and comprises about 40 species. Due to its nutritional value and flavor, *Citrus* is one of the fruits being consumed fresh or as juice. *Citrus* fruits have been shown to possess anti-inflammatory, antioxidant, antitumor, and antifungal activities (Ghafar et al., 2010). These fruits contain a variety of sugars, citric acid, ascorbic acid, carotenoids, minerals, essential oils, etc., and play an important role in human nutrition as an excellent source of antioxidants (ascorbic acid, carotenoids, and phenolic compounds) (Rekha et al., 2012). *Citrus grandis* is a crisp citrus fruit native to south and southeast Asia. Due to favorable climatic condition, the fruit is one of the major fruit crops in Sikkim. The fruit is usually pale green to yellow when ripe, with sweet white (or, more rarely, pink or red) flesh and very thick albedo (rind pith). It is the largest citrus fruit, 15–25 cm (5.9–9.8 in) in diameter, and usually weighing 1–2 kg (2.2–4.4 lb) (Growing the Granddaddy of Grapefruit, 2014). Ascorbic acid is the most important vitamin and natural antioxidant in fruit juices, especially in *Citrus* sp., and it protects the organism from oxidative stress (Ebrahimzadeh et al., 2004; Fernandez-Lopez et al., 2005).

Except human and other primates, most of the phylogenetically higher animals can synthesize vitamin C (L-ascorbate) (Stryer, 1988). More than 90% of vitamin C in human diets is supplied by fruits and vegetables. Vitamin C is defined as the generic term for all compounds exhibiting the biological activity of L-ascorbic acid (Stryer, 1988). Vitamin C is required for the prevention of scurvy and maintenance of healthy skin, gums, and blood vessels. It functions in collagen formation, absorption of inorganic iron, reduction of plasma cholesterol level, inhibition of nitrosamine formation, enhancement of the immune system, and reaction with singlet oxygen and other free radicals (Rekha et al., 2012). As an antioxidant, it reportedly reduces the risk of arteriosclerosis, cardiovascular diseases, and some forms of cancer (Lee and Kader, 2000; Sarkar et al., 2009). The consumption of fruit juices is beneficial, and the health effects of fruits are ascribed, in part to ascorbic acid, a natural antioxidant (Rekha et al., 2012). The objective of the present study was to study the content of ascorbic acid during the different stages of ripening in *C. grandis* – a major fruit crop of Sikkim.

24.2 MATERIALS AND METHOD

24.2.1 EXTRACTION OF JUICE

Ten different plants of *C. grandis* from different localities in and around Marchak, Gangtok, Sikkim, were selected for the study. The fruits from the plants were plucked at different stages. The fruits were washed thoroughly in water. The peel was removed carefully avoiding any cut or damage to the pulp. The juice was extracted by mechanical squeezing by cutting the fruits in half. The collected juice was filtered through a muslin cloth and centrifuged at 6000 rpm for 5 mins. The supernatant was collected and used for the present study. All the materials used in the preparation of the juice were first thoroughly washed by distilled water and dried in hot air oven in order to avoid dilution of the fruit juices and other unwanted chemical reaction.

24.2.2 DETERMINATION OF PH AND TOTAL ACIDITY

pH of the juices of different stages of the fruit were determined using Eutech pH meter (EC pH 1500 – 42s) after standardization with pH 4, 7, and 10. Total acidity of the juice was determined by the titration method against sodium hydroxide using phenolphthalein indicator (Khosa et al., 2011). Further, 10% fruit juices were prepared by taking 10 mL juice and making up to 100 mL using distilled water in a volumetric flask. Then, 10 mL each juice was titrated using phenolphthalein indicator against 0.1N NaOH, which was standardized using standard oxalic acid. The end point was noted when the color changed from colorless to pale pink. Total acidity was calculated in terms of citric acid using the formula: Acidity (g/100 mL) = normality of the juice × equivalent weight of citric acid (Table 24.1 and Figure 24.1).

TABLE 24.1 pH, Total Acidity, and Ascorbic Acid Content of *Citrus grandis* Fruit Juices

Maturity of fruit (weeks after flowering)	pH	Total acidity (g/100 mL)	Ascorbic acid (mg/100 mL)
12	2.06	3.992	52.0
14	2.14	3. 842	50.3
16	2.26	3.605	48.4
18	2.38	3.241	46.5
20	2.44	2.976	40.7
22	3.06	2.888	38.5
24	3.12	2.766	36.0

Data represents mean of 10 readings.

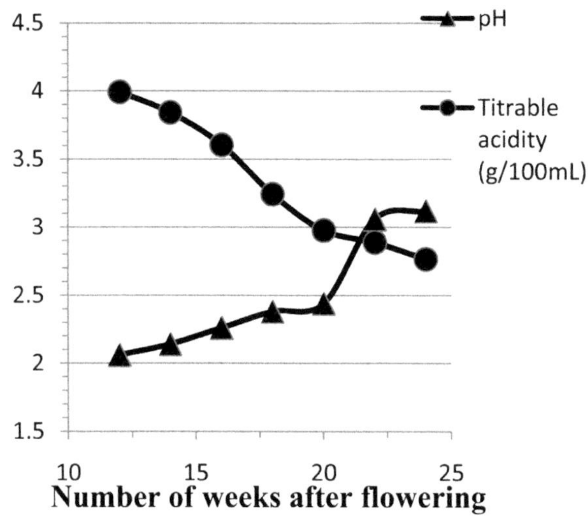

FIGURE 24.1 Changes in pH and titrable acidity during ripening of *Citrus grandis* fruit.

24.2.3 ESTIMATION OF ASCORBIC ACID CONTENT

The vitamin C content of the juices of different stages of fresh fruits was determined by the official AOAC method (1990) (AOAC. 1990) of dye-titration using 2,6-dichloroindophenols dye (DCIP). First, 5 mL of standard ascorbic acid (100 μg/mL) was taken in a conical flask containing 10 mL 4% oxalic acid and was titrated against the 2,6-dichlorophenol indophenol dye. The appearance and persistence of pink color was taken as the end point. The amount of dye consumed (V_1 mL) is equivalent to the amount of ascorbic acid. Then, 5 mL of sample (prepared by taking 5 g of juice in 100 mL 4% oxalic acid) was taken in a conical flask with 10 mL of 4% oxalic acid and titrated against the dye (V_2 mL). The amount of ascorbic acid was calculated using the formula: Ascorbic acid (mg/100 g) = (0.5 mg/V_1 mL) × (V_2/15 mL) × (100 mL/Wt. of sample) × 100 (Figure 24.2).

24.3 RESULTS AND DISCUSSION

Pomelo (the biggest citrus in the world) has a sweet and deliciously tart taste. The rind of the fruit is thick and leathery, which contains oil glands.

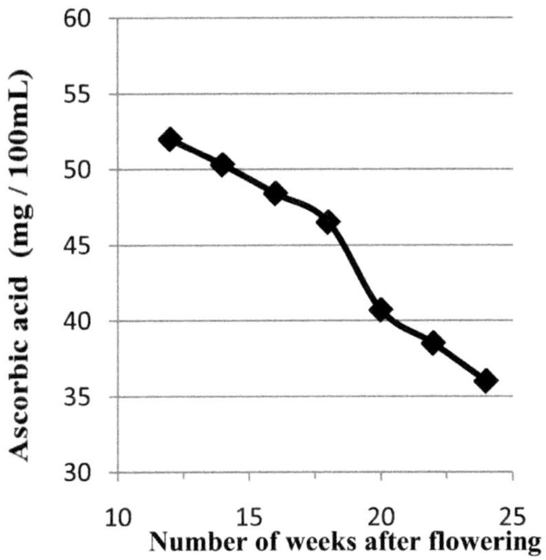

FIGURE 24.2 Changes in ascorbic acid content during ripening of *Citrus grandis* fruit.

White mesocarp covers the edible flesh. The pulp is juicy, and the color varies from pale yellow, pink or red. The pulp can be sweet, bland, or tangy but not as acidic as other citruses. Fruits for marketing are generally harvested when they begin to turn color. The dull skin of the unripe fruit brightens upon ripening as the oil glands in the rind becomes more prominent and shiny (Pomelo, 2015). With regard to the climatic condition in Sikkim (Marchak), India, the fruits are observed to mature in about 5.5–6 months after flowering. For the purpose of our study, the analysis of the fruits was performed after 3 months of flowering. The fruits were plucked after every 2 weeks starting from the third month after flowering. The fruits were analyzed for its pH, total acidity, and ascorbic acid content.

An increase in the pH was observed from unripe to ripe fruit during the various stages of the ripening process (2.06 to 3.12). The pH was found to be lesser in unripe fruits than in ripe fruits. This indicates that the unripe fruit is more acidic than the ripe fruit. The total acidity was found to higher in unripe fruit (3.992 g/100 mL) than in ripe fruits (2.766 g/100 mL). Acidity in terms of citric acid content was found to be more in the unripe fruit juice and less in ripe fruit juice. The ascorbic acid content, which was estimated by the volumetric method, was found to be higher in unripe fruits (52.0 mg/100 mL) than in ripe fruits (36.0 mg/100 mL). It was observed that the unripe fruits of *C. grandis* have higher antioxidant activity in terms of increased ascorbic acid content (52 mg/100 mL) than the ripe fruits (36 mg/100 mL). These findings are in accordance with the results obtained by Rekha et al. (2012), Gardner et al. (2000), Zvaigzne et al. (2009), and Sony Kumari et al. (2013). The antioxidant activity was found to be higher in fruit juices containing high ascorbic acid (Rekha et al., 2012; Gardner et al. 2000; Zvaigzne et al., 2009; Sony Kumari et al., 2013).

Citrus grandis

Citrus grandis

Fruit of Citrus grandis

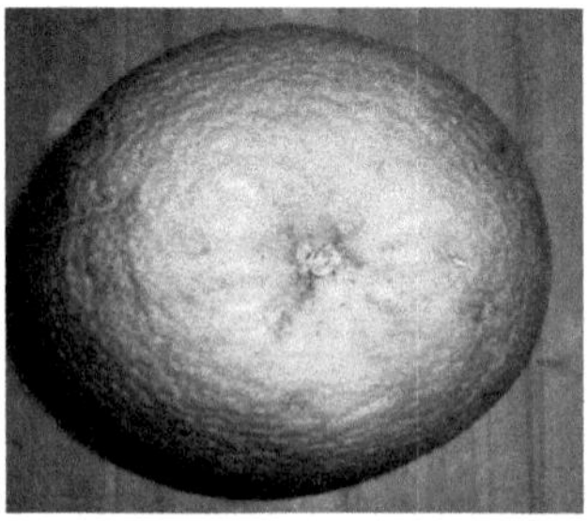

Fruit of Citrus grandis

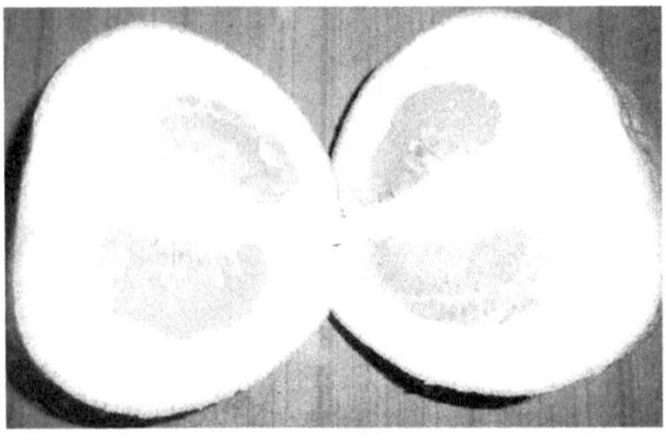

Fruit of Citrus grandis cut in half

Fruit pulp of Citrus grandis

(See color insert.)

Decrease in the ascorbic acid content during the ripening process can be attributed to the changes that occur during the process of fruit ripening. Unripe fruits have a high amount of ascorbic acid, phenolic, starch, chlorophyll, pectin, acids, and organics. During ripening, ethylene (a phytohormone) is released that activates the transcription genes for the synthesis of various enzymes that degrade the phyto-constituents involve in ripening process. During ripening, metabolism speeds up and a number of free radicals are generated. Antioxidants like ascorbic acid, phenolics, etc. intervene and detoxify the generated free radicals into harmless reduced substances. Thus, there is a reduction in the ascorbic acid, phenolic, and other antioxidant contents in the fruits while ripening, as a result of which the antioxidant activity of ripe fruit juices is comparatively less than that of unripe fruit juices (Jacob-Wilk et al., 1999).

In recent years, there is an increasing interest in finding antioxidants from natural origin (Chung et al., 2006). The present results have shown that *C. grandis* consumed for their nutritional value also have antioxidant activity in terms of ascorbic acid content. Besides its high nutritional and medicinal properties, the pomelo fruit can be consumed for its appreciably high ascorbic acid and antioxidant contents. The global consumption of pomelo is steadily increasing but is still low compared to other major citruses. Marketing and consumption of the pomelo fruit should be encouraged for nutritional food security.

ACKNOWLEDGMENT

The authors sincerely express thanks to the Dean, College of Agricultural Engineering and Post-Harvest Technology, Central Agricultural University, Ranipool, Gangtok, Sikkim, India, for providing necessary laboratory facilities.

KEYWORDS

- **acidity**
- **antioxidant**
- **ascorbic acid**
- ***Citrus grandis***

REFERENCES

AOAC, (1990). Official methods of the association of official analytical chemists. 15th edn. Washington, DC, *Association of Official Analytical Chemists.*

Chung, Y., Chien, C., Teng, K., & Chou, S., (2006). *Food Chem., 97,* 418–425.

Ebrahimzadeh, M. A., Hosseinimehr, S. J., & Gayekhloo, M. R., (2004). " Measuring and comparison of vitamin C content in citrus fruits: introduction of native variety", *Chemistry: An Indian Journal, 1*(9), 650–652.

Fernandez-Lopez, J., Zhi, N., Aleson-Carbonell, L., Perez-Alvarez, J. A., & Kuri, V., (2005). "Antioxidant and antibacterial activities of natural extracts: application in beef meatballs," *Meat Sci., 69,* 371–380.

Gardner, P. T., White, T. A. C., McPhail, D. B., & Duthie, G. G., (2000).The relative contributions of vitamin C, carotenoids and phenolics to the antioxidant potential of fruit juices. *Food Chem., 68,* 471–474.

Ghafar, M. F. A., Prasad, K. N., Weng, K. K., & Ismail, A., (2010). Flavonoid, hesperidine, total phenolic contents and antioxidant activities from Citrus species. *Afr. J. Biotechnol., 9*(3), 326–330.

Growing the granddaddy of grapefruit, http:www.sfgate.com, (accessed December 25, 2014).

Jacob-Wilk, D., Holland, D., Goldschmidt, E. E., Riov, J., & Eval, Y., (1999).Chlorophyll breakdown by chlorophyllase: isolation and functional expression of the Chlase1 gene from ethylene-treated citrus fruit and its regulation during development. *The Plant J., 20*(6), 653–661.

Khosa, M. A., Chatha, S. A. S., Hussain, A. I., Zea, K. M., Riaz, H., & Aslam, K. J., (2011). Spectrophotometric quantification of antioxidant phytochemicals in juices from four different varieties of Citrus limon, Indigenous to Pakistan. *J. Chem. Soc. Pak., 33*(2), 188–192.

Lee, S. K., & Kader, A. A., (2000). Preharvest and postharvest factors influencing vitamin C content of horticultural crops. *Postharvest Biology Technology, 20*(3), 207–220.

Pomelo, The "lucky" giant citrus. www. Itfnet.org.(accessed December 2, 2015).

Rekha, G., Poornima, M., Manasa, V., Abhipsa, J., Pavithra Devi, H. T., Vijay Kumar, & Prashith Kekuda, T. R., (2012). Ascorbic acid, total phenol content and antioxidant activity of fresh juices of four ripe and unripe citrus fruits. *Chem. Sci. Trans., 1*(2), 303–310. DOI:10.7598/cst2012.182; ISSN/E-ISSN: 2278–3458/2278–3318.

Sarkar, N, Srivastava, P. K., & Dubey, V. K., (2009). *Curr. Nutri. Food Sci., 5,* 53–55.

Sony Kumari, Neelanjana Sarmah, & Handique, A. K., (2013). Antioxidant activities of the unripe and ripe *Citrus aurantifolia* of Assam. *International Journal of Innovative Research in Science, Engineering and Technology, 2*(9), 4811–4815C.

Stryer, L., (1988). *"Biochemistry",* 3rd Ed., W. H. Freeman & Co., New York, 268.

Zvaigzne, G., Karklina, D., Seglina, D., & Krasnova, I., (2009). Antioxidants in various citrus fruit juices. *Chemine Technologija, 3*(52), 56–61.

CHAPTER 25

FRUIT SETTING, YIELD, AND ECONOMICS OF BER (*ZIZYPHUS MAURITIANA* LAMK) cv. 'BAU KUL-1' TO THE EXPOSURE OF GROWTH REGULATORS AND MICRONUTRIENTS

SAYAN SAU,[1] DEBJIT ROY,[1] BIKASH GHOSH,[1] INDRANI MAJUMDER,[1] SUBHASIS KUNDU,[1] and SUKAMAL SARKAR[2]

[1]*Department of Fruits and Orchard Management, Bidhan Chandra Krishi Viswavidyalaya, Mohanpur, West Bengal–741252, India, E-mail: hortsayan17rs@gmail.com*

[2]*Department of Agronomy, Bidhan Chandra Krishi Viswavidyalaya, Mohanpur, West Bengal–741252, India*

CONTENTS

ABSTRACT

Ber is an indigenous delicious, nursing fruit that grows widely throughout the India. Heavy fruit drops fruits due to several factors like moisture and nutritional stress, root damage, hormonal imbalance, high wind velocity, and heavy pathogenic infestation, thus showing a declining trend in *ber* production over the year. Keeping these concerns in mind, a replicated field experiment was conducted during 2013–2014 and 2014–2015 to determine the effect of foliar application of growth regulators and micronutrients on fruit setting, yield, and economics of popular *ber* cv. "BAU Kul-1" at the Horticulture Research Station, Mondouri, Bidhan Chandra Krishi Viswavidyalaya, West Bengal, India. The *ber* trees under investigation were subjected to 11 treatments combinations as follows: GA_3 @ 10 and 20 mg/L, NAA @ 10 and 20 mg/L, 2,4-D @ 5 and 10 mg/L, $ZnSO_4$ at 0.5 and 1%, H_3BO_3 @ 0.2 and 0.4%, and control (water spray only). These chemicals were thoroughly sprayed twice immediately after fruit set at 15 days interval. The results of 2 years of investigation revealed that the application of 2,4-D @ 10 mg/L recorded highest fruit set (48.80%), followed by $ZnSO_4$ at 1% and NAA at 10 mg/L. Maximum fruit retention % (42.83%) and total no. of fruits tree^{-1} (514) were obtained with the application of NAA @ 20 mg/L, which was significantly higher than rest of the treatments. The application of GA_3 @ 20 mg/L recorded significantly higher yield (30.67 kg/tree) and economic returns over the control during both years of the experiment.

25.1 INTRODUCTION

Ber (*Ziziphus mauritiana* Lamk.) is one of the hardy fruits that is suitable for cultivation mainly in arid and semi-arid condition where most of the trees fail to grow due to lack of irrigation. It is cultivated for its fresh fruits and popularly called as poor man's apple due to its high nutritional content such as higher protein (0.8 g), β-carotene (70 IU), vitamin C (50–100 mg) and medicinal value (Rai and Gupta, 1994). The total area under *ber* cultivation in India is 48,450 ha with the production of 0.66 million tons (NHB, 2014). In north India, *ber* flowers in the month of August–September, while in West Bengal, flowering occurs mainly in September–November in different varieties. It mainly produces heavy flowers in the auxiliary cymes on maturation, and current season's growth with very high fruit set percentage (Teaotia and Chauhan, 1963). But, high degree of bulk of immature fruits dropped during initial stage of fruit growth and development experiences all over India may be due to various factors like hormonal imbalance, abortion of embryo and inclement weather (Bal et al., 1988; Singh et al., 2001), nutrition (Chauhan and Gupta, 1985), moisture stress (Ghosh and Tarai, 2007), and pathogen infestation (Reddy et al., 1997) makes *ber* cultivation nonprofitable. "BAU Kul-1" is an introduced cultivar from Bangladesh released from Bangladesh Agricultural University – Germplasm Center, FTIP; it has large fruit size, round, and slightly elongate in shape; medium sweet and good flavor in taste, crispy texture, skin color is brightly yellowish with smooth and shiny surface. This cultivar is well acclimatized and a better performing one in West Bengal condition both with respect of quality and quantity parameters but facing problem of high fruit drop. Plant growth regulators and micronutrients in minute quantities play an important role in enhancing growth and development of plants to influence yield and quality, by affecting plant metabolism and bringing about a change in nutritional and hormonal status of the plant (Gadi, 2005). NAA is an important plant hormone reported to enhance the fruit set, growth, retention, yield, and market price of some fruit species (Nawaz et al., 2008); it also delayed fruit ripening and enhanced fruit formation through cell division and elongation (Dutta and Banik, 2007). Gibberellins are reported to increase fruit set, size, retention and yield as well as improve fruit physico-chemical characteristics and ripening (Rizk-Alla et al., 2011). Micronutrients (B, Fe, and Zn) have a positive effect on *ber* fruit set, yield, fruit quality and storage life (Samant et al., 2008). In this context, the present study was undertaken to investigate the effect of plant growth regulators and micronutrients in different concentrations on fruit setting, yield, and economics of *ber* cv. "BAU Kul-1" during 2013–2015.

25.2 MATERIAL AND METHODS

25.2.1 EXPERIMENTAL SITE

A field experiment was conducted during 2013–2014 and 2014–2015 at the Horticulture Research Station, Mondouri, Bidhan Chandra Krishi Viswavidyalaya, West Bengal, India (22° 43′N, 88° 30′E and 9.75 m above mean sea level) to evaluate the effect of different plant growth regulators and micronutrients in different concentrations on fruit setting, yield, and economics of ber cv. "BAU Kul-1" at the fixed site. The climate of the region is humid subtropical with hot humid summers and cool winters. The mean annual rainfall is 1,750 mm, out of which 80–90% is normally received from June to September. The experiment was conducted in a 3-years aged well-managed *ber* orchard. Soil at the experiment site was sandy clay loam (sand 64.8%, silt 10.4%, and clay 24.8%) with a pH of 6.8 and contained organic carbon of 0.56%, available N of 131.58 kg/ha, available P of 20.5 kg/ha, and available K of 170.63 kg/ha.

25.2.2 EXPERIMENTAL DESIGN AND CROP MANAGEMENT

The experiments in both years were arranged in a randomized complete block design with three replications. The *ber* trees under investigation were planted at 5 m × 5 m spacing and subjected to 11 treatments combinations as follows: GA_3 @ 10 and 20 mg/L, NAA @ 10 and 20 mg/L, 2,4-D @ 5 and 10 mg/L, $ZnSO_4$ at 0.5% and 1%, H_3BO_3 @ 0.2% and 0.4%, and control (water spray only). These chemicals were thoroughly sprayed twice along with the control (water spray) using *teepol* as a surfactant. The first spray was given immediately after fruit set (second week of October) and second spray after 15 days from the first spray (first week of November) using a knapsack sprayer having a delivery of about 5 l per plant of spray solution through a flat fan nozzle at a spray pressure of 140 kPa. All *ber* plants received RDF of NPK fertilizer with a dose of 300:200:300 g N, P_2O_5, and K_2O per plant, respectively. The chemical fertilizer was applied as basal in two splits, one before flowering (1st week of June) and the rest amount after fruit set of *ber* (second week of October) in both the years. Two to three prophylactic sprays were followed for protection against insect pest and diseases. In each plant, four main (primary) branches were kept and were pruned in April every year by removing 25% of primary shoots from the tip.

25.2.3 MEASUREMENTS AND OBSERVATIONS

Fruit setting and yield indicators of *ber* were measured as fruit set %, fruit retention %, total no. of fruits/tree, fruit weight (g), and fruit yield (kg/tree and q/ha). In order to calculate the percentage of fruit set, ten branches for each tree of all treatments were selected at random and tagged, and their flowers were counted during the full bloom. Fruitlets were also counted and recorded at the right time of fruit setting in last week of October. Fruit set % was calculated using the following formula:

$$\text{Fruit set \%} = \frac{\text{No. of developed fruitlets}}{\text{Total no. of flowers at full bloom stage}} \times 100$$

Eight bearing shoots of uniform size and vigor were randomly selected from different directions on each tree and tagged immediately after fruit set to record fruit retention. Fruit set % was calculated using the following formula:

$$\text{Fruit retention \%} = \frac{\text{No. of fruits retained upto harvest}}{\text{Total no. of fruits counted during the time of spraying}} \times 100$$

Fruit weight was calculated by taking 50 matured fruits from tagged branches and taking total weight of them, and the average fruit weight was then determined. The fruit yield was calculated by multiplying the average fruit weight to the number of fruits/plant, and the value was then converted into t/ha. Economic analysis of the treatments was calculated on the basis of total expenditure to total income or profit.

25.2.4 STATISTICAL ANALYSIS

Data were analyzed using analysis of variance (ANOVA) to evaluate the differences among treatments, while the mean values were separated using the critical difference (CD) test at the 5% level of significance using SPSS (version 18.0; SPSS, Inc., Chicago, IL, USA).

25.3 RESULTS AND DISCUSSION

25.3.1 FRUIT YIELD AND YIELD COMPONENTS

Fruit set, fruit retention, number of fruits/plant, and fruit weight are the most important yield contributing component for any fruit crops. Most of the yield components and yield of *ber* was significantly improved by the application of micronutrients and growth regulators over control. Though there was no significant variation in fruit set percentages in both the experimental years initially, fruit retention, number of fruits/plant, and fruit weight significantly increased by application of micronutrients and growth regulators (Table 25.1).

Maximum fruit retention (42.83%) and number of fruit plant (514) was measured with 20 mg/L NAA and was statistically at par with GA_3 at 20 mg/L. This finding is partially in agreement with Bal and Randhawa (2007), where

TABLE 25.1 Effect of Different Growth Regulators and Micronutrients on Fruit Set, Fruit Retention, and Yield of ber cv. "BAU Kul-1"

Treatment	Fruit set (%)	Fruit Retention (%)	No. of fruits/ plant	Fruit weight (g)	Yield (kg/ plant)	Yield (t/ ha)	Percent Increment of the yield over control
GA_3 @ 10 mg/l	47.97	39.09	469	58.4	27.39	10.95	68.63
GA_3 @ 20 mg/l	47.83	42.25	507	60.5	30.67	12.27	88.87
NAA @ 10 mg/l	48.56	38.19	458	54.8	25.11	10.04	54.61
NAA @ 20 mg/l	48.47	42.83	514	55.4	28.48	11.39	75.33
2,4-D @ 5 mg/l	48.56	35.84	430	52.9	22.75	9.09	40.05
2,4-D @ 10mg/l	48.80	36.92	443	51.1	22.64	9.05	39.37
ZnSO4 @ 0.5%	48.20	34.33	412	57.3	23.61	9.44	45.36
ZnSO4 @ 1%	48.75	36.33	436	57.9	25.24	10.09	55.41
H_3BO_3 @ 0.2%	47.75	39.75	477	55.7	26.57	10.62	63.57
H_3BO_3 @ 0.4%	47.70	40.08	481	56.8	27.32	10.92	68.22
Control	48.47	32.00	38	42.3	16.2	6.49	–
S. Em (±)	0.360	0.289	2.516	0.513	0.144	0.578	–
$LSD_{0.05}$	NS[a]	0.859	7.475	1.524	0.429	1.176	–

[a]NS, Non-significant.

they recorded minimum fruit drop with higher fruit set percent with 20–30 mg/L NAA. Singh et al. (2001) and Ghosh et al. (2009) also recorded highest fruit set in *ber* with NAA at 20 mg/L and 25 mg/L, respectively. The increase in fruit retention by the application of NAA might be attributed to reduced fruit drop by the on-going physiological and biochemical process of inhibition of abscission (Tomaszewska and Tomaszewski, 1970). Chaudhury (2006) also reported that auxin-like plant hormones (i.e., NAA) are known to stimulate cell division, cell elongation, photosynthesis, RNA synthesis, and membrane permeability, resulting higher water and nutrient uptake that finally reduced fruit drop and higher fruit retention. Application of GA_3 at 20 mg/L recorded significantly highest fruit weight (60.5 g) over the other treatments, closely followed by GA_3 at 10 mg/L and $ZnSO_4$ at 1.0% foliar spray. Similarly, highest yield (12.27 t/ha) was recorded with GA_3 at 20 mg/L, which was statistically at par with NAA at 20 mg/L. Higher fruit weight due to application of GA_3 might be due to greater photosynthesis and plant metabolism that enable higher cell growth and cell expansion, which in turn results in bigger fruit size compared to that in other treatments. Samant et al. (2008) also recorded higher fruit weight after the application of $ZnSO_4$ at 0.4%. The beneficial effect of micronutrients on fruit weight might be due to the higher availability of photosynthates, and these chemicals are also associated with hormone metabolism, which promotes synthesis of auxin necessary for fruit set and growth. Richard (2006) suggested that gibberellic acid promotes fruit growth by increasing plasticity of the cell wall, followed by the hydrolysis of starch into sugars, which reduces the cell water potential and results in the entry of water into the cell, thereby causing elongation.

A strong positive correlation between fruit retention ($R^2 = 0.843$) and fruit weight ($R^2 = 0.807$) with fruit yield (t/ha) (Figure 25.1) confirmed that both are the most important yield contributing parameters of *ber* rather than fruit set percent ($R^2 = 0.269$). This may be due to production of large number of hermaphrodite flowers and absence of pollination problem, which lead to high initial fruit set in "BAU Kul-1," and hence, even an initial high fruit percentage may not contribute to the final fruit yield.

25.3.2 ECONOMIC ASSESSMENT

Ber cultivation without growth regulator and micronutrient treatments (water spray only) had lower cost of production than the plants with growth regulator and micronutrient treatments, because in the former case no cost was involved in chemical application and resulted in less yield (Table 25.2). However, all growth regulator and micronutrient treatments yielded higher gross and net returns than control (water spray only). *Ber* with GA_3 20 mg/L gave the highest net returns (US$ 6220) as well as the highest benefit:cost ratio (3.17). Higher net returns in growth regulator and micronutrient treatments than the water spray (control) indicated the need for effective nutrient management for higher yield and economic return from *ber*.

25.4 CONCLUSIONS

From this study, it is evident that the application of growth regulators and micronutrients significantly improved fruit retention, weight, and yield as well as economic return. Thus, it can be concluded that twice spraying of GA_3 @ 20 mg/L to *ber* cv. "BAU Kul-1" at fortnight interval from fruit set

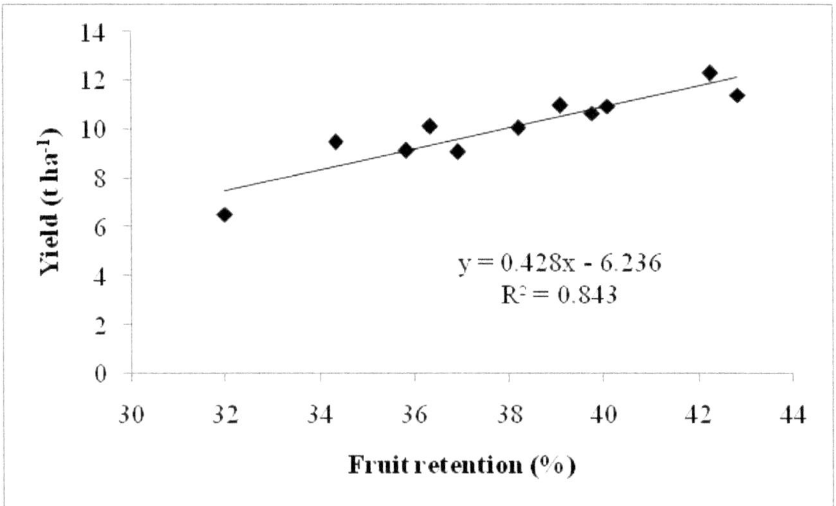

[a]

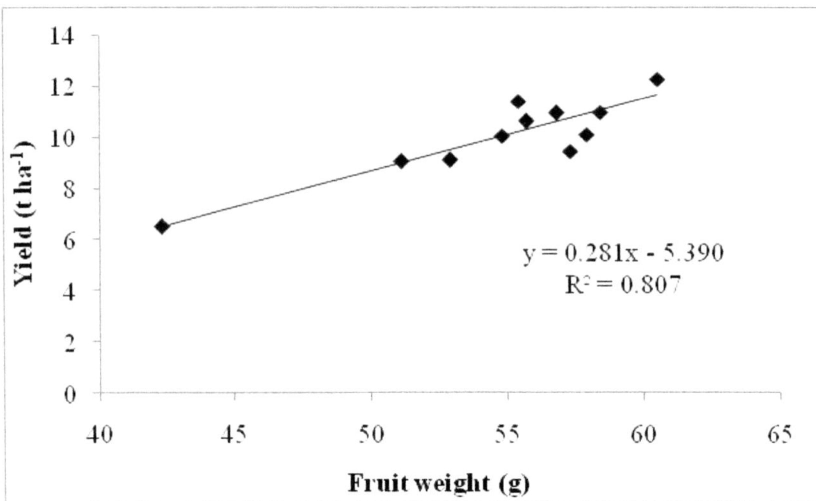

[b]

FIGURE 25.1 Relationship between ber cv. "BAU Kul-1" yield (t/ha) to fruit retention % [a] and fruit weight (g) [b].

resulted in not only more yield to the farmers but also in higher economic returns.

TABLE 25.2 Effect of Different Growth Regulators and Micronutrients on Benefit:Cost Ratio of the Cultivation of ber cv. "BAU Kul-1"

Treatment	Yield t/ ha	Total cost of production (US $/ha/year)	Gross revenue (US$/ ha / year)	Net benefit (US$/ ha/ year)[a]	Benefit : Cost ratio
GA_3 @ 10 mg/l	10.956	1925	7303	5378	2.79
GA_3 @ 20 mg/l	12.271	1960	8180	6220	3.17
NAA @ 10 mg/l	10.045	1892	6697	4805	2.54
NAA @ 20 mg/l	11.391	1893	7595	5702	3.01
2,4-D @ 5 mg/l	9.099	1891	6067	4176	2.21
2,4-D @ 10mg/l	9.055	1892	6037	4145	2.19
$ZnSO_4$ @ 0.5%	9.444	1958	6297	4338	2.22
$ZnSO_4$ @ 1%	10.097	2027	6732	4705	2.32
H_3BO_3 @ 0.2%	10.627	1918	7085	5167	2.69
H_3BO_3 @ 0.4%	10.929	1947	7285	5338	2.74
Control	6.497	1890	4330	2440	1.29

[a]Net benefit = Gross revenue – Total cost of production; 1 US$ = 60 INR (Indian rupees); Market price for ber (BAU Kul-1) US$ 666.66 per tone.

KEYWORDS

- **ber**
- **economics**
- **fruiting**
- **growth regulator**
- **micro nutrient**
- **yield**

REFERENCES

Bal, J. S., & Randhawa, J. S., (2007). Effect of NAA on fruit drop and quality of ber. *Haryana J. Hort. Sci.*, *36*(3&4), 231–231.

Bal, J. S., Randhawa, J. S., & Singh, S. N., (1988). Effect of NAA on fruit characters and quality of ber. cv. Umran. *Haryana J. Hort. Sci.*, *17*, 20–23.

Chaudhary, B. R., Sharma, M. D., Shakya, S. M., & Gautam, D. M., (2006). Effect of plant growth regulators on growth, yield and quality of chilly (*Capsicum annuum*L.) at Rampur, Chitwan. *J. Inst. Agric. Anim. Sci.*, *27*, 65–68.

Chauhan, K. S., & Gupta, A. K., (1985). Effect of foliar application of urea on the fruit drop physico-chemical composition of ber fruits under arid conditions. *Haryana J. Hort. Sci.*, *14*,9–11.

Dutta, P., & Banik, A. K., (2007). Effect of foliar feeding of nutrients and plant growth regulators on physico-chemical quality of Sardar guava grown in West Bengal. *Acta. Hort.*, *335*(6), 407–411.

Gadi, B. R., & Bohra, S. P., (2005). Effect of plant growth regulators on photosynthesis and some biochemical parameters in ber cv. Gola. *Indian J. Hort.*, *62*(3), 296–297.

Gajbhiya, V. T., Sinha, P., Gupta, S., Singh, C. P., & Gupta, R. K., (2003). Efficiency of persistence of carbendazim following pre- and post-harvest application in ber (*Zizyphus mauritiana* Lam.). *Ann. Plant Protection Sci.*, *11*, 154–58.

Ghosh, S. N., & Tarai, R. K., (2007). Effect of mulching on soil moisture, yield and quality of ber. *Indian J. Soil Cons.*, *35*, 246–248.

Ghosh, S. N., Bera, B., Kundu, A., & Roy, S., (2009). Effect of plant growth regulators on fruit retention, yield and physico-chemical characteristics of fruits in ber 'Banarasi Karka' grown in close spacing. *Acta. Hort.*, *840*, 357–362.

National Horticulture Board, (2015). Online statistical database of NHB, Ministry of Agriculture, Govt. of India. http://www.Indiastat.com. (accessed on Feb. 13, 2015).

Nawaz, M. A., Waqar, A., Saeed, A., & Mumtaz Khan, M., (2008). Role of growth regulators on preharvest fruit drop, yield and quality in kinnow mandarin. *Pak. J. Bot.*, *40*(5), 1971–1981.

Panse, V. G., & Sukhatme, P. V., (1978). Statistical methods for agricultural workers. *Indian Council of Agricultural Research.*, New Delhi, 145–152.

Rai, M., & Gupta, P. N., (1994). Genetic diversity in ber. *Indian J. Hort.*, *39*, 42–47.

Reddy, M. M., Reddy, G. S., & Madhusudan, T., (1997). Evaluation of some ber (*Zizyphus muritiana* L.) varieties and fungicides against powdery mildew. *Journal of Research ANGRAU.*, *25*, 19–26.

Richard, M., (2006). *How to Grow Big Peaches*. Dep. of Hort. Virginia Tech. Blacksburg, VA 24061. http://www. Rce.rutgers.edu. (accessed on Aug 18, 2006).

Rizk-Alla, M. S., Abd El-Wahab, M. A., & Fekry, O. M., (2011). Application of GA and NAA as a means for improving yield, fruit quality and storability of black monukka Grape. *Nature and Sci.*, *9*(1), 1–19.

Samant, D., Mishra, N. K., Singh, A. K., & Lal, R. L., (2008). Effect of micronutrient sprays on fruit yield and quality during storage in ber cv. Umran under ambient conditions. *Indian J. Hort.*, *65*(4), 399–404.

Singh, K., & Randhawa, J. S., (2001). Effect of growth regulators and fungicides on fruit drop, yield and quality of fruit in ber cv. Umran. *J. Res., Punjab Agric. Univ.*, *38*(3/4), 181–185.

Teaotia, S. S., & Chauhan, R. S., (1963). Flowering, pollination, fruit set and fruit drop studies in ber (*Zizyphus mauritiana* Lamk.). *Punjab Hort. J.*, *3*, 60–70.

Tomaszewska, E., & Tomaszewska, M., (1970). Endogenous growth regulators in fruit and leaf abscission. *Zeszyty Nauk Biol.,* Copernicus Univ. *Torun Pol., 23,* 45–53.

CHAPTER 26

IMPACT OF VARIOUS PRUNING LEVELS ON PETIOLE NUTRIENT STATUS, PHYSIOLOGY, YIELD, AND QUALITY OF GRAPE (*VITIS VINIFERA* L.) cv. ITALIA

S. SENTHIL KUMAR,[1,2] R. M. VIJAYA KUMAR,[1] K. SOORIANATHA SUNDARAM,[1] and D. DURGA DEVI[1]

[1]*Department of Fruit Crops, Horticultural College and Research Institute, TNAU, Coimbatore, Tamil Nadu – 641 003, India, Phone: +919056585411, E-mail: senthilshanmugam87@gmail.com*

[2]*School of Agriculture, Lovely Professional University, Phagwara, Punjab, India*

CONTENTS

ABSTRACT

In viticulture, pruning grape vines is considered as the most important cultural practice, as it helps to control the growth and crop load. An investigation was made to access the impact of various pruning levels on petiole nutrient status, physiology, yield, and quality in grape (*Vitis vinifera* L.) cv. Italia at the Horticultural College and Research Institute, TNAU, Coimbatore, in two seasons, summer and winter season crop during December 2012–January 2014. Different pruning levels, viz., 100%, 75%, 50%, and 25% pruning of canes for crop were imposed in two different seasons, namely summer and winter season crop. The results revealed that the treatment "T_3", i.e., pruning 50% of the canes for crop yield and 50% of the canes for vegetative growth was found to perform well in both the seasons for the above mentioned characteristics.

26.1 INTRODUCTION

Grape (*Vitis vinifera* L.) is an important commercial fruit crop in India, which is grown successfully in tropical and subtropical regions of the country. In successful viticulture, pruning the grape vines for optimum cropping is the most sensible method to maintain the balance between growth and production (Reddy and Prakash, 1990). The area under grape cultivation in the last three decades is increasing steadily with the introduction of exotic varieties. Pruning all the matured canes to 4–5 bud level for fruiting, as adopted by grape growers of Tamil Nadu, results in gradual depletion of reserved food materials, leading to loss of vigor, quality, and early setting of senility in vines. Recently, an exotic cultivar Italia, commercially under cultivation in Italy and other parts of the world, is introduced; the performance of which under Tamil Nadu conditions is unknown. The vines of this cultivar are vigorous, but tend to produce more number of bunches with variable size.

Nutrition is a key component in vineyard management. Imbalances in nutrient content of grapevine can lead to imbalance in vegetative growth and fruit production (Robinson, 1992; Christensen and Bianchi, 1994; Nair, 1998). Therefore, plant nutrient analysis acts as a tool to optimize the vine performance. Petiole analysis is widely recognized as the most reliable way to ascertain the grapevine nutritional status. Some authors relate total foliage area as a common characteristic to describe the capacity of growth and production in grape (Hunter, 2000; Naor et al., 2002). Photosynthesis is an important physiological process in plants, where the food is manufactured in leaves by trapping solar energy from direct sunlight. Adequate leaf area per shoot with good amount of chlorophyll content tends to nourish the developing bunches in a vine (Chadha and Shikhamany, 1999).

Proper canopy management in grapevine influences not only the photosynthetic activity and yield but also the quality. Enhanced N and K concentrations in the petioles improved bunch weight and TSS, TSS: Acid ratio but decreased acidity in grape cv. "Crimson Seedless" (Abd El-Razek et al., 2011).

Keeping the above facts in view, the present study was undertaken to evolve the optimum crop load management practices by means of pruning and study certain nutritional and physiological characteristics with respect to yield and quality attributes in grape cv. Italia.

26.2 MATERIALS AND METHODS

The present investigation was conducted at College Orchard, Horticultural College and Research Institute, TNAU, Coimbatore, Tamil Nadu, during December 2012–January 2014. The orchard located at an altitude of 426.6 m above mean sea level with latitude of 11°N and longitude of 77°E. The 8-year-old grape (*Vitis vinifera* L.) cv. Italia raised on Dog Ridge rootstocks were trained on the bower system. The spacing followed was 4.0 × 2.0 m. The field experiment was carried out in a randomized block design (RBD) with six treatments and replicated four times.

The pruning treatments imposed were as follows: (T_1) Pruning all (100%) the canes to 5–6 bud level for crop yield in both summer and winter season, (T_2) Pruning 75% of the canes to 5–6 bud level for crop yield and 25% of the canes to 2 bud level for vegetative growth in both summer and winter season, (T_3) Pruning 50% of the canes to 5–6 bud level for crop yield and 50% of the canes to 2 bud level for vegetative growth in both summer and winter season, (T_4) Pruning 75% of the canes to 5–6 bud level for crop yield and 25% of the canes to 2 bud level for vegetative growth in summer season; Pruning 25% of the canes to 5–6 bud level for crop yield and 75% of the canes to 2 bud level for vegetative growth during winter season, (T_5) Pruning all (100%) the canes to 2 bud level for vegetative growth in summer season; Pruning all (100%) the canes to 5–6 bud level for crop yield during winter season, and (T_6) Pruning all (100%) the canes to 5–6 bud level for crop yield in summer season; Pruning all (100%) the canes to 2 bud level for vegetative growth during winter season crop.

Leaf petioles opposite to the panicles were collected at the flowering stage in both the seasons. The influence of pruning severity on petiole nutrient status of the vine such as total nitrogen, total phosphorous, and total potassium was recorded, while the physiological parameters as chlorophyll "a," chlorophyll "b," total chlorophyll content, and total leaf area per shoot were also assessed.

Further, yield parameters as number of bunches per vine, average bunch weight, and yield per vine were also recorded. Certain quality traits like TSS, titrable acidity, and TSS:acid ratio were assessed.

26.3 RESULTS AND DISCUSSION

The results on petiole nutrient status, physiological characteristics, yield, and quality attributes as influenced by various pruning levels are presented as given in the following subsections.

26.3.1 PRUNING SEVERITY ON PETIOLE NUTRIENT STATUS OF THE VINE

In viticulture, fruiting is an exhaustive process and heavy crop load generally leads to depletion of nutrient reserves of the vine resulting in early senility. In this context, petiole analysis of the vine was taken up for major nutrients (nitrogen, phosphorus, and potassium). Nitrogen is one of the major nutrients promoting adequate growth of plants. Phosphorous is immobile in the plants and is essential to photosynthesis, respiration, and many metabolic processes (Salisbury and Ross, 1992). Potassium is an activator of enzymes that are essential for photosynthesis and respiration (Bhandal and Malik, 1988). Fallahi et al. (2005) reported that of several major nutrients, potassium has more pronounced effects on various yield and quality components of "Qermez Bidaneh" table grape.

Among the pruning levels, Pruning 50% of the canes for crop yield and 50% of the canes for vegetative growth maintained better petiole nutrient status in respect of total nitrogen, total phosphorus and total potassium at the time of flowering in both the seasons when compared to other pruning levels (see Table 26.1).

TABLE 26.1 Effect of Pruning Treatments on Petiole NPK Content (Percentage) at Flowering Stage in Grape cv. Italia

Treatments	Total Nitrogen (%)		Total Phosphorus (%)		Total Potassium (%)	
	Season I (Summer crop)	Season II (Winter crop)	Season I (Summer crop)	Season II (Winter crop)	Season I (Summer crop)	Season II (Winter crop)
T_1	1.690	1.760	0.760	0.742	2.166	2.090
T_2	2.040	2.190	0.822	0.780	2.500	2.375
T_3	2.270	2.410	0.850	0.795	2.785	2.592
T_4	2.040	2.080	0.830	0.770	2.530	2.340
T_5	1.830	1.750	0.785	0.740	2.440	2.104
T_6	1.690	2.050	0.752	0.764	2.160	2.310
S.Ed	0.025	0.030	0.011	0.010	0.039	0.030
CD (0.05%)	0.054	0.060	0.023	0.021	0.083	0.064

For total nitrogen content, in first season (summer season crop), among the treatments "T_3" recorded the maximum content (2.27%) and the minimum content (1.69%) was observed in both "T_1" and "T_6," which were on par with each other. During the second season (winter season crop) also, a similar trend was noticed. The treatment "T_3" registered the maximum petiole nitrogen content (2.41%) followed by "T_2" (2.19%) and the minimum petiole nitrogen content (1.76 and 1.75%) was recorded in "T_1" and "T_5" that were on par with each other.

Regarding total phosphorus content, for summer crop, the treatment "T_3" recorded the maximum value (0.85%) which was followed by "T_4" (0.83% percent). The minimum total phosphorus content (0.76% and 0.75%t) was registered in "T_1" and "T_6," which were on par with each other. In winter crop also, a similar trend was noticed, and the treatment "T_3," registered the maximum petiole phosphorus content (0.79%) and the minimum level (0.74%) was recorded in both "T_1" and "T_5," which were on par with each other.

The potassium content was correlated positively with yield in Thompson seedless grape (Bhujbal, 1977). For the total potassium content, during summer crop, among the treatments, "T_3" registered the maximum content (2.78%) and the minimum level (2.17% and 2.16%t) was recorded in both "T_1" and "T_6," which were on par with each other. In winter crop also, a similar trend prevailed as that of first season. The maximum petiole potassium content (2.59%) was recorded in "T_3." The minimum total potassium content (2.09% and 2.10%) was observed in "T_1" and "T_5" that were on par with each other. Relative potassium content of the petiole was less in low yielding vines as compared to high yielding ones in Thompson seedless grape (Chittiraichelvan et al., 1984).

Pruning all (100%) the canes to 5–6 bud level for crop yield in both summer and winter season exhibited lower level of nutrients in the petiole due to relatively more number of bunches per vine, competing for drawing more nutrients for the development of bunches. This finding was strongly supported by the results of Ahlawat and Yamdagni (1991) and Jeet Ram et al. (1993), indicating higher depletion of nutrients due to heavy bunch load.

26.3.2 PRUNING SEVERITY ON PHYSIOLOGICAL CHARACTERISTICS

26.3.2.1 Chlorophyll Content

The leaf chlorophyll content is a key factor in determining the rate of photosynthesis, which is considered as an index of the metabolic efficiency of plants. This pigment, responsible for harnessing solar energy and converting it into chemical energy, exhibits a differential pattern of its accumulation in response to different levels of pruning done during summer and winter seasons. A slight fluctuation in chlorophyll content is enough to trigger changes in physiological processes of plants, particularly, photosynthesis (Winkler et al., 1974).

Pruning levels significantly influenced the chlorophyll "a," "b," and total chlorophyll contents at the flowering stage (see Table 26.2).

Among the treatments, "T_3" recorded the maximum chlorophyll "a," "b," and total chlorophyll contents in both seasons. In summer season, the treatment "T_3" registered the maximum chlorophyll "a" content (1.236 mg/g) followed by "T_4" (1.082 mg/g) and "T_2" (1.074 mg/g), which were on par with each other. During winter crop also, "T_3" recorded the highest chlorophyll "a" content (0.778 mg/g). With regard to "chlorophyll "b" content," the treatment "T_3" registered the maximum (0.672 mg/g) in summer, which was followed by "T_2" (0.602 mg/g) and "T_4" (0.595 mg/g) that were on par with each other. In winter season crop, the maximum chlorophyll

TABLE 26.2 Effect of Pruning Treatments on Chlorophyll Content in Grape cv. Italia

Treatments	Chlorophyll content					
	Chlorophyll 'a' (mg/g)		Chlorophyll 'b' (mg/g)		Total Chlorophyll (mg/g)	
	Summer crop	Winter crop	Summer crop	Winter crop	Summer crop	Winter crop
T_1	0.925	0.690	0.564	0.506	1.489	1.196
T_2	1.074	0.756	0.602	0.527	1.676	1.283
T_3	1.436	0.778	0.672	0.544	2.108	1.322
T_4	1.082	0.745	0.595	0.518	1.677	1.263
T_5	1.030	0.694	0.536	0.504	1.566	1.198
T_6	0.918	0.723	0.570	0.485	1.488	1.208
S.Ed	0.014	0.009	0.008	0.007	0.022	0.016
CD (0.05%)	0.030	0.020	0.016	0.014	0.046	0.034

"b" content (0.544 mg/g) was recorded in treatment "T_3" followed by "T_2" (0.527 mg/g). Regarding total chlorophyll content, among the treatments, "T_3" registered the maximum total chlorophyll content (1.908 mg/g) in summer and (1.322 mg/g) in winter season crops.

The vegetative growth developed from 50% of shoots retained for two bud level might have produced sufficient photosynthates through enhanced chlorophyll content in these vines. The chlorophyll content during summer was found to be significantly higher than that in the winter season crop. This might be due to the prevailing of high temperature; more sunshine hours; and less relative humidity, which might have favored the synthesis of more chlorophyll during summer season than during winter season. Similar results were noticed by Slavtcheva et al. (1997) and Kumar (1999) in grape cultivar "Bangalore Blue." The decline in chlorophyll content among the treatments may be due to the increased chlorophyllase activity with limited chlorophyll synthesis (Sritharan et al., 2010).

Availability of sufficient vegetative growth in these vines with enhanced chlorophyll content due to pruning of 50% of the shoots for crop yield and remaining 50% of the shoots for vegetative growth might have accelerated the photosynthetic efficiency of the crop in both seasons.

26.3.2.2 Total Leaf Area per Shoot

Physiologically, leaf area constitutes the main photosynthetic surface and supplies most of the photosynthates required by seed, fruit, or any storage organ. Therefore, the estimation of leaf area is an essential part of classical growth analysis and is often important in physiological reasoning of variations in crop productivity. The observations on "total leaf area per shoot" exhibited significant differences among the treatments in both the seasons (see Figure 26.1). In summer season crop, the maximum total leaf area per shoot (2538.63 cm^2) was observed in treatment "T_3." The minimum leaf area per shoot (2313.40 and 2301.38 cm^2) was registered in treatments "T_1" and "$T_{6,}$" which were on par with each other. The total leaf area per shoot assessed during winter crop revealed that among the treatments, "T_3" recorded the highest total leaf area (2451.38 cm^2) and the minimum total leaf area per shoot (2120.63 cm^2) was recorded in "T_4." Higher leaf production was noticed in summer season irrespective of the pruning levels than in the winter season crop. However, comparing the effect of pruning levels on leaf production, differential response could be noticed. Pruning 50% of the canes for crop yield and 50% of the canes

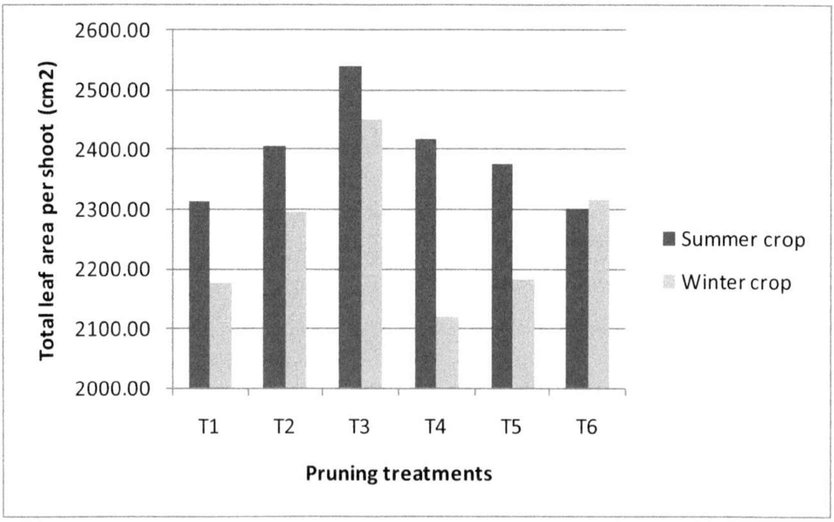

FIGURE 26.1 Effect of pruning treatments on total leaf area per shoot (cm^2) in grape cv. Italia.

for vegetative growth ("T_3") resulted in higher leaf production in terms of total leaf area per shoot. Similar results were obtained earlier by Chougule (2004).

26.3.3 SEVERITY OF PRUNING ON SEVERAL YIELD PARAMETERS

26.3.3.1 Number of Bunches per Vine

The number of bunches per vine showed wide differences in the present study between treatments in both the seasons, indicating the influence of treatments (see Table 26.3).

During summer season crop, the treatments "T_1" and "T_6" recorded the maximum number of bunches per vine (30.70 and 30.35), which were on par with each other and superior over rest of the treatments. This is normally expected as all the available canes were pruned for 5–6 bud level for fruiting and no cane was left for vegetative growth. In "T_5," although the canes were pruned at the two bud level, it was able to produce the least number of bunches per vine indicating fruit bud potential even in the lower bud level (Thatai et al., 1987). In winter crop, the treatments "T_1" and "T_5" recorded the maximum number per vine (26.45 and 26.60). This was expected because of the differential pruning methods adopted. The possible reduction of number

TABLE 26.3 Effect of Pruning Treatments on Yield Attributes in Grape cv. Italia

Treatments	Number of bunches per vine		Average bunch weight (g)		Yield per vine (kg/vine)	
	Summer crop	Winter crop	Summer crop	Winter crop	Summer crop	Winter crop
T_1	30.70	26.45	543.14	498.75	16.68	13.19
T_2	26.10	23.20	650.96	585.40	17.05	13.58
T_3	24.25	21.52	720.21	674.28	17.52	14.50
T_4	26.50	15.80	649.67	619.50	17.20	9.79
T_5	11.78	26.60	637.30	523.10	7.50	13.91
T_6	30.35	9.50	539.25	560.18	16.35	5.32
S.Ed	0.46	0.27	6.95	7.46	0.21	0.16
CD (0.05%)	0.98	0.58	14.81	15.89	0.44	0.33

of bunches per vine in severely pruned vines may be explained with the reduction in number of fruiting canes (Chougule, 2004).

Another critical observation between summer and winter seasons is that all the treatments except "T_5" had relatively higher number of bunches during summer season. This may be attributed to the buildup of vigorous growth, higher photosynthetic capacity, and better partitioning efficiency of the crop in response to climatic variations prevailing during summer season.

26.3.3.2 Average Bunch Weight

Apart from the number of bunches, the bunch weight is also very important as it decides the market value of the produce. Observations on "average bunch weight" are presented (see Table 26.3). Significant differences among the treatments were recorded during summer and winter season crops. In summer season, the maximum bunch weight (720.21 g) was recorded in "T_3" followed by "T_2" and "T_4" (650.96 and 649.67 g, respectively), which were on par with each other. During winter season crop, the treatment "T_3" recorded the maximum bunch weight (674.28 g), and it was significantly superior over the rest of the treatments. It was followed by "T_4" (619.50 g) and the minimum bunch weight (498.75 and 523.10 g) was recorded in "T_1" and "T_5." In the present study, among the treatments, "T_3" could be adjudged as the best as it has produced the maximum bunch weight in both seasons. This indicates that available photosynthates in this crop (source) was able to support the developing bunches (sink) in an efficient manner.

26.3.3.3 Yield per Vine

Usually, yield is the manifestation of morphological, growth, physiological, and biochemical parameters (Nagajothi and Jeyakumar, 2014). Yield in grape is a multiplicative factor of number of bunches per vine and average bunch weight. Pruning levels had a pronounced effect on yield per vine in both seasons in a year (see Table 26.3). In both summer and winter crop, the treatment "T_3" registered maximum yield per vine. During summer, the maximum yield per vine (17.52 kg/vine) was recorded in treatment "T_3"

followed by "T_4" and "T_2" (17.20 and 17.05 kg/vine), which were on par with each other. In winter season crop, the treatment "T_3" registered maximum yield (14.50 kg/vine) which was significantly superior over rest of the treatments.

26.3.3.4 Effect of Pruning on Quality Attributes

In any fruit crop, quality is an important criterion to fetch better price for the produce. Quality is generally judged by chemical components such as total soluble solids, acidity, sugar-acid ratio, etc.

Regarding "total soluble solids," summer crop showed no significant difference among the treatments (see Table 26.4).

However, the maximum values (14.82 °Brix) were noticed in the treatment "T_5," whereas the minimum values (14.33 and 14.35 °Brix) were observed in "T_6" and "T_1." In winter season crop, significant differences were observed among the treatments. The treatment "T_6" registered maximum TSS (13.75 °Brix), which was followed by "T_4" (13.67 °Brix) and the minimum (12.93 and 12.95 °Brix) was registered in "T_1" and "T_5."

Observations on "titrable acidity" in the berries showed significant differences in the first season (summer crop); the treatment "T_5" registered the

TABLE 26.4 Effect of Pruning Treatments on TSS (°Brix), Titrable Acidity (Percent), and TSS: Acid Ratio in Grape cv. Italia

Treatments	TSS (°B)		Titrable acidity (%)		TSS : acid ratio	
	Season I Summer crop	Season II Winter crop	Season I Summer crop	Season II Winter crop	Season I Summer crop	Season II Winter crop
T_1	14.35	12.93	0.63	0.65	22.78	19.82
T_2	14.52	13.20	0.61	0.64	23.80	20.63
T_3	14.70	13.58	0.60	0.64	24.50	21.22
T_4	14.55	13.67	0.61	0.63	23.85	21.70
T_5	14.82	12.95	0.59	0.65	25.12	19.92
T_6	14.33	13.75	0.63	0.63	22.75	21.83
S.Ed	0.19	0.17	0.01	0.01	0.30	0.27
CD (0.05%)	NS	0.36	0.02	NS	0.65	0.57

least acidity content (0.59%) followed by "T_3" (0.60%) that were on par with each other. In the case of second season (winter crop), the treatment "T_6" exhibited minimum acidity content (0.63%).

Observations on "TSS:acid ratio" was also computed in both seasons. During summer crop, the maximum TSS:acid ratio (25.12) was observed in "T_5," followed by "T_3" (24.50) and the minimum (22.75 and 22.78) was recorded in "T_6" and "T_1," respectively, which were on par with each other. During winter crop, the maximum TSS:acid ratio (21.83) was noticed in "T_6" and the minimum (19.82 and 19.92) was observed in treatments "T_1" and "T_5," respectively, which were on par with each other.

In the present study, invariably, severely pruned vines had produced bunches with better TSS, TSS:acid ratio, and lower acidity in both seasons than less severely pruned vines. This clearly indicates that crop load has a negative effect on the quality of bunches and it is vital to regulate the crop load in order to produce quality bunches. This was evident with the report of Zabadal et al. (2002) that the vines with heavy fruit loads may not ripen to desired soluble solids in climates with short growing season.

The reason for high TSS and TSS:acid ratio in severely pruned vines might be due to lesser competition for metabolites, among the limited number of bunches per vine, availability of more photosynthates consequent to better vigor and physiological activities induced in them. The predominant acids of grapes, viz., tartaric and malic acids are synthesized in leaves. These acids are translocated from leaves to bunch. This higher quantum of acids might have deposited in bunch during development, and this resulted in higher acidity in less intensive pruning treatments (Sehrawat et al., 1998; Chougule, 2004; Somkuwar and Ramteke, 2007; Harikanth, 2013).

26.4 CONCLUSION

The overall results of the study, based on petiole nutrient status, physiology, yield and quality attributes for summer and winter season crop in grape cv. Italia depicted that the treatment "T_3" (Pruning 50% of the canes to 5–6 bud level for crop yield and 50% of the canes to 2 bud level for vegetative growth) is better suited to get two crops in a year under Tamil Nadu condition.

KEYWORDS

- **grape Italia**
- **physiology**
- **pruning intensity**
- **quality**
- **vine nutrients**
- **yield**

REFERENCES

Abd El-Razek, E., Treutter, D., Saleh, M. M. S., El-Shammaa, M., Amera, A., Fouad, A., & Abdel-Hamid, N., (2011). Effect of nitrogen and potassium fertilization on productivity and fruit quality of 'Crimson Seedless' grape. *Agric. Biol. J. N. Am., 2*(2), 330–340.

Ahlawat, V. P., & Yamdagni, R., (1991). Note on the seasonal variations in nutrient contents of grapes cv. Perlette. In: *Scientific Horticulture* (Ed. S. P. Singh) Scientific Publishers, Jodhpur, *1*, 33–36.

Bhandal, J. S., & Malik, C. P., (1988). Potassium estimation, uptake, and its role in the physiology and metabolism of flowering plants. *Int. 'l. Rev. of Cytl., 110*, 205–254.

Bhujbal, B. G., (1977), PhD. *Dissertation.* MPKV, Rahuri.

Chadha, K. L., & Shikhamany, S. D., (1999). *In*: *The Grape Improvement, Production and Post Harvest Management.* Pub: Malhotra publishing house, New Delhi, pp. 340–345.

Chittraichelvan, R., Shikhamany, S. D., & Chadha, K. L., (1984). Evaluation of low yielding vines of Thompson Seedless for nutrient indices by DRIS analysis. *Ind. J. Hort., 41*, 66–79.

Chougule, R. A. M., (2004). *Sc. (Hort.) Dissertation.* MPKV, Rahuri.

Christensen, P., & Bianchi, M., (1994). Effect of nitrogen fertilizer timing and rate on inorganic nitrogen status, fruit composition, and yield of grapevines. *Am. J. Enol. and Vitic., 45*, 377–387.

Fallahi, E., Shafii, B., Stark, J. C., & Fallahi, B., (2005). Canopy growth and leaf mineral partitioning in various wine grapes. *J. Am. Poml. Soc., 59*(4), 182–190.

Harikanth, H. M., (2013). *Sc. (Hort.) Dissertation.* TNAU, Coimbatore.

Hunter, J. J., (2000). Implications of seasonal canopy management and growth compensation in grapevine. *S. Afr. J. Enol. and Vitic., 21*(2), 81–91.

Jeet Ram, D., Singh, S., Jain, P. K., & Ahlawat, V. P., (1993). Growth and nutrient composition of Perlette grapes (*Vitis vinifera* L.) as affected by nitrogen levels and pruning intensities. *Haryana J. Hort. Sci., 22*(4), 280–284.

Kumar, R. K. M., (1999). *Sc. (Hort.) Dissertation.* UAS, Bangalore.

Nagajothi, R., & Jeyakumar, P., (2014). Differential response of trifloxystrobin in combination with tebuconazole on growth, nutrient uptake and yield of rice (*Oryza sativa* L.). *Int. 'l. J. Agrl. Environ. and Biotech., 6*(1), 87–93.

Nair, J., (1998). Studies on grape nutrition in India. *Ind. J. Agron., 51*, 23–34.

Naor, A., Gal, Y., & Bravdo, B., (2002). Shoot and cluster thinning influence vegetative growth, fruit, yield and wine quality of 'Sauvignon Blanc' grapevines. *J. Am. Soc. Hort. Sci., 127*(4), 628–634.

Reddy, N. N., & Prakash, G. S., (1990). Effect of rootstock on bud fruitfulness in Anab-e-Shahi grape. *J. Mah. Agric. Univ., 15*, 218–220.

Robinson, J. B., (1992). Grapevine nutrition. In: *Viticulture: Practices*. Eds. BG Coombe and PR Dry (Winetitle, Adelaide), vol. *2*, pp. 178–208.

Salisbury, F. B., & Ross, C. W., (1992). *Plant Physiology, 4th ed.,* Wadsworth Publishing Company, USA.

Sehrawat, S. K., Daulta, B. S., Dahiya, D. S., & Bharadwaj, R., (1998). Effect of pruning on growth, yield and fruit quality in grapes (*Vitis vinifera* L.) cv. Thompson Seedless. *Int.'l. J. Trop. Agric., 16*(1–4), 185–188.

Slavtcheva, T., Poni, S., Iacono, F., & Intrieri, C., (1997). Effect of cultivation practices on leaf area, photosynthetic rate and grape yield. *Acta. Hort., 427*, 209–213.

Somkuwar, R. G., & Ramteke, S. D., (2007). Effect of bunch retention, quality and yield in Sharad Seedless. *Annual Report 2006–07,* NRC Grapes, Pune.

Sritharan, N., Vijayalakshmi, C., & Selvaraj, P. K., (2010). Effect of micro-irrigation technique on physiological and yield traits in aerobic rice. *Int'l. J. Agrl. Environ. and Biotech., 3*(1), 26–28.

Thatai, S. K., Chohan, G. S., & Kumar, H., (1987). Effect of pruning intensity on yield and fruit quality in Perlette grape trained on head system. *Ind. J. Hort., 44*(1), 66–71.

Winkler, A. J., Cook, J. A., Kliewer, W. M., & Lider, L. A., (1974). *General Viticulture*. Univ. of Calif. Press, Berkeley, pp. 633.

Zabadal, T. J., Vanee, G. R., Dittmer, T. W., & Ledebuhr, R. L., (2002). Evaluation of strategies for pruning and crop control of Concord grapevines in southwest Michigan. *Am. J. Enol. Viticult., 53*, 204–209.

CHAPTER 27

GREEN SPACE FOR SUSTAINABLE DISASTER MANAGEMENT

JENNY ZOREMTLUANGI,[1] VANLALRUATI,[1] and VANLALREMRUATI HNAMTE[2]

[1]Department of Floriculture and Landscaping, Faculty of Horticulture, BCKV, Mohanpur–741252, Nadia, West Bengal, India, E-mail: jennyzhorti@gmail.com

[2]Department of Spices and Plantation Crops, Faculty of Horticulture, Bidhan Chandra Krishi Viswavidyalaya, Mohanpur–741252, Nadia, West Bengal, India

CONTENTS

ABSTRACT

Greenspace is an important part of complex urban ecosystems and benefits urban communities environmentally, esthetically, recreationally, and economically. Green space systems play important roles in improving the ecological environment and diversity of the urban landscape. The shortage of disaster reduction and prevention function is one of the key points that leads to security issues of living environment, especially in the hill region. Focusing on large-scale botanical gardens and ecology parks for green space development meets the requirements of long-term and temporary shelter needs of local residents. Protective green space, the main components of temporary shelters, are reduced when they are changed to lands for construction. Attached and other green spaces, the main components of emergency shelters, could be destroyed by urban expansion and old urban reconstruction. Green space systems should be in right proportion with larger area of protective and attached green space to effectively function for disaster prevention.

27.1 INTRODUCTION

Urbanization has become the worldwide choice of achieving modernization in every country, and high-speed development has affected the natural ecosystem and social environment. The rapid and often unplanned expansion of cities has also exposed a greater number of people and economic assets to the risk of disasters and the effects of climate change. India is one of the most disaster-prone countries of the world, and the reasons for this increasing trend in losses due to natural hazards is the exponential increase in the vulnerability of the region and the society to disasters (Satendra, 2003). For city governments, increased climate variability imposes additional challenges to effective urban management and the delivery of key services, while for residents, it increasingly affects their lives and livelihoods due to a greater frequency floods, landslides, heat waves, droughts, or fires. Natural disasters also exert an enormous toll on the development of society and pose a significant threat to prospects for achieving the "millennium development goals" in particular (Shields et al., 2009). Disaster reduction has thus become a key component to create comprehensive disaster preparedness shelters in cities and offer a shelter and emergency rescue for victims and the homeless when disaster occurs (Jerome et al., 1986). Public facilities such as schools, hospitals, and stadiums are generally suitable for use as emergency shelters in Western countries (Shaw, 2001), but in the case of earthquake, such facilities have failed and collapsed (Liu et al., 2011). Thus, facilities like urban green spaces (UGS) have become more and more important due to their potential role in disaster prevention and reduction. UGS form an integral part of any urban area and quantity and quality of UGS is of prime concern for planners and city administrators (Gupta et al., 2012). Greenspace is an important part of complex urban ecosystems and benefits urban communities environmentally, esthetically, recreationally, and economically (Zhang, 2005) and includes highly modified parks, gardens, and recreation venues (Stephen et al., 2004).

27.2 GREENSPACE BENEFITS

The multiple benefits that greenspace provides can be categorized into ecological, social, and economic benefits and are mentioned as under:

27.2.1 ECOLOGICAL BENEFITS

Parks and other greenspaces provide many ecosystem benefits, such as regulating ambient temperatures, filtering air, reducing noise, and sequestering carbon and attenuating storm water (Bolund and Hunhammar, 1999). Aside from these human benefits, carefully designed UGS can also protect habitats and preserve biodiversity (Wilby and Perry, 2006).

27.2.2 SOCIAL BENEFITS

UGS offer urban residents solace from their stressful lives, hasten recovery from disease or illness, and can foster active living, combating sedentary lifestyles associated with obesity, heart disease and several types of cancer (Maller et al., 2006). They also foster closer community ties (Armstrong, 2000).

27.2.3 ECONOMIC BENEFITS

Seeland et al. (2009) reported that parks and provide promotes tourism and lessen environmental impacts. UGS help in adapting cities to the anticipated impacts of climate change such as higher temperatures, increased flooding, increased storminess, and the like (Carthey et al., 2009).

27.2.4 A TYPOLOGY OF URBAN GREEN/OPEN SPACES

Urban open space and greenspace can also be classified based on its size, its use, its intended function, and its location (viz. parks, plazas, urban trails, and streets) though other typologies that include cemeteries, rail reserves, roof-tops, and the like are also possible.

27.2.5 PARKS

Typically, classification schemes are based upon the size of the park, its deemed function, it geographic location, and the types of facilities present within the park and sometimes the degree of naturalness of the park. National park features large greenspace, catchments, or ranges, whereas regional parks are typically not as large as national parks and serve smaller populations and areas that typically comprise several municipalities (State of Queensland Department of Communities, 2009). District parks served several neighborhoods, whereas community parks serve a single neighborhood. Local parks consist of a bench, trees and a small patch of grass and serve a few blocks and pocket parks a single street. Though nature strips and traffic islands tend to be the most common type of UGS, they are not considered as parks.

27.2.6 PLAZAS

Plazas are traditional open spaces – often acting as civic focal points. Typically, they are paved spaces in between or completely surrounded by buildings, and function as meeting places. Plazas often take the form of a public square, but their shapes can vary widely (Marcus, Francis, and Russell, 1998). Some plazas contain no vegetation, whereas others are richly planted (Childs, 2006).

27.2.7 URBAN TRAILS/GREENWAYS

Urban trails or greenways are linear corridors used for walking, cycling, jogging, skating, etc. Some trails are intentionally designed for this purpose, whereas other have been converted from disused rail corridors or retrofitted to easements like power transmission corridors (Lindsey, Maraj, and Kuan, 2001). Urban trails traverse varied landscapes including flood-plains, river banks, lakefronts, woodlands, and sea-sides and also cross a wide range of land uses from industrial areas and vacant land to residential areas and nature reserves (Lindsey and Nguyen, 2004; Hellmund and Smith, 2006). Urban trails can increase the physical activity levels of residents, thus combating chronic diseases associated with sedentary lifestyles, especially where trails are connected to other greenspaces (Troped et al., 2001).

27.2.8 STREETS

Street types may range from small alleys to large expressways but could not be all time considered as green or open spaces as they are major traffic arteries. But some, like boulevards, lanes, and pedestrian-only streets (malls), can perform functions other than acting as transport corridors, viz., the presence of street trees, comfortable places to sit, wider footpaths to accommodate more pedestrians, shelter, and shade (Mehta, 2007).

27.3 URBAN GREENS FOR DISASTER REDUCTION

When disaster occurs, UGS can be a safe venue for various kinds of emergency services such as provision of relief supplies as well as for setting up a security command center and medical aid stations (Liu, 2011). Most cities still focus on the construction of a single emergency shelter and mitigation park or street and lack an overall perspective on green space for disaster shelter and mitigation, which limited the effective function of disaster shelter and mitigation on green space (Liu, 2011). The UGS system could be used to slow down the spread of natural disaster owing to its wide distribution and ease in accessibility. It provides a safe shelter for providing emergency medical rescue as well as a shelter for the homeless and the displaced (Greene, 1992). UGS can also be divided into park green space, productive green space, protective green space, and attached green space (The Ministry of Construction of China, 2002). Big parks can be used as long-term disaster shelters because of their large area and their location, thus being aloof from the city. Camps and temporary buildings are set up in such parks for providing long-term refuge to disaster victims. Medium-sized center parks and protective green spaces serve as temporary shelters or refuge. Such places can provide temporary asylum and treatment for the wounded, typically 10 and 20 min after disaster strikes. Emergency shelters mainly include small green spaces of central parks and attached green spaces close to residential districts. They can provide emergency shelter and escape immediately (3 to 7 min) after unexpected disasters strike (Fan et al., 2012).

UGS also help in climate change mitigation by moderation of urban temperatures as well as linking green spaces and storm water management. It also influences the growth conditions of urban vegetation, and as beneficial effects, depend on vital and lush plants, establishment and management practices of urban plantings have to adapt to climate change challenges as well. UGS such as forests or parks, can the ameliorate urban heat island effect (anthropogenic heat produced from heating and cooling processes in buildings and vehicles) by preventing incoming solar radiation from heating the surrounding buildings and surfaces, cooling the air by evapotranspiration, and reducing wind speed (Akbari et al., 2001). Besides urban green management, conservation of natural forest and increasing forest cover by plantation of appropriate species along with sustainable and integrated management of natural resources, proper land use, watershed management, and good management of land water bodies and coastal areas may be the appropriate step in reducing the adverse impact of natural hazards. These activities, on the one hand, can reverse the current trend of environmental degradation and decrease the intensity and frequency of these hazards, and on the other hand, can make the society economically and physically strong enough so that it is resilient to the adverse impact of these hazards.

Cities can better prepare for natural disasters through preparation and publicizing of hazard maps and evacuation routes, as well as through development of early warning systems. Anticipation of likely natural disasters that may occur on a repeated basis, such as tropical cyclones and hurricanes, necessitate coordinated evacuation drills and the development of special shelters for citizens. Besides, integrated and coordinated use of space-based and terrestrial technologies and their applications can play a useful role in supporting disaster risk management by providing accurate and timely information and communication support through improved risk assessment, early warning and monitoring of disasters. Improving access to geographical information and geospatial data, and building capacities to use this data in areas such as climate monitoring, land use planning, water management, disaster risk reduction, health, and food security, will allow for more accurate environmental and social impact assessments that could lead to better informed decision-making. Space-derived and in situ geographic information and geospatial data are also of benefit during times of emergency response and reconstruction, particularly in large urban areas with a high population density and especially after the occurrence of major events such as earthquakes or floods. Using geographic data and information collected before the occurrence of major disasters in combination with post-disaster data could yield important ideas for improved urban planning, especially in disaster-prone areas and highly-populated regions. Availability of structured and easily-accessible, shared geographic information is also essential for disaster management activities, such as identifying access corridors or establishing the optimal location for essential public institutions such as hospitals or emergency shelters. Such geographic data and related resources and capacities are part of what is known as "spatial data infrastructure" (SDI) (www.uncosa.unvienna.org).

27.4 EFFECTIVE SPACE FOR DISASTER SHELTER

Different sizes and patterns of green space are suitable for different aspects of disaster mitigation and shelters. UGS can be divided into three regions based on information from actual disasters: disaster area, calamity buffer and safety zone (Pursals, 2009; Saadatseresht, 2009; Shendarkar, 2008). The usefulness of green spaces for disaster mitigation and shelter mainly relies on the safety zone of greenbelt patches. When UGS patches have an area less than 1.0 ha, their internal safety area is 0 ha making them invalid for use as disaster shelter viz., small accessory green spaces of streets and residential areas (Chen et al., 2009). Minimum space requirement for making shelter for sit, stand and basic activities under emergency shelter category was reported as 1.2 m^2/ person, whereas for making temporary shelter for medical aid and rest was 2.5 m^2/ person and long-term shelter for making tents and transitional housing for long-term living was 9 m^2/ person (Fan et al., 2012).

27.5 CONCLUSION

Focusing on large-scale botanical gardens and ecology parks for green space development meets the requirements of long-term and temporary shelter needs of local residents. Protective green space, the main components of temporary shelters, are reduced when they are changed to lands for construction. Attached and other green spaces, the main components of emergency shelters, could be destroyed by urban expansion and old urban reconstruction (Fan et al., 2012). Green space systems should be in right proportion with larger area of protective and attached green space to effectively function for disaster prevention.

KEYWORDS

- **disaster**
- **greenspace**
- **shelter**
- **urban**

REFERENCES

Armstrong, D., (2000). A survey of community gardens in upstate New York: implications for health promotion and community development', *Health and Place* 6, pp. 319–27.

Bolund, P., & Hunhammar, S., (1999). Ecosystem services in urban areas', *Ecological Economics, 29*, pp. 293–301.

Carthey, J., Chandra, V., & Loosemore, M., (2009). Adapting Australian health facilities to cope with climate-related extreme weather events', *Journal of Facilities Management, 7*, pp. 36–51.

Chen, H. G., Liang, T., & Zhang, H. W., (2009). Study on the methodology for evaluating urban and regional disaster carrying capacity and its application. *Safety Sci., 47*(1), 50–58 [InChin.].

Childs, M. C., (2006). *Squares: a Public Place Design Guide for Urbanists,* University of New Mexico Press, Albuquerque.

Fan, L., Sha Xue, L., & Guobin, L., (2012). Patterns and its disaster shelter of urban green space: Empirical evidence from Jiaozuo city, China. *African J. Agril. Res.,7*(7), 1184–1191.

Fryd, O., Pauleit, S., & Buhler, O., (2011). The role of urban green space and trees in relation to climate change. *Perspectives in Agriculture, Veterinary-Science, Nutrition and Natural-Resources, 6*(053), 1–18.

Greene, M., (1992). Housing recovery and reconstruction: Lessons from recent urban earthquakes. In: *Proceedings of the 3rd U. S./Japan Workshop on Urban Earthquakes*, Oakland, CA, Earthquake Engineering Research Institute (EERI) Publication No. 93-B.

Hellmund, P. C., & Smith, D. S., (2006). *Designing Greenways: Sustainable Landscapes for Nature and People,* Island Press, Washington, D. C.

Lindsey, G., & Nguyen, D. B. L., (2004). 'Use of Greenway Trails in Indiana', *Journal of Urban Planning and Development, 130*, pp. 213–217.

Liu, Q., Ruan, X. J., & Shi, P., (2011). Selection of emergency shelter sites for seismic disasters in mountainous regions: Lessons from the 2008 Wenchuan Ms 8.0 Earthquake, China. *J. Asian Earth Sci., 40*(4), 926–934.

Maller, C., Townsend, M., Pryor, A., Brown, P., & St Leger, L., (2006). 'Healthy nature healthy people: 'contact with nature' as an upstream health promotion intervention for populations', *Health Promotion International, 21*, pp. 45–54.

Marcus, C. C., Francis, C., & Russell, R., (1998). 'Urban Plazas', In. *People Places: Design-Guidelines for Urban Open Space*, Clare Cooper Marcus & Carolyn Francis (eds.), John Wiley and Sons, New York.

Mehta, V., (2007). 'Lively streets: determining environmental characteristics to support social behavior', *Journal of Planning Education and Research, 27*, pp. 165.

Satendra, (2003). Forestry and disaster mitigation. *Journal of Rural Development Hyderabad, 22*(2), 181–197.

Seeland, K., Dübendorfer, S., & Hansmann, R., (2009). 'Making friends in Zurich's urban forests and parks: The role of public green space for social inclusion of youths from different cultures', *Forest Policy and Economics, 11*, pp. 10–7.

Shaw, R., (2001). Role of schools in creating earthquake-safer environment. *Disaster Management and Educational Facilities*, Greece. pp. 1–7.

State of Queensland Department of Communities, (2009). 'Capalaba Regional Park All Abilities Playground', Queensland State Government.

Troped, P. J., Saunders, R. P., Pate, R. R., Reininger, B., Ureda, J. R., & Thompson, S. J., (2001). 'Associations between self-reported and objective physical environmental factors and use of a community rail-trail', *Preventative Medicine, 32*, pp. 191–200.

Wilby, R. L., & Perry, G. L. W., (2006). 'Climate change, biodiversity and the urban environment: A critical review based on London, UK', *Progress in Physical Geography, 30*, pp. 73.

CHAPTER 28

EFFECT OF ORGANIC AND INORGANIC SOURCE OF NPK ON PLANT GROWTH, SPIKE YIELD, AND QUALITY OF GLADIOLUS (*GLADIOLUS GRANDIFLORUS*) cv. JESTER

V. M. PRASAD, DEVI SINGH, and BALAJI VIKRAM

Department of Horticulture, Sam Higginbottom Institute of Agriculture, Technology and Sciences, Allahabad, U.P., India, E-mail: devisinghaaidu@gmail.com

CONTENTS

ABSTRACT

A field experiment was conducted at the Department of Horticulture, Sam Higginbottom Institute of Agriculture, Technology and Science, Allahabad,

India during 2014–2015 *rabi* season. The experiment was carried out in a randomized block design with 13 treatments in three replications on different levels of inorganic and organic fertilizers (RDF 75%, 50%, 25%) with different sources; the results revealed that maximum plant height (104.39 cm), number of leaves (8.53), number of shoots per corm (2.73), days to spike initiation(53.93 days), days for opening of first florets (70.00), diameter of the florets (9.38 cm), length of spikes (93.31 cm), number of florets per spike (20.60), first floret durability (9.53 days), number of spike per plant (2.73), number of spike per hectare (150706.47), number of corm per plant (3.53), number of corm per hectare (246861.40), and benefit:cost ratio (2.35:1) were obtained for treatment T_{11}(75% RDF + 25% vermicompost) under Allahabad agro-climatic conditions.

28.1 INTRODUCTION

Gladiolus (*Gladiolus grandiflorus) is* a perennial bulbous flowering plants belongs to the family Iridacea. It is distributed in Mediterranean Europe, Asia, Tropical Africa, and South Africa. The center of diversity of the genus is located in the Cape Floristic Region. Gladiolus, popularly called sword lily, takes its name from the Latin word Gladius because of sword-shaped leaves. It is one of the most important ornamentals for cut flower trade in India and abroad. Among the different cut flowers, gladiolus stands at the fourth place in the international trade, after rose, carnation, and chrysanthemum. The yield and quality of flowers and corms can be improved by adopting integrated nutrient management practices, which include the judicious and combined use of organic, inorganic (Singh et al., 2006). It is one of the major cut flowers in national and international markets, and it is grown commercially to an extent of 1,500 ha in India. It is mainly cultivated in Karnataka, West Bengal, Maharashtra, Punjab, Haryana, Uttar Pradesh, Tamil Nadu, Jammu and Kashmir, Uttarakhand, Delhi, Sikkim, and Himachal Pradesh. It is widely used in flower arrangement, bouquets, bunches, baskets, and indoor decorations. Gladiolus has gained much importance as it is the "Queen of bulbous flowers" (Ramach, Rudu, and Thangam, 2009). Integrated application of inorganic fertilizers, organic manures, and biological sources of nutrients is an efficient way for improving the fertility, productivity, and physical conditions of the soil. The success of gladiolus cultivation depends upon many factors like soil fertility, irrigation, planting time, planting density, plant protection measures, plant growth regulators, some chemicals, etc.; these may play major role in increasing production and quality of gladiolus. They are cost-effective, inexpensive, and eco-friendly source of nutrient and do not require nonrenewable source of energy during their production (Swaroop and Janakiram, 2010). Gangadharan and Gopinath (2000) reported the effect of organic and inorganic fertilizers on growth, flowering, and quality of gladiolus (*Gladiolus grandiflorus*) cv. White prosperity. Atta-Alla et al. (2003) reported using different organic manure (cattle manure, chicken manure, sewage sludge and compost) and NPK slow release fertilizer on the vegetative growth, flowering, and chemical composition of gladiolus (*Gladiolus grandiflorus*) cultivars Eurovision, Novolux, Peter Pears, and Rose Supreme. Sharma et al. (2003) was reported the application of N (up to 400 kg/ha) and P (up to 200 kg/ha) increased floret size and number of florets per spike. Saurabhjha et al. (2012) reported the application of 75% RDF+FYM 10 t ha^{-1} showed better in days to sprouting, number of sprouts, number of leaves $plant^{-1}$, girth of plant base, width of leaf, height of the plant, days to spike emergence, diameter of corm, weight $corm^{-1}$, total corm weights $plot^{-1}$, and number of corms $plant^{-1}$.

28.2 MATERIALS AND METHODS

The research work was carried out the under Allahabad agro climatic conditions at the experimental field of the Department of Horticulture, Allahabad School of Agriculture, Sam Higginbottom Institute of Agriculture, Technology, and Sciences, Deemed-to-be-University (formerly known as Allahabad Agricultural Institute AAI-DU) during the year 2014–2015. The experimental design was randomized block design (RBD) and included three replications and 13 treatments, with a planting space of 30 × 30 cm^2 and a plot size 1.5 × 1.5 m^2 (Table 28.1).

28.3 RESULT AND DISCUSSION

Among the different treatments studied with respect of plant height, the highest plant height was recorded in T_{11} (104.39 cm, 75% RDF + 25% vermicompost) followed by T_{12} (100.18 cm, 75% RDF + 25% poultry manure) and

TABLE 28.1 Treatment Combinations

Treatments	Combinations
T_0	Control
T_1	25% RDF
T_2	25% RDF + 75% FYM
T_3	25% RDF + 75% Vermicompost
T_4	25% RDF + 75% Poultry manure
T_5	50% RDF
T_6	50% RDF + 50% FYM
T_7	50% RDF + 50% Vermicompost
T_8	50% RDF + 50% Poultry manure
T_9	75% RDF
T_{10}	75% RDF + 25% FYM
T_{11}	75% RDF + 25% Vermicompost
T_{12}	75% RDF + 25% Poultry manure

the minimum was recorded for T_0 (79.22 cm).Similar results were obtained by Gaur et al. (2006) in Gladiolus.

The highest number of leaves per plant was recorded in T_{11} (8.53, 75% RDF + 25% vermicompost) followed by T_{12} (8.20, 75% RDF + 25% poultry manure) and the minimum was recorded for T_0 (5.93). Similar results were obtained by Chaitra et al. (2007) in China aster.

The maximum number of shoots per plant (2.73) was observed with treatment T_{11} (75% RDF + 25% vermicompost) followed by T_{12} (2.60, 75% RDF + 25% poultry manure) and the minimum number of shoots per plant was observed with T_0 (1.07, control).

Early spike initiation (53.93) was observed with treatment T_{11} (75% RDF + 25% vermicompost) followed by T_{12} (55.87, 75% RDF + 25% poultry manure) and the maximum days to spike initiation was observed with T_0 (73.13, control). Similar findings were reported by Anitha et al. (2013) in marigold.

The minimum days for opening of first florets (70.00) was observed with treatment T_{11} (75% RDF + 25% vermicompost) followed by T_{12} (72.80, 75% RDF + 25% poultry manure) and the maximum days for opening of first florets was observed with T_0 (89.93, control). Similar findings were reported by Narendra et al. (2013) in Gladiolus.

TABLE 28.2 Effect of Treatments on Plant Height; Number of Leaves, Shoots; Spike Initiation and Florets Opening

Treatments	Plant height (cm)			Number of leaves per plant			No. of shoots per corm	Day to spike initiation	Days for opening of 1st florets
	30 DAP	60 DAP	90 DAP	30 DAP	60 DAP	90 DAP			
T_0	20.60	40.35	79.22	2.93	5.00	5.93	1.07	73.13	89.93
T_1	23.13	45.28	85.32	3.40	5.93	6.73	1.27	69.53	87.07
T_2	27.20	48.63	88.20	3.80	6.20	7.00	1.67	66.00	82.80
T_3	29.73	50.92	91.73	4.13	6.67	7.53	1.80	63.07	80.40
T_4	27.87	49.56	90.46	3.93	6.33	7.20	1.73	65.20	82.13
T_5	25.93	47.40	87.37	3.67	6.13	6.93	1.53	66.80	85.00
T_6	31.73	54.39	94.37	4.33	6.80	7.73	2.00	61.07	77.73
T_7	34.87	58.21	97.31	4.53	7.13	8.00	2.53	58.40	75.00
T_8	32.87	55.24	95.40	4.47	6.93	7.80	2.13	59.93	77.07
T_9	30.93	53.10	93.31	4.20	6.73	7.67	1.87	61.07	78.87
T_{10}	32.47	58.73	99.03	4.44	7.20	8.13	2.47	56.87	74.67
T_{11}	29.84	64.39	104.39	4.42	7.53	8.53	2.73	53.93	70.00
T_{12}	30.59	61.41	100.18	5.00	7.33	8.20	2.60	55.87	72.80
F- test	NS	S	S	NS	S	S	S	S	S
S.Ed ±	4.95	0.20	0.30	0.53	0.12	0.14	0.10	0.75	0.31
CD (5%)	2.18	0.42	0.61	1.09	0.25	0.28	0.20	1.54	0.63

TABLE 28.3 Effect of Treatments on Flower Characters, Corm Yield and Production Economics

Treatments	Diameter of the florets (cm)	Length of Spikes (cm)	Number of florets per spike	1st floret durability (days)	No. of spike per plant	No. of spike per hectare	No. of corm per plant	No. of corm per hectare	Benefit Cost Ratio
T_0	7.03	75.32	9.33	6.73	1.07	109230.93	1.13	130868.93	1.42:1
T_1	7.38	78.20	10.60	7.67	1.27	116782.27	1.40	161682.73	1.67:1
T_2	7.74	81.73	12.13	8.27	1.67	117058.60	1.80	176859.40	1.72 :1
T_3	8.11	84.37	13.53	8.67	1.80	121292.67	2.27	197181.93	1.90:1
T_4	7.83	82.32	12.47	8.47	1.73	118623.40	2.00	185329.40	1.81:1
T_5	7.63	80.46	11.53	8.20	1.53	115891.60	2.53	175391.67	1.73:1
T_6	8.31	86.37	14.87	8.80	2.00	126022.67	2.67	207232.00	1.96:1
T_7	8.74	89.37	16.13	9.13	2.53	133832.93	3.00	213696.20	2.06.:1
T_8	8.43	87.37	15.13	8.93	2.13	127785.80	2.87	207813.27	1.99:1
T_9	8.26	85.32	14.07	8.73	1.87	123510.73	2.47	205176.33	1.95:1
T_{10}	8.83	91.73	17.07	9.20	2.47	137782.80	3.13	216605.53	2.08:1
T_{11}	9.38	93.31	20.60	9.53	2.73	150706.47	3.53	246861.40	2.35:1
T_{12}	9.07	92.25	18.07	9.27	2.60	143028.93	3.33	227939.20	2.19:1
F- test	S	S	S	S	S	S	S	S	-
S.Ed ±	0.15	0.25	0.36	0.11	0.10	492.36	0.12	3782.41	-
CD (5%)	0.31	0.52	0.75	0.24	0.20	1016.19	0.25	7806.52	-

The maximum diameter of the florets (9.38 cm) was observed with treatment T_{11} (75% RDF + 25% vermicompost) followed by T_{12} (9.07 cm, 75% RDF + 25% poultry manure) and the minimum diameter of the florets was observed with T_0 (7.03 cm, control). Similar result was reported by Gaur et al. (2006) in Gladiolus.

The maximum length of spike (93.31 cm) was observed with treatment T_{11} (75% RDF + 25% vermicompost) followed by T_{12} (92.25 cm, 75% RDF + 25% poultry manure) and minimum length of spike (cm) was observed with T_0 (75.32 cm, control). Similar findings were reported by Kadu et al. (2003) in Gladiolus.

The maximum number of florets per spike (20.60) was observed with treatment T_{11} (75% RDF + 25% vermicompost) followed by T_{12} (18.07, 75% RDF + 25% poultry manure) and the minimum number of florets per spike was observed with T_0 (9.33, control).

The maximum first floret durability (9.53 days) was observed with treatment T_{11} (75% RDF + 25% vermicompost) followed by T_{12} (9.27, 75% RDF + 25% poultry manure) and the minimum first floret durability (days) was observed with T_0 (6.73, control).

The maximum number of spike per plant (2.73) was observed with treatment T_{11} (75% RDF + 25% vermicompost) followed by T_{12} (2.60, 75% RDF + 25% poultry manure) and the minimum number of spike per plant was observed with T_0 (1.07, control).

The maximum number of spike per hectare (150706.47) was observed with treatment T_{11} (75% RDF + 25% vermicompost) followed by T_{12} (143028.93, 75% RDF + 25% poultry manure) and the minimum number of spike per hectare was observed with T_0 (109230.93, control). Similar findings were reported by Nagaraju et al. (2007) and Pooja et al. (2012).

The maximum number of corms per plant (3.53) was observed with treatment T_{11} (75% RDF + 25% vermicompost) followed by T_{12} (3.33, 75% RDF + 25% poultry manure) and the minimum number of corms per plant was observed with T_0 (1.13, control). Similar findings were reported by Gangadharam et al. (2000) and Barman et al. (2005) in Gladiolus.

The maximum number of corms per hectare (246861.40) was observed with treatment T_{11} (75% RDF + 25% vermicompost) followed by T_{12} (227939.20, 75% RDF + 25% poultry manure), and the minimum number of corms per hectare was observed with T_0 (130868.93, control). Similar findings were reported by Gangadharam et al. (2000) and Barman et al. (2005) in Gladiolus.

28.4 CONCLUSION

The present investigation on Gladiolus concluded that the application of (75% RDF+ 25% vermicompost) T_{11} treatment gave maximum plant height (105.60 cm) and highest spike yield (140848.84 no./ha) that is effective for enhancing corm (241482.28 no./ha.) production. The treatment T_{11} (75% RDF + 25% Vermicompost) was found to be most economically viable in terms of gross return (719782.51), net return (413256.51), and benefit:cost ratio (2.35:1).

KEYWORDS

- **gladiolus *(Gladiolus grandiflorus)***
- **inorganic**
- **NPK**
- **organic**
- **plant growth quality**
- **spikes yield**

REFERENCES

Anita Mohanty, C. R., Mohanty, P. K., & Mohapatra, (2013). Studies on the response of integrated nutrient management on growth and yield of marigold (*Tagetes erecta* L). *Research Journal of Agricultural Sciences, 4(3), 383–385.*

Atta-Alla, H. K., Zaghloul, M. A., Barka, M., Hashish, K. H., (2003). Effect of organic manure and NPK fertilizers on the vegetative growth, flowering and chemical composition of some gladiolus cultivars. *Annals of Agricultural Science, Moshtohor, 41*(2), 889–912.

Barman, D., Rajni, K., Rampal, & Upadhyaya, R. C., (2005). Corm multiplication of gladiolus as influenced by application of potassium and spike removal. *J. Orna. Hort., 8*(2), 104–107.

Chaitra, R., (2006). *Effect of Integrated nutrient Management on Growth, Yield and Quality of China Aster* (*Callistephus chinensis* L.). University of Agricultural Sciences, Dharwad.

Gangadharan, G. D., & Gopinath, G., (2000). Effect of organic and inorganic fertilizers on growth, flowering and quality of gladiolus Cv. White prosperity. *Karnataka Journal of Agricultural Sciences, 13*(2), 401–405.

Gaur, A., Misra, R. L., Kumar, P. N., & Sarkar, J., (2006). Studies on nutrient management in gladiolus. Paper presented in the *"National Symposium on Ornamental Bulbous Crops"* held on 5–6 December, 2006 at S.V.B.P.U. of Ag. & T., Modipuram, Meerut (U.P.), 107.

Kadu, A. P., & Sable, A. S., (2003). Commercial flower production of tuberose cv. Single as influenced by various level of nitrogen and phosphorus. *National Symposium on Recent Advances in Indian Floriculture*.

Nagaraju, H. T., Narayanagowda, J. V., Rajanna, M. P., & Venkatesha, J., (2003). Influence of different levels of N, P and K on growth, flowering and shelf life of tuberose (*Polianthus tuberosa* L.) (cv. Double). *National Symposium on Recent Advances in Indian Floriculture, 18*.

Narendra Chaudhary, Kishan Swaroop, Janakiram, T., Biswas, D. R., & Geeta Singh, (2013). Effect of integrated nutrient management on vegetative growth and flowering characters of gladiolus. *Indian J. Hort., 70*(1), 156–159.

Pooja Gupta, Sunila Kumari, & Dikshit, S. N., (2012). Response of African marigold (*Tagetes erecta* L.) to integrated nutrient management. *Annals of Biology,. 28*(1), 66–67.

Ramachandrudu, M., & Thangam, M., (2009). *Production Technology of Gladiolus in Goa.* Technical Bulletin No. 20.

Saurabh Jha, Sharma, G. L., Dikshit, S. N., Patel, K. L., Tirkey, T., & Sarnaik, D. A., (2012). Effect of vermicompost and FYM in combination with inorganic fertilizer on growth, yield and flower quality of gladiolus (*Gladiolus hybridus*).

Sharma, J. R., Gupta, R. B., Panwar, R. D., & Kaushik, R. A., (2003). Growth and flowering of gladiolus as affected by N and P levels. *Journal of Ornamental Horticulture (New Series), 6*(1), 76–77.

Singh, P. V., Kumar, V., & Kumar, R., (2006). Effect of organic manures and inorganic fertilizers on flowering of gladiolus cv. American Beauty. "*National Symposium on Ornamental Bulbous Crops*". Held on 5–6 December, 2006 at S.V.B.P.U. of Ag.& T., Modipuram, Meerut (U.P.), 110–111.

CHAPTER 29

EFFECT OF PLANT GROWTH REGULATORS ON GROWTH, FLOWERING, AND CORM PRODUCTION OF GLADIOLUS (*GLADIOLUS GRANDIFLORUS*) CV. JESTER

V. M. PRASAD, DEVI SINGH, and BALAJI VIKRAM

Department of Horticulture, Sam Higginbottom Institute of Agriculture, Technology and Sciences, Allahabad, U.P., India, E-mail: devisinghaaidu@gmail.com

CONTENTS

ABSTRACT

The field experiment entitled "Effect of Plant Growth Regulators on Growth, Flowering and Corm Production of Gladiolus (*Gladiolus grandiflorus*) cv. Jester" was planned and conducted during the *rabi* season of 2014–2015 at the Research Farm of Department of Horticulture at Sam Higginbottom Institute of Agriculture Technology and Sciences (Deemed-to-be-University), Allahabad, Uttar Pradesh, India, with two plant growth regulators (GA_3 and CCC) at different concentrations on growth and spike yield of gladiolus. The experiment consisted of nine treatments with three replications carried out in a randomized block design. The results obtained from the experiment show that the combination of different hormone source of plant growth regulators significantly affected the growth parameters of gladiolus, such as plant height (124.46 cm), number of sprouts (2.75), number of leaves per plant (9.5), days to spike initiation (52.28 days), days to opening of the first florets (8.02 days), first florets durability (10.95 days), spike length (90.62 cm), number of florets/spike (19.02), number of spike/plant (2.68), spike yield/ha (1,50,000 lakh), corm per plant (2.75), corm per plot (30.65), and corm yield/ha (1,50,002.95 lakh). The spike yield attributes of gladiolus were also influenced significantly by combination of different plant growth regulators. The maximum value of yield and yield attributes parameters, viz., maximum number of spike per plant, spike yield/ha, number of corms/plant, and corm yield/ha was found to be higher under the treatment T_2 (GA_3 @ 150 ppm) and T_6 (CCC @ 750 ppm).

29.1 INTRODUCTION

Flowers are one of god's most beautiful boon to mankind that bring joy and happiness to one and all; they are symbol of beauty, love, and tranquility. They form the soul of garden and convey the message of nature to man. The gladiolus crop that grows in the subtropical condition is more popular among the flower growers because of its easy cultivation and higher value for cut flowers.

Botanically, Gladiolus (*Gladiolus grandiflorus*) belongs to the family Iridaceae with chromosome number n=15 and originated from South Africa. A thick stem with blossoms is located on one side of the stem. Blossoms may be plain, fringed, or ruffled, and forms may vary from trumpet shaped to rosebud-like to tulip shaped.

Plant growth regulators are the organic chemical compounds that modify or regulate physiological processes in an appreciable measure in plants when used in small concentrations. They are readily absorbed and move rapidly through tissues when applied to different parts of the plant.

It has generally been accepted that many plant processes including senescence are controlled through a balance between plant hormones interacting with each other and with other internal factors (Mayak et al., 1980). Although growth retarding chemicals did not increase the number of flowers, they produced flowers with compact shape, developed short stalk, and flowers remained fresh for a longer period, and they suppressed the height of the plant (Bhattacharjee et al., 1974). It is known fact that the application of growth regulators such GA3, NAA, CCC and MH had positive effects on growth and development of gladiolus plants at different concentrations. The reports indicate that the growth and yield of gladiolus was enhanced by the application of GA3 (Peanav et al., 2005; Vijai et al., 2007), NAA (Kumar et al., 2008), CCC (Patel et al., 2010; Ravidas et al., 1992), and MH (De et al., 2002).

29.2 METHOD AND MATERIALS

A field experiment entitled "Effect of Plant Growth Regulators on Growth, Flowering, and Corm Production of Gladiolus (*Gladiolus grandiflorus*) cv. Jester" was carried out in the experimental field of the Department of Horticulture, Sam Higginbottom Institute of Agricultural, Technology and Sciences (formerly known as Allahabad Agriculture Institute Deemed University, AAI-DU) during the winter season of 2014–2015. Details of the materials and methods adopted are discussed in this chapter.

29.2.1 GEOGRAPHICAL LOCATION OF THE EXPERIMENTAL SITE

The experimental site is situated at a latitude of North and longitude of 60° 3′ East and at altitude of 78 m above mean sea level (MSL).

29.2.2 CLIMATIC CONDITION OF THE EXPERIMENTAL AREA

The area of Allahabad district comes under the sub-tropical belt in the southeastern Uttar Pradesh, which experience extremely hot summer and fairly cold winter. The maximum temperature of the location reaches up to 46°C–48°C and seldom falls as low as 4°C–5°C. The relative humidity ranged between 20% and 94%. The average rainfalls in this area are around 900 mm annually. The meteorological data (October 2014 to March 2015) with respective to total rainfall, maximum and minimum temperature, relative humidity are presented in Table 29.2.

29.3 EXPERIMENTAL SITE

The experiment was carried out at the horticulture research farm, Department of Horticulture, Allahabad School of Agriculture, Sam Higginbottom Institute of Agriculture, Technology and Sciences, Allahabad, Uttar Pradesh.

29.4 TREATMENTS DETAILS

Recommended dose of fertilizer (RDF) for Gladiolus is 100-120-80 kg of nitrogen (N), phosphorus (P), and potassium (K) per hectare (Table 29.1).

29.4.1 ANALYSIS OF VARIANCE (ANOVA)

In the present experiment, randomized block design (RDB) was applied. The analysis of variance technique was applied for drawing conclusions from the data. The calculated value of F was compared with tabulated value at 5% level of probability for the appropriate degree of freedom.

TABLE 29.1 Treatment Details

No of treatments	Details
T_0 (Control)	RDF
T_1	GA3 @ 100 ppm
T_2	GA3 @ 150 ppm
T_3	NAA @ 100 ppm
T_4	NAA @ 150 ppm
T_5	CCC @ 500 ppm
T_6	CCC @ 750 ppm
T_7	IBA @ 100 ppm
T_8	IBA @ 150 ppm

29.5 RESULTS AND DISCUSSION

Result presented in Table 29.2 revealed that the growth and flowering parameter of gladiolus plants were significantly altered due to the application of growth regulators.

The plant height, number of leaves, number of shoots per plant, number of shoots per corm, number of florets per spike, number of spike per plant, and yield of corms were significantly increased due to GA_3, NAA, CCC, and IBA application. CCC and IBA application significantly reduced these parameters when compared with control. The maximum plant height (70.39 cm) was observed with treatment T_2 (GA_3 @ 150 ppm) followed by T_1 (GA_3 @ 100 ppm) (69.77 cm); these growth parameters might be due to the fact that GA_3 promotes vegetative growth by inducing active cell division in the apical meristem. These findings are in consonance with the reports of Sharma et al. (2004), Kumar et al. (2008), Chopde et al. (2012), and Awasthi et al. (2012) in gladiolus. The data showed that the application of GA_3 and NAA significantly hastened flowering as compared to control. The application of GA_3 hastened the flower for abate 10 days hastening of flowering by 10 days earlier with GA_3 application might be attributed to the enhanced vegetative growth early phase due to increased photosynthesis and CO_2 fixation. Further exogenous application of GA_3 would have favored the convenience of factors influencing floral initiation, i.e., the carbohydrate pathway and photo-periodic pathway with the GA_3 pathway. The quality parameters of flowers, such as member of florets per spike and flower length, were significantly increased by the application of all growth regulators. The highest number of florets per spike (19.02) and flower length (7.25 cm) recorded in GA_3 @ 150 ppm was at par with that for GA_3 @ 100 ppm. The findings of the present study agree with those of Barman et al. (2004), Pal and Chowdhary (1998), Das et al. (1992), Patel et al. (2010), and Chopde et al. (2012) with GA_3 in gladiolus. The yield attributes related to corms and cormels are significantly increased by the application of growth retardants like CCC and IBA in all the concentration when compared to control and other growth regulators. The maximum corm yield per ha. was recorded in T_6 (CCC @ 750 ppm) (150002.95) followed by T_5 (CCC @ 500 ppm) (125002.95). These might be due to influence of growth retardants in delaying floral initiation, which would have enhanced source to sink ratio by reducing the partition of carbohydrates to floral spike; this is evident from the reduction in spike length due to CCC application when compared to control. These results are

TABLE 29.2 Effect of Plant Growth Regulators on Growth, Flowering and Corm Production of Gladiolus (*Gladiolus grandiflorus*) cv. Jester

Treatments		Plant height (cm)	No. of leaves per plant	No. of shoots per corm	No. of florets per spike	No. of spike per plant	Yield of corm (no.) per ha
T_0	RDF	62.52	5.62	1.05	13.88	1.80	80002.95
T_1	GA3 @ 100 ppm	69.77	7.82	1.45	18.22	1.95	85003.95
T_2	GA3 @ 150 ppm	70.39	8.75	2.01	19.02	2.02	90002.95
T_3	NAA @ 100 ppm	68.96	6.42	2.15	17.48	2.28	95002.95
T_4	NAA @ 150 ppm	64.44	6.59	2.23	17.05	1.98	115002.95
T_5	CCC @ 500 ppm	66.72	6.82	2.45	14.79	2.35	125002.95
T_6	CCC@ 750 ppm	64.84	6.75	2.75	15.15	2.68	150002.95
T_7	IBA @ 100 ppm	68.83	7.12	1.25	15.68	2.25	110003.95
T_8	IBA @ 150 ppm	68.07	7.28	1.95	16.48	2.28	100002.95
F-test		S	S	S	S	NS	S
S.Ed. (±)		0.768	0.786	1.130	0.575	1.058	1.254
C.D. (P = 0.05)		1.586	1.623	2.332	1.186	2.185	2.589

in accordance with the findings of Ragaa et al. (2012) in Irish plant and Patel et al. (2012) in gladiolus. From the above result, it could be concluded that the foliar application of GA_3 @ 150 ppm on 30th and 60th days after planting was most effective to obtain early flowering and highest yield of good quality spikes. The application of CCC @ 750 ppm was most effective to obtain highest yield of corms and carmels.

29.6 CONCLUSION

On the basis of the present investigation in gladiolus, it is concluded that the application of GA_3 @ 150 ppm and CCC @ 750 ppm at 30, 60, and 90 DAS had a significantly direct effect on growth and spike yield (105003.95 lakh/ha). Application of different plant growth regulator may be adopted in the agro-climatic conditions of the Allahabad region of Uttar Pradesh for obtaining the maximum yield of gladiolus. However, these results are based on 1 year of experimentation; hence, they need validation through further experimentation before formulating a recommendation.

KEYWORDS

- **CCC**
- **GA_3**
- **gladiolus**
- **yield**

REFERENCES

Arun Awasthi, Yadaw, A. L., & Singh, A. K., (2012). Effect of GA_3 on growth, flowering and corn production of gladiolus (*G. grandiflorus*) cv. Red Beauty. *Plant Archives, 12*(2), 853–855.

Barman, D., & Rajni, K., (2004). Effect of chemicals on dormancy breaking, growth, flowering and multiplication in gladiolus. *J. Orna. Horti, 7*(1), 38–44.

Bhattacharjee, S. K., Bose, T. K., & Mukhopdhyay, T. P., (1974). Experiments with growth retardants on dahlia. *Indi. J. Horti., 35*(2), 85–90.

Chopde Neha, Gonge, V. S., & Dalal, S. R., (2012). Growth flowering and corm production of gladiolus as influenced by foliar application of growth regulators. *Plant Archives, 12*(1), 41–46.

Jinegh Patel, Patel, H. C., Chavda, J. C., & Saiyad, M. Y., (2010). Effect of plant growth regulators on flowering and yield of gladiolus *(G. grandiflorus* L.)cv. American beauty. *Asian Journal of Horticulture, 5*(2), 483–485.

Kumar, S., Bhagawati, R., Kumar, R., & Ronya, T., (2008). Effect of plant growth regulators on vegetative growth, flowering and corm production of gladiolus. *J. Orna Horti, 11*(4), 265–270.

Leena, R., Rajeevan, P. K., & Valsalakumari, P. K., (1992). Effect of plant growth regulators on the growth, flowering and corm yield of gladiolus cv. Friendship. *South India Horti., 10*(6), 329–335.

Lepcha, B., Nautiyal, M. C., & Rao, V. K., (2007). Variability studies in gladiolus under mid hill conditions of Uttarakh. *J. Orna. Horti., 10*(3), 169–172.

Neha Chopde, & Gonge, V. S., (2014). *Influence of Varieties and Growth Regulators on Growth, Yield and Quality of Gladiolus Horticulture Section*, College of Agriculture, *15*(1), 215–219.

Pal, P., & Chowdhary, T., (1998). Effect of growth regulators and duration of soaking on sprouting growth, flowering and corm yield of gladiolus cv. '*Tropic Sea*'. *Horti. J., 11*(2), 69–77.

Patel Jinegh, Patel, H. C., Chavda, J. C., & Saiyad, M. Y., (2010). Effect of plant growth regulators on flowering and yield of gladiolus *(Grandiflorus L.)* cv. American beauty. *Asian J. Horti, 5*(2), 483–485.

Prakash Ved, & Jha, K. K., (1998). Effect of GA_3 on the floral parameters of gladiolus cultivars. *J. Applied Biology, 8*(2), 24–28.

Ragaa, A., Taha, (2012). Effect of some growth regulators on growth, flowering, bulb productivity and chemical composition of Irish plant. *Journal of Horticulture Science and Ornamental Plants, 4*(2), 215–220.

Rana Peanav, Kumar Jitendra, & Kumar Mukesh, (2005). Response of Ga3, plant spacing and planting depth on growth, flowering and corm production in gladiolus. *Journal of Ornamental Horticulture 8*(1), 41–44.

Sharma, D. P., Chattar, Y. K., & Gupta, N., (2006). Effect of gibberellic acid on growth, flowering and corm yield in three cultivars of gladiolus. *J. Orna. Horti, 9*(20), 106–109.

Suresh Kumar, P., Bhagawati, R., Rajiv Kumar, & Ronya, T., (2008). Effect of plant growth regulators on vegetative growth, flowering and corm production of gladiolus. *Journal of Ornamental Horticulture, 11*(4), 265–270.

Umrao Vijai, K., Singh, R. P., & Singh, A. R., (2007). Effect of gibberellic acid and growing media on vegetative and floral attributes of gladiolus. *Indian Journal of Horticulture, 64*, 73–76.

CHAPTER 30

EFFECT OF IBA AND DIFFERENT ROOTING MEDIA ON FIG CUTTINGS

MEGHA H. DAHALE, G. S. SHINDE, S. G. BHARAD, P. K. NAGRE, and D. N. MUSKE

Department of Agricultural Botany, Dr. Panjabrao Deshmukh Krishi Vidyapeeth, Akola–444104, Maharashtra, India, E-mail: meghadahale13@gmail.com

CONTENTS

ABSTRACT

Ficus is a genus of the family_Moraceae that is collectively known as figs; they are native of the tropics with a semi-warm temperate zone. The common fig (*F. carica*) is a temperate species native to southwest Asia and has been widely cultivated from ancient times for its fruit; it is also referred to as figs grown in India. In a 100-gram serving, raw figs provide 74 calories, but no essential nutrients in significant content, with all having less than 10% of the daily value. When dried (uncooked), however, 100 grams of figs supply 60 calories with the dietary mineral, manganese, and rich content of minerals and vitamin K. Figs are propagated by cuttings. It is grown in tropical and semi tropical zone so it is having scarcity of water it is directly impact on the survival of cuttings. To improve the survival rate of cuttings, the present study was carried out by using different combinations of rooting media with indole butyric acid (IBA). Here, factorial completely randomized design (FCRD) was followed with 24 different combinations comprising three treatments of IBA concentration, i.e., 500, 1000, 1500 ppm with control, replicated thrice. From the experiment, it was observed that among different concentration, 1500 ppm IBA gave maximum shoot, root growth, percentage of rooted cuttings survival percentage of rooted cutting, and root:shoot ratio over the remaining treatments. In terms of using different rooting media, soil + farm yard manure (FYM) + Cocopeat recorded better performance than using soil as a rooting media. The combined effect of IBA with rooting media was found to be significant with the maximum number of sprouts, percentage of rooted cuttings, survival percentage of rooted cuttings, and root:shoot ratio. Thus, this method can be used to develop nursery with better survival rate, and the number of plants per cuttings will be more.

30.1 INTRODUCTION

The common fig (*Ficus carica)* is subtropical and large, deciduous shrub or a small tree of 3–9 m height, with short and twisted trunk with leaves deeply 3 to 5 lobed. It contains copious milky latex. The fig is a native of Asia Minor and spread early to the Mediterranean region. It is a plant of extremely ancient cultivation and reported to be under cultivation from 3000 to 2000 BC in the eastern Mediterranean region. It has been used as a principle food in Mediterranean countries for thousands of years. It is believed that the leaves of this plant provided the first clothing. In the Book of Genesis in the Bible, Adam and Eve clad themselves with fig leaves. In India, commercial cultivation of common (edible) fig is mostly confined to western parts of Maharashtra, Gujarat, Uttar Pradesh (Lucknow and Saharanpur), Karnataka (Bellary, Chitradurga, and Srirangapatna), and Tamil Nadu (Coimbatore).

In India, "Poona fig" is the most popular cultivar grown for consumption as fresh fruit. Recently, a variety "Dinkar," an improvement over Daulatabad variety for yield and fruit quality is gaining commercial importance. Asexual propagation is the best way to maintain some species. Fig is propagated by asexual propagation, which is very useful for replicating true-to-type clonal planting material for multiplication of elite plants for plantation purpose, germplasm conservation, and introduction of fast growing species. Thus, it is the most important, fast, convenient economic propagation technique to raise superior planting material. In comparison to other methods of asexual propagation in fig, propagation by stem cutting has been the modern commercial nursery practices because as a rule fig is hard to root. Although fig can strike roots, rooting is not appreciable. Growth regulators are to be used to improve its rooting ability. Plant growth regulators usually auxin have an important role in stimulation and initiation of roots to cutting. Auxin induces root formation by breaking root apical dominance induced by cytokinin (Cline, 2000). There exists lot of contradiction with regard to optimum concentration of growth regulator treatments. Hence, it is possible that optimum use of growth regulators would help for rapid multiplication in propagating of fig cuttings.

Different concentrations of indole butyric acid (IBA) and rooting media have been studied under Vidharbha condition for quicker multiplication in nursery. Very less research work has so far been done on the propagation of fig by cuttings using plant growth regulators. Therefore, it is felt necessary to undertake the study on propagation of fig by using different concentrations of plant growth regulators under Akola condition for quicker multiplication in nursery. Considering these circumstances, the present study was carried out to investigate the effect of IBA and different rooting media on fig cutting.

30.2 MATERIAL AND METHODS

The experiment was conducted at the Main garden of Department of Horticulture, Dr. Panjabrao Deshmukh Krishi Vidyapeeth, Akola, Maharashtra, India. Dinkar cultivar was selected for the experiment with factorial completely randomized design (FCRD) with three replications of 24 treatments. The details of treatments are given in Table 30.1.

The experiment was carried out by planting cuttings in 22.0 cm × 12.5 cm-size polythene bags. The polythene bags were punctured to improve the drainage and filled with different rooting media that were prepared according to treatments. In order to prevent infestation of insect pest and infection of diseases, the plant protection scheduled followed during the course of investigation. For the cuttings of fig cv. Dinkar, figs were selected from 6-year-old healthy mother plants grown at commercial fruit nursery, Saswad, Pune. The hardwood cuttings were taken from the mature wood aged

TABLE 30.1 Treatment Combinations

T1–I0 M1	Control + Silt (alone)
T2–I0 M2	Control + Soil + FYM (1:1)
T3–I0 M3	Control + Soil + FYM + Sand (1:1:1)
T4–I0 M4	Control + Soil + FYM + Cocopeat (1:1:1)
T5–I0 M5	Control + Soil + FYM + Cocopeat + Sand (1:1:1:1)
T6–I0 M6	Control + Soil + FYM + Sand (2:1:1)
T7–I1 M1	IBA 500 ppm + Silt (alone)
T8–I1 M2	IBA 500 ppm + Soil + FYM (1:1)
T9–I1 M3	IBA 500 ppm + Soil + FYM + Sand (1:1:1)
T10–I1 M4	IBA 500 ppm + Soil + FYM + Cocopeat (1:1:1)
T11–I1 M5	IBA 500 ppm + Soil + FYM + Cocopeat + Sand (1:1:1:1)
T12–I1 M6	IBA 500 ppm + Soil + FYM + Sand (2:1:1)
T13–I2 M1	IBA 1000 ppm + Silt (alone)
T14–I2 M2	IBA 1000 ppm + Soil + FYM (1:1)
T15–I2 M3	IBA 1000 ppm + Soil + FYM + Sand (1:1:1)
T16–I2 M4	IBA 1000 ppm + Soil + FYM + Cocopeat (1:1:1)
T17–I2 M5	IBA 1000 ppm + Soil + FYM + Cocopeat + Sand (1:1:1:1)
T18–I2 M6	IBA 1500 ppm + Soil + FYM + Sand (2:1:1)
T19–I3 M1	IBA 1500 ppm + Silt (alone)
T20–I3 M2	IBA 1500 ppm + Soil + FYM (1:1)
T21–I3 M3	IBA 1500 ppm + Soil + FYM + Sand (1:1:1)
T22–I3 M4	IBA 1500 ppm + Soil + FYM + Cocopeat (1:1:1)
T23–I3 M5	IBA 1500 ppm + Soil + FYM + Cocopeat + Sand (1:1:1:1)
T24–I3 M6	IBA 1500 ppm + Soil + FYM + Sand (2:1:1)

about 1 year, and the suckers that arise from the base of the stem according to the suggestion of Purohit and Shekharappa (1985) and Saroj et al. (2008). The cutting was 4–6 bud, having approximate size 20–25 cm with 1.0–1.2 cm diameter as reported by Saroj et al. (2008) in October.

30.2.1 TREATMENT APPLICATION

Treatment-wise IBA solutions were prepared according to Hartman and Kester (1989). The required quantities of IBA powder was dissolved in 5 mL of ethyl alcohol (50%) and then required quantity of distilled water was added to make the solutions of desired concentrations. The basal 1.5–2.0 cm portion of the cuttings was dipped in different concentrations of IBA formulation for 12 hours and immediately planted in the polythene bags to a depth of 4.5–5.0 cm. After the cuttings were planted, light irrigation was applied gradually in the morning and evening with the help of a water can.

Five sprouted cuttings were selected randomly from each treatment of each replication. These five cuttings were labeled for recording observation through the study. Observations such as days required to sprouting recorded at 45 days and percentage of rooted cuttings were noted after 60 days of planting; the remaining observations regarding shoot and root parameter were recorded 150 days after planting. Survivals of rooted cuttings were recorded 150 days after planting, while the leaf area was measured at 150 days after planting.

The data collected on various observations during the course of investigation were statistically analyzed by FCRD as suggested by Panse and Sukhatme (1961).

30.3 RESULTS AND DISCUSSION

The results of the investigation were based on the various observations of fig cutting for shoot growth, root growth, percentage of rooted cuttings, survival percent of rooted cuttings, and root:shoot ratio.

30.3.1 EFFECT OF IBA ON FIG

The data presented in Table 30.1 indicated that days required to sprouting per cutting was significantly influenced by IBA. Significantly early sprouting was observed when hardwood cuttings of fig were treated with treatment I3 (18.89) followed by I2 (21.06). Treatment Io showed late sprouting as compared to other treatments (24.73). Data presented in Table 30.1 indicate that treatment 1500 ppm IBA showed early sprouting. This might be due to the fact that external application of auxins promotes growth and produce more favorable condition for sprouting of dormant buds on the cutting (Hartman and Kester, 1989). These findings are conformity with Kumar et al. (1995) in lemon cv. Baramasi when the cutting was treated with IBA 2000 ppm.

30.3.2 EFFECT OF ROOTING MEDIA

The data for days required to sprouting per cutting as influenced by different rooting media were recorded and presented in Table 30.1. Significantly early sprouting was observed when hardwood cuttings of figs were treated with treatment M4 (19.50) followed by M3 (20.75). Treatment M1 (24.17) showed late sprouting as compared to other treatments. The above results clearly indicates that hardwood cutting of fig performance in M4 Soil + farm yard manure (FYM) + Cocopeat showed minimum days required sprouting. This might be due to the fact that cocopeat is a highly porous material, and therefore, it absorbs large volume of water; thus, Cocopeat has great oxygenation properties that makes cocopeat rooting media highly suitable for fig cutting survival. Further, cocopeat can store and release nutrients to cuttings for long period of time, and therefore, it may help to make available nutrient to the cutting of fig. Cocopeat also has excellent water retention property, i.e., it requires less water and thus helps to avoid loss of nutrient through leaching (Yahya et al., 2009). With a carbon:nitrogen ratio of 104:1, cocopeat also releases nutrients to plants when mixed with soil.

30.3.3 EFFECT OF IBA AND ROOTING MEDIA ON NUMBER OF SPROUTS PER CUTTINGS

The data regarding number of sprouts per cutting of fig as influenced by different IBA concentration and different rooting media which was recorded and presented in Table 30.2 and depicted in Figure 30.1 indicated that the number of sprout was significantly influenced by the different IBA concentrations and rooting media. Increase in the number of leaves under the 1500 ppm IBA treatment might be due to the absorption of more nutrients along with moisture as compared to cuttings in all other treatments, which in turn increase the production of more number of leaves. Among the plant growth regulator treatments, 1500 ppm IBA showed significantly higher number of leaves. Similar results were observed by Chauhan and Maheshwari (1970) in peach cutting, Shukla and Bist (1994) in pear cutting, Singh and Kumar (1970) in phalsa, Upadhyay and Badyal (2007) in pomegranate, and Kucuk et al. (2010) in fig cutting when treated with IBA 2000 ppm (Figure 30.2).

TABLE 30.2 Effect of IBA and Rooting Media on Number of Sprouts in Fig

Treatments	**Number of sprouts**				**Mean**
IBA					
Rooting Media	I0	I1		I3	I3
M1	1.55	1.78	2.33	2.90	2.14
M2	1.76	2.01	2.63	3.28	2.42
M3	2.27	2.60	3.41	4.24	3.13
M4	2.92	3.34	4.38	5.45	4.02
M5	2.58	2.96	3.87	4.82	3.56
M6	1.94	2.22	2.91	3.62	2.67
Mean	2.17	2.49	3.25	4.05	-
Interaction (IBA x Rooting Media)					

I	**M**			**I x M**
F-test	**Sig.**	**Sig.**	**Sig.**	**Sig.**
SE (m) +	0.05	0.07	0.07	0.14
CD at 5%	0.16	0.21	0.21	0.41

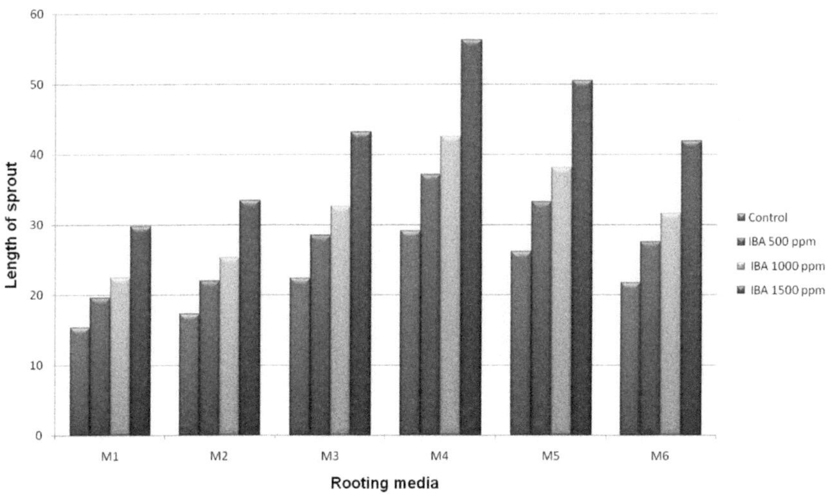

FIGURE 30.1 **(See color insert.)** Effect of IBA and rooting media on length of sprout (cm) in fig.

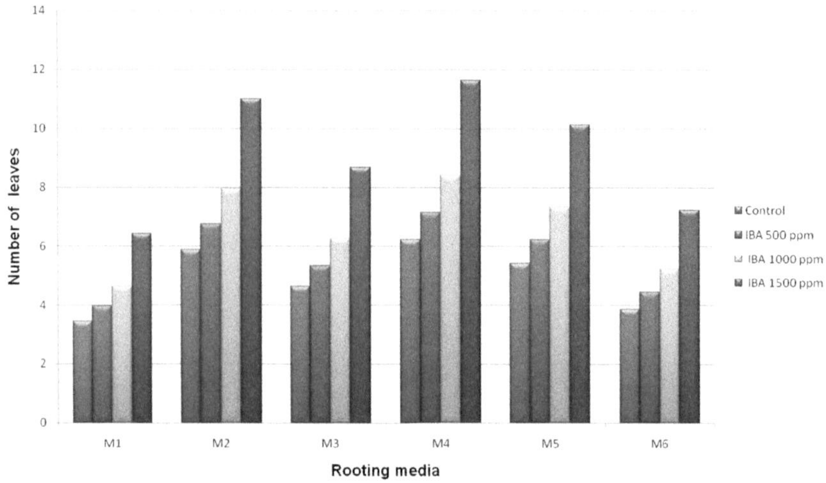

FIGURE 30.2 **(See color insert.)** Effect of IBA and rooting media on number of leaves in fig.

An experiment was conducted with FCRD with 24 treatment combinations comparing three treatments of IBA combinations, i.e., 1500, 1000, and 500 ppm IBA and control. Six treatments of rooting media The combined effect of IBA concentrations and different rooting media indicate that hardwood cuttings treated with treatment combination 1500 ppm IBA with Soil + FYM + Cocopeat as a rooting media showed the maximum shoot growth in terms of length of sprouts, number of sprouts, number of root, survival percentage of rooted cuttings, and root:shoot ratio.

KEYWORDS

- **cuttings**
- **fig**
- **growth regulators**
- **IBA**

REFERENCES

Chahuan, K. S., & Maheshwari, L. D., (1970). Effect of certain plant growth regulators, seasons and types of cutting on root initiation and vegetative growth in stem cutting of peach variety sarbati. *Indian J. Hort.*, *27*, 136–40.

Cline, M. G., (2000). Execution of the auxin replacement apical dominance experiment in temperate woody species. *American J. Bot.*, *87*(2), 182–190.

Hartman, H. T., & Kester, D. E., (1975). *Plant Propagation : Principles and Practices*, Prentice Hall Pub., pp. 623.

Kucuk, S. M., Temirkaynak, R. C., & Namal, H., (2010). *Determination of Effects of Different IBA Doses on Rooting of the Hardwood Cuttings of Some Fig Cultivars*. Batı Akdeniz Agricultural Research Institute, Antalya, Turkey.

Kumar, R., Gill, D. S., & Kaushik, R. A., (1995). Effect of indole butyric acid, P-Hydroxybenzoic acid and season on the propagation of lemon Cv. Baramasi from cuttings. *Haryana J. Hort. Sci.*, *24*(1), 13–18.

Panse, V. G., & Sukhatme, P. V., (1961). *Statistical Method of Agricultural Workers*, ICAR Publication, New Delhi, pp. 381.

Purohit, A. C., & Shekharappa, K. E., (1985). Effect of type of cuttings and IBA on rooting of hardwood cuttings of pomegranate. *Indian J. Hort.*, *42*(1–4), 30–36.

Saroj, P. L., Awasthi, O. P., Bhargava, R., & Singh, U. V., (2008). Standardization of pomegranate propagation by cutting under mist system in hot arid region. *Indian J. Hort.*, *65*(1), 0972–8538.

Shukla, G. S., & Bist, L. D., (1994). Studies on the efficiency of IBA and NAA on clonal propagation by cutting in low chilling pear rootstock. *Indian J. Hort.*, *51*(4), 351–357.

Singh Kumar, (1970). Response of some growth substance on the propagation of Phalsa. *Indian J. Hort.*, *27*, 163–164.

Upadhyay, S. D., & Badyal, J., (2007). Effect of growth regulators on rooting of pomegranate (Punica granatum L.) cutting. *Haryana J. Hort. Sci.*, *36*(1–2), 58–59.

Yahya, A., Shararom, A. S., Mohammad, B. R., & Selamet, A., (2009). Chemical and physical characteristics of cocopeat based media mixture and their effects on the growth and development of celosia cristata. *American Journal of Agricultural and Biological Sciences*, *4*(1), 63–71.

PART III

CROP IMPROVEMENT AND BIOTECHNOLOGY

CHAPTER 31

NEW DYNAMICS IN BREEDING OF HORTICULTURAL CROPS

THANESHWARI and C. ASWATH

Division of Ornamental Crops, ICAR – IIHR, Bangalore,India, E-mail: aswath@iihr.ernet.in

CONTENTS

ABSTRACT

Horticultural crops have been domesticated and further bred to satisfy human's special dietary, medicinal, or esthetic needs. Their breeding, an interdisciplinary science, is moving toward a new horizon. Improvement of the unique qualities of horticultural crops has always been a major objective in breeding of these crops. In spite of the vast diversity of horticultural crops, most of the breeding goals are the same, for example, increasing productivity, concentrations of secondary metabolites, plant tolerance to pest, shelf-life, and altering the plant architecture of fruit trees and ornamentals. Recent advancements in the field of biotechnology have created a new domain to complement the methods of horticulture crop breeding.

31.1 BREEDING OF HORTICULTURAL CROPS IN PRE-MENDELIAN ERA

One of the greatest achievement of plant breeding dated back to 10,000 years ago when people started domesticating crop plants. Among all the domesticated crop plants, the majority of crops by number are horticultural crops. Earlier, the plant breeders used to select plant from wild species having unique characteristics that allow them to survive in the wild. The main trait used in selection was adaptation, which was based on the goal of the breeder. Even today, plant breeders occasionally seek for important characteristics (e.g., biotic and abiotic stress tolerance) from the progenitors of our cultivated crop species. Luther Burbank experimented with a variety of techniques such as selection, grafting, hybridization, and cross-breeding during the last half of the nineteenth century and the starting of the twentieth century. He developed more than 800 varieties of horticultural plants, including a spineless cactus (now extinct), plumcot, Shasta daisy, July Elberta peach, Santa Rosa plum, Freestone peach, Flaming Gold nectarine, and Burbank and Russet Burbank potatoes.

31.2 PLANT BREEDING IN THE 20TH CENTURY

The rediscovery of Mendel's paper in 1905 had resulted in a genetic revolution on plant improvement. It was the time when the breeders had unconsciously been using the emerging science of genetics, especially the fusion of Mendelian and quantitative genetics. There are a number of breeding protocols that can exemplify the relationship between genetics and post-Mendelian plant breeding, such as hybrid breeding, backcross breeding, and disease resistance breeding. Many of the horticultural crops are developed by hybridization and selection. However, crop hybridization breeding has limitations that are very difficult to overcome. Hybridization cannot be conducted successfully between unrelated species. When plants are hybridized, many undesirable traits are transferred along with trait of interest. The hybridization breeding of perennial horticultural crops can take as many as 20–30 years for their improvement. New cultivars selected from natural mutations are especially productive in perennial horticultural crops. Mutagens were rapidly applied to horticulture plant breeding, and a number of varieties have been generated. However, there are some limitations with mutation breeding, such as frequency of desirable mutation is very low, and most of the mutations are recessive, deleterious and chimeric. The advances of molecular genetics have paved a wide road for modern biotechnological breeding. By knowing the details of how traits are inherited and genetically controlled, molecular biologists can create novel phenotypes through DNA recombinant technologies. Through transgenic technology, breeders can introduce desirable genes into plants, even from distant plant species or unrelated genus. A number of important varieties have been developed by transgenic approaches in horticultural crops. The first transgenic food "FlavrSavr tomato" was approved for sale in the USA in 1994 (Bruening and Lyons, 2000). GM papaya (*Carica papaya* L.), resistant to papaya ringspot virus (PRSV) has been cultivated in the USA for more than a decade (Azad et al., 2014). Although transgenic technology has achieved great success in supplementing horticultural crop breeding, this technology faces some technical and social challenges. RNA interference (RNAi) technology leads to post-transcriptional gene silencing caused by double-stranded RNA (dsRNA) molecules to prevent the expression of specific genes (Kim and Rossi, 2007). If judiciously used, RNAi may go a long way in the production of pest-resistant and nutritionally rich crops. Through RNA interference, Pons et al. (2014) enhanced the β-carotene content of orange fruit by blocking the expression of an endogenous β-carotene hydroxylase gene (Csb-CHX). Mumbanza et al. (2013) studied the in vitro antifungal activity of synthetic dsRNA molecules against *Fusarium oxysporum* f. sp. *cubense* (causing Panama disease in banana) and *Mycosphaerella fijiensis* (causing black sigatoka in banana). All the tested synthetic dsRNAs successfully stimulated the silencing of target genes and exhibit varying degrees of inhibition for spore germination of both tested banana pathogens. The field of RNAi is moving at an impressive pace and generating exciting results. As it offers a great potential in understanding gene functions and their utilization in improving crop quality and production, it is a matter of time before we see the products of this RNAi research in the farmers' fields around the world.

31.3 NEW DYNAMICS IN HORTICULTURAL CROP BREEDING

Sequencing technologies has resulted in a significant increase in the availability of genomic information for numbers of horticulture crop species. Molecular information on horticultural crops is drastically improving and many horticultural crops have been whole-genome sequenced including grapevine, papaya, strawberry, sweet orange, etc. (Bolger et al., 2014). Presently, genome-editing technologies is an advanced biotechnological tool for crop improvement. Genome editing started with the technologies like chemical mutagenesis by ethyl methane sulfonate (EMS), ionizing radiation, and use of *Agrobacterium*. These approaches lead to randomly distributed modification in the genome. In 1980s and 1990s, nuclease enzyme was discovered for targeted editing. Few agricultural plants were precisely modified using "zinc finger nucleases" (ZFNs) in 2006. "Transcription activator-like effector nucleases (TALENs)" became more famous in 2009. TALENs were found to be useful in generating heritable mutations in tomato (Lor et al., 2014). The novel biotechnological tool "clustered regulatory interspaced short palindromic repeat (CRISPR)/Cas-9" is a latest revolution in the field of genome editing. ZFN and TALENs are based on protein-DNA interactions, whereas CRISPR (Liu and Fan, 2014) is an RNA-guided DNA endonuclease system. These tools allow the precise insertion of specific genes for modification at specific genomic location without the use of foreign DNA. This may help to increase consumer acceptance of novel GM plant products. Presently, out of these three gene-editing tools, CRISPR has seen a rapid rise due to its straightforward construct design and its implementation to bacterial, animal, human, and plant systems. This tool utilizes guide RNA (gRNA) to guide Cas-9 endonuclease to generate a double-stranded break (DSB) with in a preplanned genomic sequence. This subsequently triggers homologous recombination or chromosomal mutation near the DSB site through cellular DNA repair mechanism. So far, CRISPR have been successfully demonstrated in *Arabidopsis* (Li et al., 2013), tobacco (Nekrasov et al., 2013), sorghum (Jiang et al., 2013), rice (Zhang et al., 2014), wheat (Wang et al., 2014), and tomato (Brooks et al., 2014). Brooks et al. 2014 demonstrate the CRISPR/Cas9-induced mutations in stable transgenic lines of tomato and also their heritability to next generations. Off targeting is still a major concern in plant improvement. However, advances in CRISPR/Cas technology have reduced off targeting up to 5,000-fold in nonplant systems (Tsai et al., 2014). Finally, this refinement in technology will be used in the plant system, providing more stimuli to adopt CRISPR/Cas9 for plant genome editing.

Genome editing technologies will be an advanced tool for the development of new variety with specific traits in horticultural crop plants. Studies in horticultural crops have revealed the anthocyanin biosynthetic and regulatory pathway. MYB-bHLH-WD repeat (MBW) complexes regulate the transcription of anthocyanin genes (Albert et al., 2014). Therefore, new crop varieties with high concentrations of anthocyanin can be generated by modifying some of these genes through these tools. Studies have shown the close relationship between plant architecture and phytohormones. For example, strigolactonem (SL) influences shoot branching, while gibberellin (GA) affects plant height. Thus, new horticultural crops with more branches or semi-dwarf phenotypes could be generated by the interference of the functions of these genes using these new technologies. The plant hormone ethylene is correlated with flower wilting and fruit ripening. Therefore, horticultural crops with improved shelf life could be developed by disrupting the main genes involved in the biosynthesis of ethylene. Studies in barley by Buschges et al. (1997) showed that the gene Mildew-Resistance Locus (MLO) encodes a protein that inhibits defenses against powdery mildew disease. Loss-of-function MLO alleles in barley, *Arabidopsis*, tomato, and pea lead to resistance to the powdery mildew disease (Wang et al., 2014). This highlights the application of genome-editing technologies in the modification of the MLO alleles to generate disease resistant varieties in horticultural crops. The latest genome-editing technologies, especially CRISPR/Cas, promise to be more precise to edit genes when the genome sequences for target genes are known. These technologies could be as direct and efficient as transgenic methods and could be used to generate new varieties without introducing foreign genes. Therefore, new crop varieties generated using these methods might be acceptable in countries where transgenic crops are refused by the public.

31.4 NEW THRUST AREAS FOR BREEDING AND BIOTECHNOLOGY IN HORTICULTURE CROPS

- Development of efficient transformation technologies for economically important crop plants where the source of resistance is not available in the existing germplasm.

- Discovery and utilization of new promoters for spatiotemporal expression like tissue specific, specific developmental stage, expression in response disease, insect, and environmental stimuli.
- Development of targeted gene-insertion techniques to control the site of integration.
- Development of a generally recognized as safe (GRAS) set of methodologies that would not require characterization and registration of individual genetic-insertion "events."
- Development of products with clear and significant benefits for consumers.
- Development of a collaborative public-technology and intellectual-property resource.
- Development of technology and trait-licensing packages to enable public and entrepreneurial commercialization of specialty and subsistence crops.

31.5 POSSIBLE BREAKTHROUGHS FOR NORTHEAST INDIA IN PARTICULAR (RESEARCH)

- Breeding for resistance to anthracnose, *furke*, and *chhirke* disease in large cardamom.
- Biosynthetic pathway for curcumin and capsaicin synthesis in Lakadong and Naga Miricha.
- Research on gametophytic incompatibility in pineapple and development of hybrids.
- Breeding fragrant subtropical Cymbidiums.
- Exploitation of wild and indigenous cucurbit germplasm for improvement.
- Documentation of indigenous tribal medicines of plant origin and identify the molecules using transcripts.
- Specially designed genetic reserves, gene sanctuaries and/or genetic garden should be earmarked for wild species of food value and other economic importance.
- Greater emphasis needs to be given on in situ conservation of endangered species, ex situ conservation of base collections in field gene banks, and in vitro storage and cryopreservation of important germplasm, where tissue culture protocols are available.

- Germplasm screening for processing and diversified use should be encouraged.

KEYWORDS

- **alleles**
- **breeding**
- **crops**
- **genome**
- **germplasm**
- **horticulture**

REFERENCES

Albert, N. W., Davies, K. M., Lewis, D. H., Zhang, H., Montefiori, M., & Brendolise, C. A., (2014). Conserved network of transcriptional activators and repressors regulates anthocyanin pigmentation in eudicots. *Plant Cell, 26*, 962–980.

Azad, M. A., Amin, L., & Sidik, N. M., (2014). Gene technology for papaya ringspot virus disease management. *The Scientific World J.*, 768038.

Bolger, M. E., Weisshaar, B., Scholz, U., Stein, N., Usadel, B., & Mayer, K. F. X., (2014). Plant genome sequencing – applications for crop improvement. *Curr. Op*In. *Biotech.*, *26*, 31–37.

Brooks, C., Nekrasov, V., Lippman, Z. B., & Eck, J. V., (2014). Efficient gene editing in tomato in the first generation using the clustered regularly interspaced short palindromic repeats/CRISPR-Associated9 System. *Plant Physiol.*, *166*, 1292–1297.

Bruening, G., & Lyons, J., (2000). The case of the flavr savr tomato. *Calif. Agr.*, *54*, 6–7.

Buschges, R., Hollricher, K., Panstruga, R., Simons, G., Wolter, M., & Frijters, A., (1997). The barley MLO gene: a novel control element of plant pathogen resistance. *Cell*, *88*, 695–705.

Jiang, W., Zhou, H., Bi, H., Fromm, M., Yang, B., & Weeks, D. P., (2013). Demonstration of CRISPR/Cas9/sgRNA-mediated targeted gene modification in Arabidopsis, tobacco, sorghum and rice. *Nucleic Acids Res.*, *41*,188.

Kim, D. H., & Rossi, J. J., (2007). Strategies for silencing human disease using RNA interference. *Nat. Rev. Genet.*, *8*(3), 173–84.

Li, J. F., Norville, J. E., Aach, J., McCormack, M., Zhang, D., Bush, J., Church, G. M., & Sheen, J., (2013). Multiplex and homologous recombination-mediated genome editing in *Arabidopsis* and *Nicotiana benthamiana* using guide RNA and Cas9. *Nat. Biotechnol.*, *31*, 688–691.

Liu, L., & Fan, X. D., (2014). CRISPR-Cas system: a powerful tool for genome engineering. *Plant Mol Biol., 85*, 209–218.

Lor, V. S., Starker, C. G., Voytas, D. F., Weiss, D., & Olszewski, N. E., (2014). Targeted mutagenesis of the tomato PROCERA gene using TALENs. *Plant Physiol., 166*, 1288–1291.

Mumbanza, F. M., Kiggundu, A., & Tusiime, G., (2013). *In vitro* antifungal activity of synthetic dsRNA molecules against two pathogens of banana, *Fusarium oxysporum f.sp. cubense* and *Mycosphaerella fijiensis*. *Pest Manag. Sci., 69*, 1155–1162.

Nekrasov, V., Staskawicz, B., Weigel, D., Jones, J. D., & Kamoun, S., (2013). Targeted mutagenesis in the model plant *Nicotiana Benthamiana* using Cas9 RNA-guided endonuclease. *Nat Biotechnol., 31*, 691–693.

Pons, E., Alquezar, B., Rodrıguez, A., Martorell, P., Genoves, S., Ramon, D., Rodrigo, M. J., Zacarıas, L., & Pena, L., (2014). Metabolic engineering of β-carotene in orange fruit increases its *in vivo* antioxidant properties. *Plant Biotech. J., 12*, 17–27.

Tsai, S. Q., Wyvekens, N., Khayter, C., Foden, J. A., Thapar, V., Reyon, D., Goodwin, M. J., Aryee, M. J., & Joung, J. K., (2014). Dimeric CRISPR RNA-guided FokI nucleases for highly specific genome editing. *Nat. Biotechnol., 32*, 569–576.

Wang, Y., Cheng, X., Shan, Q., Zhang, Y., Liu, J., Gao, C., & Qiu, J. L., (2014). Simultaneous editing of three homoeoalleles in hexaploid bread wheat confers heritable resistance to powdery mildew. *Nat. Biotechnol., 32*, 947–951.

Zhang, H., Zhang, J., Wei, P., Zhang, B., Gou, F., Feng, Z., Mao, Y., Yang, L., Zhang, H., & Xu, N., (2014). The CRISPR/Cas9 system produces specific and homozygous targeted gene editing in rice in one generation. *Plant Biotechnol. J., 12*, 797–807.

CHAPTER 32

PEPPERS (*CAPSICUM* SPP.): DOMESTICATION AND BREEDING FOR GLOBAL USE

S. KUMAR, H-C. SHIEH, S. W. LIN, R. SCHAFLEITNER, L. KENYON, R. SRINIVASAN, J. F. WANG, A. W. EBERT, and Y. Y. CHOU

AVRDC – The World Vegetable Center, P.O. Box 42, Shanhua, Tainan 74199, Taiwan, E-mail: sanjeet.kumar@worldveg.org

CONTENTS

ABSTRACT

The genus *Capsicum* ($2n = 24$) comprises a diverse group of plants producing pungent (hot pepper syn. chili pepper) or non-pungent (sweet pepper) fruits valued as spices and vegetables. The genus originated in the arid regions of the Andes Mountains (present-day Peru and Bolivia) and later spread to the tropical lowlands of the Americas. The domestication syndromes of *C. annuum* are a set of favorable traits selected by early domesticators from wild plants with small-round, erect, pungent, deciduous, and soft-fleshed fruits. In the past decade, highly pungent *Capsicum* landraces (e.g., Bhut Jolokia and variants, Trinidad Moruga Scorpion and variants) have been discovered, which need characterization and ex situ conservation. The availability in 2014 of full genome sequences of wild and cultivated species is expected to accelerate the pepper breeding progress. Pepper breeding at AVRDC – The World Vegetable Center – focuses on improvement of hot pepper and sweet pepper of *C. annuum* species. Between 2001 and 2013, AVRDC's genebank has provided more than 33,000 pepper seed samples (79% improved lines and 21% genebank accessions) in more than 120 countries spread across all continents. Since 2005, 12 germplasm accessions and 18 improved lines have been released as 30 open pollinated varieties for commercial cultivation after further evaluation in Armenia, Bangladesh, Ghana, India, Kazakhstan, Mali, and Uzbekistan. In addition, 34 hybrids developed using AVRDC materials as parental lines were commercialized, mostly by private seed companies in China, India, Senegal, South Africa, Sri Lanka, and Taiwan.

32.1 INTRODUCTION

Peppers [hot pepper (syn. chili or chilli) and sweet pepper (syn. capsicum or bell pepper)] are group of plants that belong to the genus *Capsicum* ($2n = 24$) of the nightshade (Solanaceae) family. As a spice and a vegetable, pepper fruit is valued for the color, flavor, and nutrition it contributes to cuisines in most tropical and subtropical regions in the world. The genus *Capsicum* is believed to have originated in the arid regions of the Andes mountains (current-day Peru and Bolivia) and later spread to the tropical lowlands of the Americas. Chili pepper is one of the oldest domesticated crops in the Western Hemisphere, perhaps dating back 10,000 years (Bosland and Votova, 2012). Columbus has been given credit for introducing peppers into Europe, Africa, and Asia. There are about 25 wild and five cultivated species of *Capsicum* (*C. annuum, C. frutescens, C. chinense, C. baccatum, C. pubescens*) originating from at least three domestication events. Except for *C. pubescens*, wild forms of the remaining four cultivated species are known.

These five cultivated species may have plants bearing pungent or nonpungent fruits with interchangeable fruit morphology, but can be distinguishable by some key features (Figure 32.1). *C. annuum* is most widely cultivated worldwide for its pungent and non-pungent fruits. Over the past 25 years, the pepper breeding program at AVRDC – The World Vegetable Center – has focused on improving both chili pepper and sweet pepper of this species by incorporating traits such as disease and insect resistance, male sterility, and heat tolerance.

This article, based on an invited presentation, broadly consists of two sections. The first section describes the versatile uses, domestication events, and recent discoveries of highly pungent pepper germplasm and the availability of the full pepper genome sequence. The second half briefly describes AVRDC's pepper breeding strategies and accomplishments, and their relevance in strengthening the global vegetable seed sector and improving the livelihoods of smallholder farmers.

32.2 DIVERSE MARKET TYPES AND USES

Fruits of various types of peppers have very high concentrations of vitamins C and A. They also contain other vitamins such as P (citrin), B1 (thiamine), and B2 (riboflavin) (Bosland and Votova, 2012). Pepper fruits contain more than 30 carotenoids including capsanthin, capsorubin, zeaxanthin, beta- carotene (the precursor of vitamin A), ascorbic acid (vitamin C), etc. More than

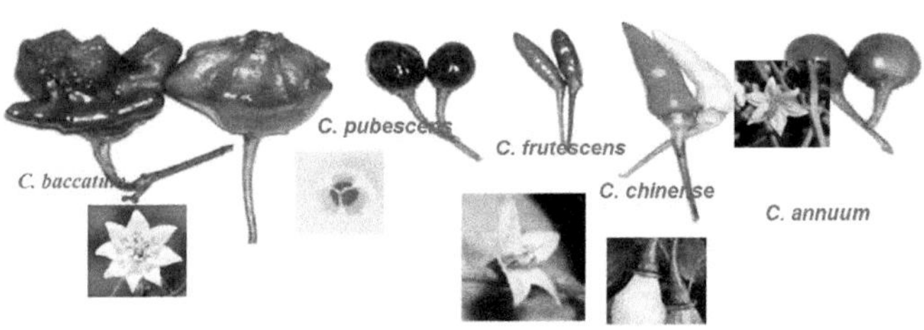

Cultivated species	Distinguishable morphology
C. annuum	White corolla and white filaments
C. frutescens	Yellow/greenish corolla and purple filaments
C. chinense	Annular constriction on pedicel attachment and yellow/greenish corolla
C. baccatum	Yellow or greenish yellow spots on corolla
C. pubescens	Hairy stems/leaves and black/brown seeds

FIGURE 32.1 (See color insert.) Distinguishable morphology of five domesticated species of *Capsicum*.

15 capsaicinoids (alkaloids) produced in the placenta of chili fruits are unique to the genus *Capsicum*. These compounds (capsaicin and dihydrocapsaicin constitute about 70% of total capsaicenoids) are responsible for the pungency (heat sensation) in pepper fruit. Vitamin C was first purified from *Capsicum* fruits in 1928 by Hungarian biochemist Albert Szent Gyorgyi, a process for which he received the 1937 Nobel Prize in physiology and medicine.

Tremendous morphological variability exists for plant type, flower morphology, fruit color, size, shape, and pungency. Based on fruit size, shape, and degree of pungency, more than 20 market types (e.g., cayenne, ancho, bell, jalapeño, pasilla, yellow wax, jwala, etc.) are commercially cultivated in different parts of the world. The breeding objectives for the quality traits of peppers can be categorized into five market types: (i) fresh market (green, red, multicolored whole fruits); (ii) fresh processing (sauce, paste, canning, pickling); (iii) dried spice (whole fruits and powder); (iv) oleoresin and capsaicin extraction; and (v) ornamental (plants and/or fruits) (Poulos, 1994).

Versatile and innovative food and nonfood uses of peppers include: (a) fresh use, of immature green fruits, mature red fruits and leaves; (b) fresh processing, for sauces, pastes, pickles, beer etc.; (c) dried spices, from mature whole fruits and powder; (d) coloring and flavoring agents, from oleoresins (carotenoids) extracts or powder; (e) ethno-botanical/traditional medicine, from fruit extracts and powders (pungent fruits); (f) modern medicine/pharmaceuticals, from extracts of capsaicinoids and carotenoids; (g) insecticides/repellents, from capsaicinoids extracts; (h) spiritual, using whole fruits, e.g. "ristras"; (i) ornamental, using whole plants or fruits; and (j) defense/punishment, using capsaicin extracts/or powder (Kumar et al., 2006).

32.3 DOMESTICATION SYNDROMES

Domestication is defined as the selection of wild species for cultivation and human use and the domestication syndrome refers to a set of desirable/favorable traits which were, or are, being selected by the domesticators (Gepts, 2014). Usually after acquiring such favorable traits, domesticated species lose the ability to grow naturally (wild) and thus need human care to complete their life cycle. Some of the most common domestication syndromes in crop species are loss of dispersal (e.g. non-shattering), increase in size (mostly of the usable part of the plant), loss of seed dormancy, loss of chemical or mechanical protection against herbivores, change in plant type (ideotypes to maximize yield), and photoperiod insensitivity (to obtain predictable yield year-round).

The wild *C. annuum* chili (bird's eye chili) typically has small-round, erect, pungent, deciduous, and soft-fleshed fruits (Figure 32.2). These traits facilitate easy seed dispersal by birds. In a well-studied case of directed deterrence of plant seed dispersal in wild chili (Tewksbury and Nabhan, 2001), secondary metabolites (capsaicinoids) in fruit function to deter fruit consumption by vertebrates that do not disperse seeds. In immature fruits/seeds, the function is simple—to deter consumption by granivores and herbivores—but in ripe fruits, the function is complex, as consumption can be either beneficial or detrimental, depending on whether the consumer disperses or destroys seeds (Levey et al., 2006). Video monitoring of wild *C. annuum* and *C. chaconese* plants revealed fruit removal only by bird species specializing in lipid-rich fruits—both chili species had fruit with unusually high lipids (35% in *C. chaconese*, 24% in *C. annuum*). These results supported Tewksbury and Nabhan's (2001) directed deterrence hypothesis in chili and suggest that through secondary metabolites (in this case, capsaicinoids) fruiting plants can distinguish between seed predators and seed dispersers (Levey et al., 2006). The major domestication syndromes of chili peppers are: (i) non-deciduous fruits (remain on the plant until harvested manually (Figure 32.2); (ii) pendent fruits (associated with size increase, better protection from sun exposure, and predation by birds); and (iii) fruit appearance and varied degree of pungency (associated with consumer preference). Peppers with diverse market types have evolved through evolution

FIGURE 32.2 (See color insert.) Fruits of wild (deciduous) and cultivated (non-deciduous) *C. annuum*.

and controlled plant breeding. It is interesting to note that due to a shortage of manual labor, breeding activities needed to develop cultivars suitable for mechanical harvesting are ongoing in several countries; breeders are now manipulating wild traits, such as deciduous and erect fruits, for easy machine harvesting.

32.4 HIGHLY PUNGENT LANDRACES

A number of highly pungent *Capsicum* landraces recently have been discovered and reported from the northeast Himalayan and Caribbean regions. In 2000, Naga Jolokia, a landrace from Assam, India was reported to be India's hottest (Mathur et al., 2000). Later, a variant of this (Bhut Jolokia) was found to be world's hottest, with a possible origin through natural hybridization between *C. chinense* and *C. frutescens* (Bosland and Baral, 2007) of a cross compatible *C. annuum* species complex (Figure 32.1). This instigated a search for highly pungent genotypes. Trinidad Scorpion, a landrace from Trinidad and Tobago, was then found to be even hotter than Bhut Jolokia, followed by Butch T Scorpion (from Australia) and most recently Trinidad Moruga Scorpion (Bosland et al., 2012). A number of close variants of Trinidad Scorpion are known from Trinidad and Tobago, e.g. Trinidad 7-Pot Jonah and Douglah Trinidad Chocolate, which are genetically distinct from Bhut Jolokia (Bosland et al., 2012). Highly pungent variants (landraces) of Bhut Jolokia are also known (Sanatombi et al., 2010; Kumar et al., 2011) and they are believed to have originated from sympatric domesticated species (Rai et al., 2013). These landraces clustered with non-*C. annuum* (*C. frutescens, C. chinense, C. baccatum*) genotypes (Rai et al., 2013) and are reproductively isolated (pre-hybridization barrier and post-hybridization barrier of hybrid breakdown) from *C. annuum* genotypes (Rai et al., 2014). The natural variability within and between these landraces (Bhut Jolokia and its variants, Trinidad Scorpion, and its variants) is currently maintained by local communities and warrants ex-situ conservation.

32.5 COMPLETE PEPPER GENOME SEQUENCE

The availability of full genome sequences of cultivated *C. annuum* CM334, Zuhla-1 and their wild progenitor Chiltepin (*C. annuum* var. *glabriusculum*) species (Kim et al., 2014; Qin et al., 2014) has provided physical maps of the large and repetitive *Capsicum* genomes, and facilitated the discovery and use of large quantities of markers for trait mapping and marker-assisted breeding. Reduced complexity sequencing approaches making use of whole genome pepper sequences as references are underway to produce high density genetic maps for tracking disease resistance loci. Comparative analyses between the whole genome sequences of pepper and other *Solanaceae* species have pinpointed key genes involved in evolution and domestication of pepper (Qin et al., 2014). Whole genome pepper sequences allow for systemic approaches to improve the horticultural, nutritional, and medicinal values of *Capsicum* species. These genome sequences have conferred unprecedented insight into the gene content of the crop, enabling the rapid identification of genes implicated with metabolic pathways, disease resistance, and other traits. With this enhanced information, we expect to make faster progress with our ongoing efforts on validation and use of markers for anthracnose resistance, leaf curl resistance, cytoplasmic male sterility, and fertility restoration traits.

32.6 PEPPER IMPROVEMENT AT AVRDC

Peppers are among AVRDC's principal breeding crops. The center's international pepper breeding program seeks to improve chili and sweet peppers with respect to resistance to anthracnose, Phytophthora blight, bacterial wilt and bacterial spot, a number of viruses, sucking insects, and broad mites.

32.6.1 DISEASE RESISTANCE

In the late 1990s, the center identified anthracnose resistance sources in two less cultivated *Capsicum* species, i.e., *C. baccatum* (PBC80/VI046804, PBC81/VI046805) and *C. chinense* (PBC932/VI047018). PBC-932 was used to develop several *C. annuum* introgression lines. Marker assisted breeding to combine resistance genes from two sources are under progress (Sowr et al., 2015). Likewise, sources resistant to various pathogens and viruses were identified and used to develop improved resistant lines (Table 32.1, Gniffke et al., 2013).

Research is ongoing to understand the temporal and spatial distribution of pathogen variations, both in genotype and virulence. Fruit anthracnose

TABLE 32.1 List of Disease-Resistant Improved Pepper Lines Developed at AVRDC (Gniffke et al., 2013)

Name of disease	Name of line
Anthracnose	HP – AVPP0205, AVPP0412, AVPP0706; AVPP0513, AVPP0514, AVPP9813, AVPP0805, AVPP0803, AVPP0903, AVPP0906, AVPP0908
Bacterial wilt	HP – AVPP9702, AVPP9703, AVPP9705; AVPP0102, AVPP0103, AVPP0104, AVPP0201, AVPP0205, AVPP0206, AVPP0307, AVPP0511
Phytophthora blight	HP – AVPP9703, AVPP9803, AVPP0302; SP – AVPP9809, AVPP0117, AVPP0504, AVPP0601, AVPP0407
Cucumber mosaic virus (CMV)	HP – AVPP0105; SP – AVPP0602
Chili venial mosaic virus (ChiVMV)	HP – AVPP0105; SP – AVPP0602
Potato virus Y (PVY)	HP – AVPP9813, AVPP9905, AVPP0012, AVPP0105; SP – AVPP9807, AVPP0006, AVPP0204, AVPP0408, AVPP0502
Leaf curl begomoviruses	HP – AVPP1127, AVPP1128, AVPP0715, AVPP0716, AVPP0906, AVPP0813, AVPP0717

HP = hot pepper; SP = sweet pepper.

caused by *Colletotrichum* species is commonly present in chili pepper production areas worldwide. Based on phenotypic and genotypic characteristics, *C. acutatum* was identified to be the predominant species causing pepper anthracnose in Taiwan, although *C. gloeosporioides*, *C. truncatum*, and *C. boninense* were present as well. *C. acutatum*, which can infect green fruit, was the predominant pathogen associated with chili pepper anthracnose in Indonesia and Fiji based on our studies (unpublished data). Moreover, two pathotypes of *C. acutatum* have been identified in Taiwan (unpublished data). The new pathotype, named CA2, could break down resistance in PBC932-derived lines. Distribution of the CA2 pathotype was limited before 2000. However, since 2000, it has been present in major pepper production areas, and has become the major pathogen in southern Taiwan. The distribution of virulence variation of *C. acutatum* and other species of *Colletotrichum* remains to be studied systemically in order to develop durable resistant lines.

Viruses of the genus *Potyvirus* are generally the most prevalent across south and southeast Asia; *Potato virus Y* (PVY), *Tobacco etch virus* (TEV) and *Chilli veinal mottle virus* (ChiVMV) have been present in Taiwan for a long time, while *Chilli ringspot virus* (ChiRSV), *Pepper veinal mottle virus* (PVMV) and *Pepper mottle virus* (PepMoV) are perhaps more recent introductions to the region. *Cucumber mosaic virus* (CMV, Cucumovirus) and the tobamoviruses, *Pepper mild mottle virus* (PMMV) and *Tomato mosaic virus* (ToMV) are present in all the pepper-growing areas of southeast Asia, though their incidences vary from season to season. As in most other vegetable crops, the whitefly-transmitted Geminiviruses (Begomoviruses) have emerged in recent years as major constraints to pepper production in many countries in south and southeast Asia, though the begomovirus species involved and the level of damage caused is generally different in each country. In Taiwan, the only begomovirus commonly infecting peppers is *Tomato yellow leaf curl Thailand virus* (TYLCTHV). Although TYLCTHV generally does not cause severe symptoms in peppers, it was used at AVRDC as part of an initial screen to identify some pepper germplasm accessions with potential begomovirus resistance for screening in the field in areas where begomovirus-induced disease is more severe (Gniffke et al., 2013). *Tomato spotted wilt virus* (TSWV) has been a sporadic pathogen of peppers in Asia for many years. However, Tospoviruses, including *Capsicum chlorosis virus* (CaCV) and *Tomato necrotic ring virus* (TNRV), have started to emerge as important in pepper production areas of several countries. In contrast, the poleroviruses *Pepper vein yellows virus* (PeVYV) and *Pepper yellow leaf curl virus* (PepYLCV) have probably both been present across south and east Asia for many years, but have been largely overlooked until now because the symptoms resemble those of nutrient deficiency and early senescence.

32.6.2 INSECT RESISTANCE

Thrips are one of the most damaging pests of pepper in the tropics. They can cause damage directly by feeding on leaves, fruits, or flowers, and also indirectly by transferring viruses, especially *Tomato spotted wilt virus* and *Capsicum chlorosis virus*. Aphids also cause severe damage in peppers. They occur in large numbers on the tender shoots and lower leaf surfaces, and suck the plant sap. Severe aphid infestations cause young leaves to curl and become deformed. Aphids also produce honeydew, which leads to the development of sooty mold that reduces the photosynthetic efficiency of the plants. Broad mite, a persistent pest of peppers, mostly causes damage on the undersides of leaves. Their feeding results in downward cupping and narrowing of the leaves. Screening of several hundred pepper accessions earlier resulted in three accessions [PBC081/VI046805 (*C. baccatum*), PBC151/VI037435 (*C. baccatum*), and PBC272 (*C. chinense*)] that are resistant to cotton aphid (*Aphis gossypii*) infestations. Additional germplasm screening resulted in four other pepper accessions with appreciable aphid resistance: PBC880/VI047003 (*C. baccatum*), and PBC18/VI046790, PBC30/VI046792, and PBC84/VI041264 (all *C. annuum*). Recently, AVRDC identified sources resistant to sucking insects including thrips and broad mite (Gniffke et al., 2013). Two accessions, VI041283/PBC145 (from India) and VI014295/C00069 (from Costa Rica) that are resistant to broad mite (*Polyphagotarsonemus latus*), aphid (*Myzus persicae*), and thrips (*Thrips palmi*) were used in breeding programs, and a set of potential improved inbred lines have been developed; several are currently being screened and evaluated in different countries.

32.6.3 MALE STERILITY

The center initiated the development of a cytoplasmic male sterility (CMS) program in chili pepper lines in 1996. To date more than a dozen pairs (with different genetic backgrounds) of CMS (Peterson's cytoplasm), maintainer and restorer inbred lines have been developed. We have also developed sweet pepper CMS and maintainer pairs. Our current focus is restorer breeding in sweet pepper, as most of the sweet pepper lines are maintainers or poor restorers. This has been a limitation for the successful use of CMS in sweet pepper hybrid development and seed production. A sequenced characterized amplified region (SCAR) marker associated with the *Rf* gene (Gulyas et al., 2006) has been validated and successfully used in marker-assisted backcrossing for the introgression of the strong *Rf* allele from hot pepper into sweet pepper genotypes (Lin et al., 2015).

32.7 GERMPLASM DISSEMINATION

The AVRDC Genebank currently has 8235 *Capsicum* spp. accessions in ex situ conservation—the largest collection in the public domain. The center disseminates seed of its germplasm accessions and improved pepper lines to public and private sector breeders, most of them working in developing countries. One of the strategies of AVRDC's pepper breeding program is to disseminate improved pepper lines through the distribution of sets of the International Chili Pepper Nursery (ICPN) and the International Sweet Pepper Nursery (ISPN). As of 2014, 23 ICPN and 12 ISPN nurseries have been released, which included 164 newly developed improved chili (90) and sweet pepper (74) lines. More than 33,000 germplasm materials were distributed from 2001–2013, comprising 81% improved breeding lines and 19% genebank accessions. The top 10 recipient countries were: India (15.6%), Republic of Korea (8.5%), Thailand (10.1%), China (7.6%), USA (7.5%), Vietnam (5.8%), Taiwan (3.8%), Indonesia (3.3%), the Netherlands (3.3%), Tanzania (1.5%), and others (33%). Except Korea and the Netherlands, all these countries imported more improved lines than genebank accessions (Figure 32.3). In addition to new elite ICPN and ISPN lines, seed samples of CMS and maintainer lines of both chili and sweet peppers were also distributed.

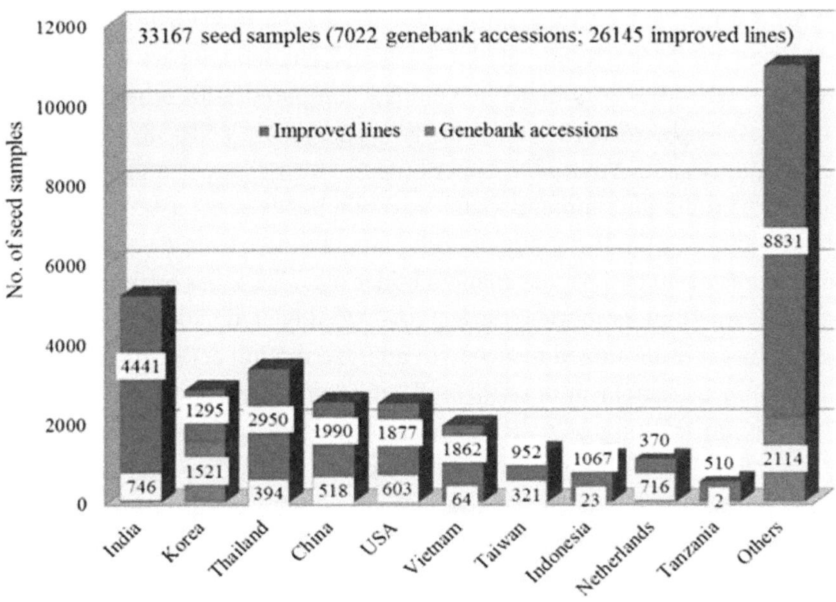

FIGURE 32.3 (See color insert.) Seed samples distributed during the period 2001-2013 (top ten recipient countries).

32.8 GERMPLASM UTILIZATION

National Agricultural Research and Extension System (NARES) and private seed companies have used AVRDC-developed improved lines and genebank accessions by: (i) directly releasing them as open-pollinated varieties through national varietal release procedures; (ii) subjecting them to selection (in cases of segregating germplasm) according to local trait preferences and subsequently releasing them as new varieties; (iii) using them (possibly after further selection and purification) as parental lines in hybrid development; or (iv) used them as sources of traits in crosses to develop new breeding lines (Lin et al., 2013).

Since 2005, 12 germplasm accessions and 18 improved lines have been released as 30 open pollinated varieties (18 chili pepper, 11 sweet pepper, 1 paprika) for commercial cultivation after further evaluation in Armenia, Bangladesh, Ghana, India, Kazakhstan, Mali, and Uzbekistan. In addition, 34 hybrids developed using AVRDC materials as parental lines were released/commercialized in China, India, Senegal, South Africa, Sri Lanka, Taiwan, and Vietnam, mostly by seed companies (Reddy et al., 2014). Exploitation of CMS lines and other improved lines for commercial hybrid development has been much higher in the private seed sector, especially in Asian countries like India (Reddy et al., 2015) and Thailand, where the vegetable hybrid seed industry is better developed compared with other developing countries. Chili cultivars such as Charmee, Jyothi, VNR-38, Rani, Vidya, Ulka, Masaya 315, Yuvraj IN, Super F_1, and Muria F_1 are examples of hybrid cultivars bred and commercialized by VNR Seeds and East-West Seeds in India and Sri Lanka using one or both parental line/s from AVRDC. An estimated area of 30,000 ha was planted under these varieties by approximately 100,000 smallholder farmers during 2012 (Reddy et al., 2014). Recently, the center has commissioned a survey to analyze the impact of AVRDC's supplied pepper germplasm in south and southeast Asian countries.

Informal tracking of AVRDC germplasm distribution and use suggests that a major multiplying impact of the center's improved lines and genebank accessions has been achieved in developing countries. Training has enhanced breeders' skills and improved seed distribution systems. Breeding and seed dissemination efforts, along with capacity building of human resources, will remain AVRDC's priority for increasing the reach and impact of the center's improved pepper lines.

ACKNOWLEDGMENTS

The authors thank the donor community for its continuous support of the center's pepper research for development.

KEYWORDS

- **chili**
- **disease resistance**
- **germplasm**
- **hot pepper**
- **male sterility**
- **sweet pepper**

REFERENCES

Bosland, P. W., & Baral, J. B., (2007). 'Bhut Jolokia'—the world's hottest known Chile pepper is a putative naturally occurring inter specific hybrid. *Hort. Science, 42*, 222–224.

Bosland, P. W., & Votava, E. J., (2012). *Peppers, Vegetable and Spice Capsicum. 2nd ed.;* CABI, Wallingford, UK.

Bosland, P. W., Coon, D., & Reeves, G., (2012). 'Trinidad Moruga Scorpion' pepper is the world's hottest measured chile pepper at more than two million scoville heat units. *Hort. Technology, 22*, 535–538.

Gepts, P., (2014). The contribution of genetic and genomic approaches to plant domestication. *Current Opinion in Plant Biology, 18*, 51–59.

Gniffke, P. A., Shieh, S. C., Lin, S. W., Sheu, Z. M., Chen, J. R., Ho, F. I., Tsai, W. S., Chou, Y. Y., Wang, J. F., Cho, M. C., Roland, S., Kenyon, L., Ebert, A. W., Srinivasan, R., & Kumar, S. Pepper research and breeding at AVRDC – The World Vegetable Center. In Proc. XV EUCARPIA Meeting on Genetics and Breeding of Capsicum and Eggplant, Turin, Italy, pp. 305–311.

Gulyas, G., Pakozdi, K., Lee, J. S., & Hirata, Y., (2006). Analysis of fertility restoration by using cytoplasmic male-sterile red pepper (*Capsicum annuum* L.) lines. *Breeding Science, 56,* 331–334.

Kim, S., Park, M., Yeom, S. I., Kim, Y. M., Lee, J. M., Lee, H. A., et al., (2014). Genome sequence of the hot pepper provides insights into the evolution of pungency in *Capsicum* species. *Nature Genetics, 46,* 270–278.

Kumar, S., Kumar, R., & Singh, J., (2006). Cayenne/American pepper (*Capsicum* species). In: Peter, K. V., (Ed.), *Handbook of Herbs and Spices*, Woodhead Publishing: Cambridge, UK. vol. *3,* pp. 299–312.

Kumar, S., Kumar, R., Kumar, S., Singh, M., Rai, A. B., & Rai, M., (2011). Incidences of leaf curl disease on *Capsicum* germplasm under field conditions. *Indian Journal of Agricultural Science, 81,* 187–189.

Levey, D. J., Tewksbury, J. J., Cipollini, M. L., & Carlo, T. A., (2006). Field test of the directed deterrence hypothesis in two species of wild chili. *Oecologia, 150*, 61–68.

Lin, S. W., Chou, Y. Y., Shieh, H. C., Ebert, A. W., Kumar, S., Mavlyanova, R., Rouamba, A., Tenkouano, A., Afari-Sefa, V., & Gniffke, P. A., (2013). Pepper (*Capsicum* spp.) germplasm dissemination by AVRDC – The World Vegetable Center: An overview and introspection. *Chronica Horticulturae, 53,* 21–27.

Mathur, R., Dangi, R. S., Dass, S. C., & Malhotra, R. C., (2000). The hottest chilli variety in India. *Current Science*, *79,* 287–288.

Poulos, J. M., (1994). Pepper breeding (*Capsicum* spp.): achievements, challenges and possibilities. *Plant Breeding Abstracts, 64*, 143–155.

Qin, C., Yu, C., Shen, Y., Fang, X., Chen, L., Min, J., et al., (2014). Whole-genome sequencing of cultivated and wild peppers provides insights into *Capsicum* domestication and specialization. *Proceedings of the National Academy of Sciences, 111*, 5135–5140.

Rai, V. P., Kumar, R., Kumar, S., Rai, A., Kumar, A., Singh, M., Singh, S. P., Rai, A. B., & Paliwal, R., (2013). Genetic diversity in *Capsicum* germplasm based on microsatellite and random amplified microsatellite polymorphism markers. *Physiology and Molecular Biology of Plants, 19*, 575–586.

Rai, V. P., Kumar, R., Singh, S. P., Singh, S., Kumar, S., Singh, M., & Rai, M., (2014). Monogenic recessive resistance to pepper leaf curl virus in *Capsicum. Scientia Horticulturae, 172*, 34–38.

Reddy, M. K., Kumar, S., Kumar, R., Srivastava, A., Chawda, N., Ebert, A. W., & Vishwakarma, M., (2014). Chilli (*Capsicum annuum* L.) breeding in India: an overview. *SABRAO J. Breeding and Genetics*, *46*, 160–173.

Reddy, M. K., Srivastava, A., Lin, S. W., Kumar, R., Shieh, H. C., Ebert, A. W., Chawda, N., & Kumar, S., (2015). Exploitation of AVRDC's chili pepper (*Capsicum* spp.) germplasm in India. *J. Taiwan Society Horticultural Sciences*, *61*, 1–9.

Sanatombi, K., Sen-Mandi, S., & Sharma, G. J., (2010). DNA profiling of *Capsicum* landraces of Manipur. *Scientia Horticulturae, 124*, 405–408.

Suwor, P., Thummabenjapone, P., Sanitchon, J., Kumar, S., & Techawongstien, S., (2015). Phenotypic and genotypic responses of chili (*Capsicum annuum* L.) progressive lines with different resistant genes against anthracnose pathogen (*Colletotrichum* spp.). *European J. Plant Pathology*, *143*, 725–736.

Tewksbury, J. J., & Nabhan, G. P., (2001). Directed deterrence by capsaicin in chillies. *Nature, 412*, 403–404.

CHAPTER 33

DETECTION OF *PHYTOPHTHORA* USING MOLECULAR AND IMMUNOLOGICAL APPROACHES IN *CITRUS* SPP.

S. J. GAHUKAR,[1] DEEPA N. MUSKE,[1] AMRAPALI A. AKHARE,[1] R. G. DANI,[1] and R. M. GADE[2]

[1]*Biotechnology Center, Department of Agricultural Botany, Dr. Panjabrao Deshmukh Krishi Vidyapeeth, Akola–444104, Maharashtra, India.*

[2]*Associate Dean Vasantrao Naik College of Agricultural Biotechnology, Yavatmal, Maharashtra, India, E-mail: sj_gahukar@yahoo.com*

CONTENTS

ABSTRACT

Citrus is one of the most popular fruit crops of the world. It is cultivated in around 140 countries, including many in the Asia-Pacific. During recent years, there has been a significant increase in the citrus production mainly on account of its increased use as a nutritious and health drink; however, with increasing area, its production has not increased because of several reasons including diseases and pests. Among all diseases, *Phytophthora* causes 80% to 90% economic losses to grower. It has an extensive host range infecting different crops and weeds. For better control management, there is a need of correct diagnosis. The present study was designed to develop correct diagnosis methods. Soil samples were collected from an infected field from different locations. From the samples, *Phytophthora* was isolated and purified using PARPH-CMA medium. The rDNA was amplified using the universal primer, i.e., ITS, and the amplicons were sequenced, and further primers were designed using primer-3 and validated using the PCR method; we found a set of primer that can detect only *Phytophthora* from soil and infected root samples. Antiserum was raised using rabbit and used for standardization of ELISA. The titer was fixed at 1:50 of antigen, 1:16000 of primer antibody, and 1:2000 of secondary antibody dilution. While proceeding with work, we were able to diagnosis lowest density of *Phytophthora* in the field. As well as using this approach field soil had been tested for presence of *phytophthora*. This diagnosis method helps in the identification and screening of diseased planting material from nursery. For better orchard, there is a need for soil as well as planting material testing for disease control management. Thus, this is very quick, accurate and sensitive method for the detection of *Phytophthora*, which is a national problem for citrus growing industry.

33.1 INTRODUCTION

The genus *Citrus* is economically very important and is known for its juice and pulp throughout the world. It is an important group of plants that has influenced the life of the people throughout the world. They have become a part of our daily food, and different fruits of the citrus group are used in different parts of the world.

India ranks 6th position in the production of citrus fruit cultivation in the world. From the last 10 years, the citrus-growing area has increased about 20.3% and production by 20.9%. Mandarin sweet oranges, lemons, and lime are major citrus fruits taken in India. At present, citrus production is about 11.14 million MT with the maximum share of sweet oranges, followed by mandarin and acid lime. In India, there are 26 states involved in citrus production, but 9 states cover more than 90% of area with 89% of total production. Maharashtra ranks first in the growing area, but its production is fourth and its productivity is far below. The Vidarbha region of Maharashtra has the largest area and production under Nagpur mandarin plantation in India (CCRI, 2015). The low productivity is because of diseases and pest attack. Citrus diseases can be broadly categorized as fungal, bacterial, viral, and mycoplasma-like disease. These include fungal diseases like gummosis (causal organism: *Phytophthora species*), pink disease, powdery mildew, citrus scab, anthracnose or wither tip, twig blight or citrus, and sooty mold; bacterial diseases like citrus canker; and viral diseases. The most damaging ones among these are Gummosis and Tristeza virus that cause serious losses in production (Francesco et al., 2015).

An extensive survey of citrus orchards and citrus nurseries in India has been conducted under the National Network Project on Phytophthora disease of Horticulture Crops (Citrus) funded by Indian Council of Agricultural Research during 1997–2002. Citrus cultivation belt of Vidarbha and Marathwada region of Maharashtra, Punjab, Madhya Pradesh, Andhra Pradesh, and northeastern states of India were surveyed to assess the impact of *Phytophthora* diseases. Two mating type (A1 and A2) of *Phytophthora parasitica* (*nicotianae*) have been recorded from the orchards of Nagpur mandarin in Nagpur district (Naqvi, 2000b, 2002b). *P. parasitica* Dastur has been a common species associated with citrus disease in Assam (Chowdhury, 1951) and in northwest India (Bajwa, 1941; Paracer and Chahal, 1962). Kapoor and Bakshi (1967) reported that a 14–18% decline in sweet orange was due to foot rot in Abohar (Punjab).

An extensive roving survey was conducted in different districts of Marathwada region of Maharashtra state to isolate the pathogen associated with the gummosis of sweet orange. In all 103 sweet orange orchards surveyed, the average disease incidence of 38.83% had been observed. Highest disease incidence and severity noticed in Nanded district (63.38%), followed by Jalna (58%), Parbhani (50.66%), Hingoli (46.66%), Aurangabad (39.11%), and Latur (14%). The lowest disease incidence was noticed in Osmanabad district (10%). The peak period of disease expression was August–September that was concomitant with heavy rainfall, high humidity percentage, and temperature range of 18°C–35°C.

There are number of host and pathogen factors which, together with features of their interactions, make phytophthora diseases so troublesome in the wet tropics. One of the important factors to consider is that the genus *Phytophthora* does not belong to the fungal kingdom. It is an Oomycete, closely related to diatoms, kelps and golden-brown algae in the Kingdom Stramenopila (Beakes, 1998). These organisms thrive in the environments found commonly in the wet tropics. There are a number of additional reasons why *Phytophthora* diseases cause so much damage in the tropics. All *Phytophthora* species need high humidity for sporulation and the germination of sporangiospores and zoospores to initiate infections. Frequent or seasonal heavy rainfall, and high levels of humidity, are common throughout the tropical lowlands. Tropical highlands have the added problem of heavy mist and dew during the morning and/or late afternoon, producing free water throughout the night and providing almost daily opportunities for sporangiospores to be formed, transported, and start new infections (Morris and Gow, 1993).

The most important species include *Phytophthora parasitica* Dastur (*P. nicotianae*), *P. citrophthora*, and *P. palmivora. P. parasitica* has been reported to cause citrus diseases in India (Kamat, 1927; Uppal and Kamat, 1936; Kumbhare and Moghe, 1976, 1982; Naqvi, 1988; Naqvi, 2000b) and Florida (Zitko and Timmer, 1994); Widmer et al., 1998; Graham and Menge, 2000). Other *Phytophthora* spp. viz. *P. boehmeriae, P. cactorum, P. cinnamon, P. citricola, P. citrophthora, P. dreschleri, P. hibernalis, P. megasperma, P. palmivora*, and *P. syngiae* have been reported pathogenic on citrus from different Citrus growing areas of World. There are about 60 species in the genus *Phytophthora*, all of them plant pathogens. *Phytophthora*, the "plant destroyer," is one of the most destructive genera of plant pathogens in temperate and tropical regions, causing annual damages of billions of dollars. Losses due to this disease vary in kind and quality from place to place and time to time or season to season, resulting in increased cost of production and reduced yield of citrus fruits very drastically. More than 20% plants die due to this pathogen in citrus nurseries of central India where 7–8 million citrus plants are being propagated every year. Rest of the apparently healthy plants from such infected nurseries becomes the primary source of spread of this pathogen to new areas (Das, 2009).

For control management, there is need of appropriate diagnosis of disease. However, for a healthy status, it is better to prevent disease than controlling it. Recently, many methods are developed to detect *Phytohphthora*. In this sense, Martin et al. (2002) reported a comparison among hybridization probes, RT-PCR, and serological techniques, concluding that ELISA was the most sensitive and universal method for pathogen detection.

It is also important to recognize that *Phytophthora* populations are not uniform across the orchard. For routine propagule density determinations, samples should be collected at random in the orchard. Propagule densities vary considerably between sample times and populations may show seasonal trends of abundance. For detection and quantification of *Phytophthora parasatica, P. citrophthora*, and *P. palmivora*, many mediums had been used, viz., Modified Kerr's medium (Hendrix and Kuhlman, 1965), MacCain medium (McCain et al., 1967), Masago medium (Masago et al., 1977), PVPH medium (Tsao and Guy, 1977), PARPH medium (Mitchell et al., 1986), and BHMPVR medium (Bist and Nene, 1988). The number of selective agar media containing antibacterial antibiotics and selective fungal agents has been proved successful in *Phytophthora* isolations from soil and plant tissues (Eckert and Tsao, 1960; Kuhlman and Hendrix, 1965; Flowers and Hendrix, 1969; Tsao and Ocana, 1969; Sneh, 1972; Fugisawa and Masago, 1975; Masago et al., 1977; Tsao and Guy, 1977; Kueh and Khew, 1982;). Leaflets of *Albizzia falcatarea* were used as baits to isolate *Phytophthora palmivora*.

Many species of *Phytophthora* can now be identified using their unique DNA sequences in the internal transcribed spacer (ITS) regions of their nuclear ribosomal DNA (rDNA) repeat region. Recently, polymerase chain reaction (PCR) has been used extensively for the detection of plant pathogens. There are several advantages of PCR-based detection methods over the traditional diagnosis methods; for example, no culturing is required and it is more reliable, sensitive, and rapid. The PCR-based methods include conventional PCR and quantitative real-time PCR. These methods have been reported as being effective for a number of *Phytophthora* species. Nested PCR, a two-step system in which the first round PCR product is subjected to a second PCR amplification, is widely used to detect fungal plant pathogens due to its greater specificity. The sensitivity of this technique also allows the detection of the target pathogen in minute amounts of infected material. This method, however, is time consuming, expensive, and labor intensive as it requires separate PCR reactions and additionally increases the risks of false positive due to cross-contamination. Multiplex PCR approach is a tool to detect several pathogens simultaneously and reduce the time spent, costs, and contamination risk. Multiplex PCR utilizes a single reaction with several primer pairs, each generating an amplicon that can be separated and visualized through electrophoresis. This technique has been extensively applied in plant pathology as it permits the identification of more than one target in a single reaction (Chirapongsatonkul et al., 2015).

Phytophthora is a major problem for growing citrus industry not only in Maharashtra but worldwide; thus, there is a need to develop an appropriate diagnosis method for control management. *Phytophthora* is a pseudofungus that requires different management practices than other fungi. For the control of *Phytophthora*, there is a need of correct diagnosis in early stage with accurate and less expensive method. Keeping this objective as priority, this study was carried out using the below-mentioned methodology.

33.2 MATERIALS AND METHODS

33.2.1 SOURCE OF ISOLATES AND DNA EXTRACTION

Phytophthora spp. isolated from soil samples were collected from various citrus growing orchards in the Vidarbha region of Maharashtra.

Soil samples were collected from rhizosphere of each plant at 5 to 30 cm depth along with feeder roots. The samples were diluted by adding 20 cc soil in 80 mL of sterile distilled water and plated on a selective medium plate containing pimaricin, ampicillin, rifampicin, penta-chloronitrobenzene, and himexazol (PARPH) at a specific concentration of each antibiotic into corn meal agar medium (Kannwischer and Michell, 1978). Typical muddy colonies of *Phytophthora* were observed on specific medium, and these colonies were then purified on PARPH-CMA selective plate. *Phytophthora* colonies were maintained on corn meal agar plate and in sterile distilled water vials for long-term storage.

For DNA extraction, *Phytophthora* were grown in 20 mL culture of a sucrose/asparagine/mineral salts broth containing 30 μg ml^{-1} β-sitosterol (Elliott et al., 1966). After vacuum filtration, the mycelium was freeze-dried for extended storage at −20°C. To extract total DNA, 10–20 mg of dry mycelia were crushed with liquid nitrogen and then suspended in 1:2 of extraction buffer (200 mM Tris-HCl [pH 8], 250 mM NaCl, 25 mM EDTA, 0.2% CTAB), vials were kept in water bath for 60 min at 65°C and centrifuged at 13,000 rpm. The supernatant precipitated with 800 μL of phenol/chloroform/isoamyl alcohol (25:24:1) and centrifuged at 13,000 g for 5 min. The aqueous phase was extracted twice with 800 μL of phenol/chloroform/isoamyl alcohol (25:24:1) and 700 μL of chloroform/isoamyl alcohol (24:1), respectively. DNA was precipitated with an equal volume of isopropanol for 1 h at 4°C, and the pellet was then washed with 70% cold ethanol (−20°C), dried, resuspended in sterile distilled water, and stored at −20°C. For routine amplifications, DNA was diluted to 50 ng/µL and maintained at 4°C deep freeze.

33.2.2 PCR AMPLIFICATION

To ensure the quality of template DNA, all extracts were amplified by PCR with universal primers, i.e., ITS1-ITS4 combination. PCR reactions were performed in a total volume of 25 µL containing 50 ng of genomic DNA, 10 mM Tris–HCl (pH 9), 50 mM KCl, 0.1% Triton X-100, 100 µM dNTPs, 1 mM $MgCl_2$, 1 unit of *Taq* polymerase (Taq DNA polymerase, Promega Corporation, WI, USA) and 2 µM of primers. The PCR reaction was incubated in a programmable thermal cycler starting with 5 min denaturation at 95°C, followed by 35 cycles at 95°C for 30 s, annealing at 56°C for 45 s, and extension at 72°C for 1 min. A negative control (no template DNA present in PCR reaction) was included in all experiments. Amplicons were analyzed by electrophoresis in 2% agarose gels in TBE buffer (Sambrook et al., 1989) and visualized by staining with ethidium bromide. The amplified products were sent for sequencing, and the sequenced products were analyzed using a sequencer analyzer, with specific primers designed for *Phytophthora* diagnosis using Primer-3 software.

33.2.3 ANTISERUM PRODUCTION

New Zealand white rabbit was injected with five 0.5 mL virus injections at weekly intervals. The first was with equal volume of Freund's complete adjuvant and the subsequent ones with Freund's incomplete adjuvant. Test bleed was done after the fourth injection, and after a 10-day gap, the final injection was given, and final bleeding was done at an interval of 10 days. The serum was separated after overnight incubation at room temperature, followed by centrifugation, and antiserum was separated (van Regenmortel, 1982; Muske et al., 2014).

33.2.4 STANDARDIZATION OF INDIRECT AND DIRECT DOT ENZYME-LINKED IMMUNOSORBENT ASSAY (DIBA)

Indirect and Direct Dot-ELISA was standardized using 2 μL of suitably diluted antigen (crude mycelia extract), following the procedure described by Timmer (1993) with few modifications. The antiserum dilutions used for both direct and indirect DIBA were 1:500, 1:1000, 1:2000, 1:4000, 1:8000, 1:16,000, and 1:32,000 and the dilutions of antigen used were 1:10, 1:20, 1:50, and healthy control; the secondary antibody was used at a dilution of 1:1000 and 1:2000.

33.2.5 STANDARDIZATION OF DIRECT AND INDIRECT PLATE ENZYME-LINKED IMMUNOSORBENT ASSAY (ELISA)

Plate ELISA was standardized on polystyrene plates by following a modified protocol described by Clark and Adams (1977). The antiserum dilutions used for both direct and indirect plate ELISA were 1:500, 1:1000, 1:2000, 1:4000, 1:8000, 1:16000, and 1:32000, and the dilutions of antigen used were 1:10, 1:20, 1:50, and healthy control with secondary antibody dilutions of 1:1000 and 1:2000. The optical density (OD) readings were recorded at 405 nm.

33.3 RESULTS AND DISCUSSION

For disease management purposes, it is frequently desirable to know whether *Phytophthora* spp. are present and if so, at what level. Fruit and leaf baiting methods were developed many years ago for the detection of *Phytophthora spp.* (Klotz et al., 1958; Grimm and Alexander, 1973). These are usually considered qualitative tests, but propagule densities can be quantitized by use of more complex techniques. They are relatively simple and require minimal equipment and supplies.

The *Phytophthora* species isolated from different locations from diseased citrus orchard from symptomatic plants are shown in Figure 33.1; they were confirmed with morphological test, and it was found that *Phytophthora nicotianae, P. palmivora,* and *P. parasitica* are present in selected locations.

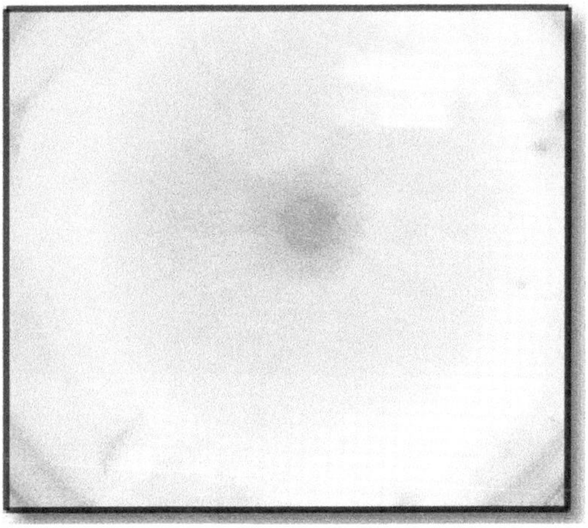

FIGURE 33.1 **(See color insert.)** Culture plate of *Phytophthora* spp.

PCR was carried using ITS1-ITS4 primer and the PCR profile is shown in Figure 33.2. The PCR products were sequenced by Eurofines, Hyderabad. On the basis of 98% sequence similarity with NCBI database, specific primers were designed from the available sequence and further validated with *Phytophthora* as shown in Figure 33.3. The designed primer can detect only *Phytophthora* spp. at minimal density.

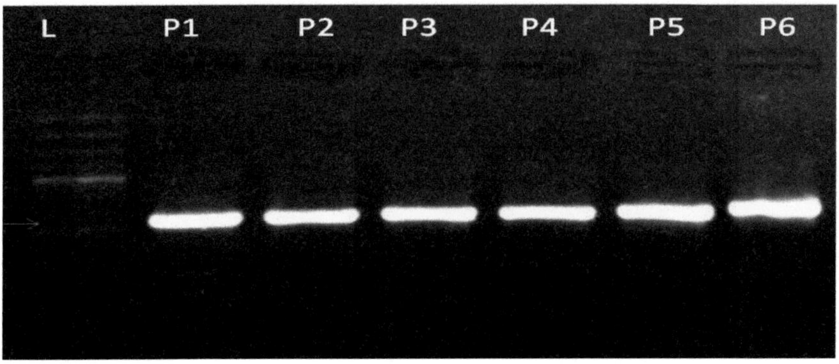

FIGURE 33.2 Amplification profiling of isolates of Phytophthora from different locations using ITS1-ITS4 primer (L-100 bp Ladder, P1-Nagpur, P2-Varud, P3- Katol, P4-Amravati, P5-Achalpur, P6-Akola).

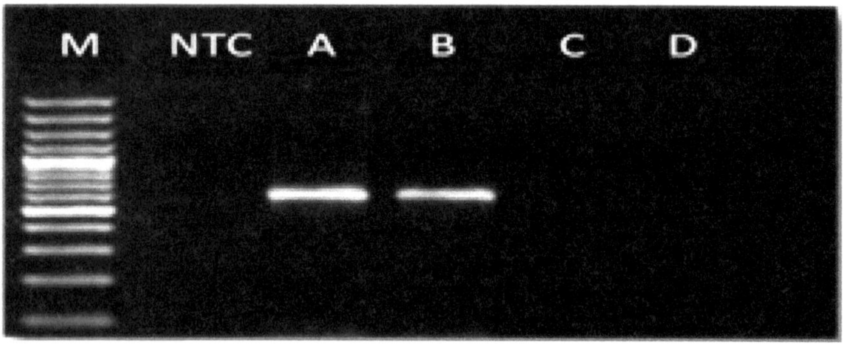

FIGURE 33.3 PCR products amplified with a specific primer pair designed using previous sequence of amplified product, i.e., ITS1-ITS4 (Lane 1-marker, 2-non-template control, A-pure culture of P. nicotianae B-infected soil sample, C-infected soil sample with Fusarium, D-Sterile soil sample).

Direct and indirect Dot-ELISA was standardized by dotting suitably diluted primary antibody and antigen in coating buffer onto nitrocellulose membrane, respectively. A titer of 1:32000 of primary antibody, 1:50 of antigen, and 1:2000 of secondary antibody was found appropriate to detect *Phytophthora* in direct and indirect dot ELISA with no color seen in control as shown in Figure 33.4.

In the plate ELISA standardization experiment, color development was observed in all combinations of the primary antibody, antigen, and secondary antibody dilutions. A titer of 1:50 dilution of antigen, 1:16000 of primary antibody and 1:2000 of secondary antibody dilution was fixed for indirect plate ELISA with an optimum OD reading of 1.016 at 405 nm. No color was observed in control as shown in Figure 33.5.

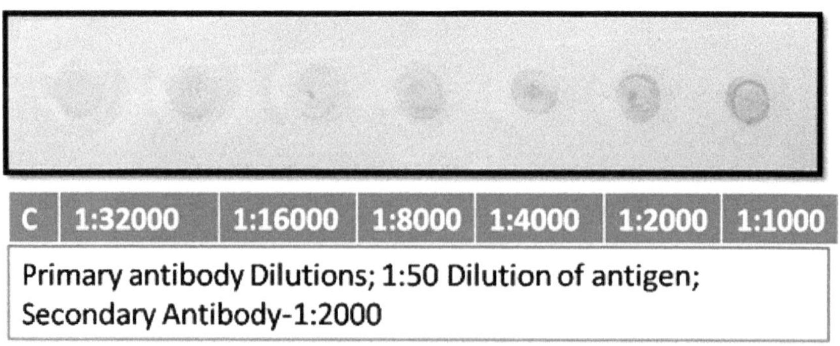

FIGURE 33.4 **(See color insert.)** Standardized DOT-ELISA for *Phytophthora* detection.

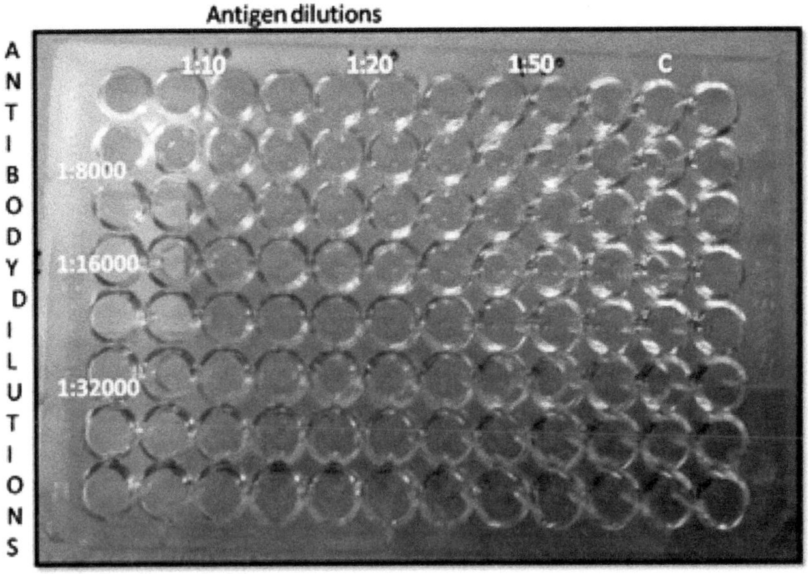

FIGURE 33.5 (See color insert.) Standardized DOT-ELISA for *Phytophthora* detection.

Genus-specific polyclonal and monoclonal antibodies were produced against *Phytophthora nicotianae*. Diagnostic kits using enzyme-linked immunosorbent assay (ELISA) were developed for the detection of *Phytophthora* spp. that enables to detect a broad range of *Phytophthora* belonging to this family in roots and in soil debris. The ELISA method is highly sensitive and can detect the presence of *Phytophthora* at lower population densities than dilution plating onto selective media (Miller and Martin, 1988; Skaria and Miller, 1989; Miller et al., 1990). Theoretically, the tissue baiting and the ELISA soil assay should be the most sensitive because they utilize larger volumes of soil. However, in one study, the leaf baiting and selective media were about equally sensitive (Timmer et al., 1993). Although it is possible to quantify *Phytophthora* spp. populations, the significance of those populations remains yet to be determined. Populations may vary considerably during a season, suggesting a need to sample repeatedly through the year as well as from year-to-year.

Future work will be focused on characterizing and grouping of strains and development of immunodiagnostic strips for the rapid detection of *Phytophthora* in the field and in nursery that controls spread of inoculum to healthy orchard.

KEYWORDS

- **citrus diseases**
- **ELISA (enzyme linked immuno-sorbent assay)**
- **gummosis**
- **ITS (internal transcribed spacer)**
- **PCR**
- ***Phytophthora* spp.**

REFERENCES

Bajwa, B. S., (1941). Gummosis in Fruit trees. *Punjab Fruit J.,* 889–891.

Bist, V. S., & Nene, Y. L., (1988). A selective medium for Phytophthora causing pigeon pea blight. *Pigeon Pea Newletter*, *8*, 12–13.

CCRI, (2015). Souvenir, brain storming dialogue on viral and greening diseases of citrus: *Challenges and Way Forward.* Central Citrus Research Institute, Nagpur, India.

Chirapongsatonkula, N., U-taynapunb, K., Chanwunc, T., & Churngchowa, N., (2015). Development of a multiplex PCR assay for rapid and simultaneous detection of rubber tree pathogens *Phytophthora* spp. and *P. palmivora,* [Online early access]. doi: 10.2306/scienceasia1513–1874.2015.41.170. Published Online, *Science Asia*, *41*, 170–179.

Chowdhary, S., (1951). Gummosis of Citrus in Assam. *Indian J. Agric. Sci.*, *16*, 570–571.

Clark, M. F., & Adams, A. N., (1977). Characteristics of the microplate method of enzyme-linked immunosorbent assay for the detection of plant viruses. *J. Gen. Virol.*, *34*, 475–483.

Das, A. K., (2009). *Fungal Disease in Citrus and its Management.* Integrated orchard management in citrus. NRCC, Nagpur, 53–57.

Eckert, J. W., & Tsao, P. H., (1960). A preliminary report on the use of pimaricin to the isolation of *Phytophthora* spp. from root tissues. *Plant Dis. Rep.*, *44*, 660–661.

Elliott, C. G., Hendrie, M. R., & Knights, B. A., (1966). The sterol requirement of *Phytophthora cactorum*. *Journal of General Microbiology*, *42*, 425–435.

Flowers, R. A., & Hendrix, J. W., (1969). Gallic acid procedure for the isolation of *Phytophthora parasitica* var. *nicotianae* and *Pythium* spp. from soil. *Phytopathology.*, *59*, 725–731.

Francesco, A. D., Costa, N., Plata, M. I., & Marıa, L. G., (2015). Improved detection of citrus psorosis virus and coat protein-derived transgenes in citrus plants: comparison between RT-qPCR and TAS-ELISA. *J. Phytopathol,* doi: 10.1111/jph.12392 1–10.

Fugisawa, Y., & Masago, H., (1975). Investigations on selective medium for the isolation of *Phytophthora* spp. *Ann. Phytopathol. Soci. Japan.*, *41*, 267.

Graham, J. H., & Menge, J. L., (2000). *Phytophthora* induces disease In: Timmer, L. W., Garnesy, S. M., Graham, J. H., editors) "Compendium of Citrus diseases", *American Phytopathological Society, St. Paul, M. N.*, pp. 12–15.

Grimm, G. R., & Alexander, A. F., (1973). Citrus leaf pieces as traps for *Phytophthora parasitica* from soil slurries. *Phytopathology*, *63*, 540–541.

Hendrix, F. F., & Kuhlman, E. G., (1965). Factors affecting direct recovery of *Phytophthora cinnamomi* from soil. *Phytopathology, 55*, 1183–1187.

Jagtapa, G. P., Dhavalea, M. C., & Deya, U., (2012). Symptomatology, survey and surveillance of citrus gummosis disease caused by *Phytophthora* spp. *Scientific Journal of Agricultural*, *1*(1), 14–20.

Kamat, M. N., (1927). The control of mosambi gummosis. *Agric. J. India, 22*, 176–179.

Kannwischer, M. E., & Mitchell, D. J., (1978). The influence of a fungicide on the epidemiology of black shank of tobacco. *Phytopathology*, *68*, 17–60.

Kapoor, S. P., & Bakshi, J. C., (1967). Foot rot or gummosis a serious disease of Citrus orchards. *Punjab Hort. J.*, *7*, 85–89.

Klotz, L. J., De Wolfe, T. A., & Wong, P. P., (1958). Decay of fibrous root rot. *Phytopathology, 48*, 616–622.

Kueh, T. K., & Khew, K. L., (1982). Survival of *Phytophthora palmivora* in soil and after passing through alimentary canals of snails. *Plant Dis.*, *66*, 897.

Kuhlman, E. G., & Hendrix, E. F., (1965). Direct recovery of *P. cinnamomi* from the field soil (Abst.). *Phytopathology, 55*, 500.

Kumbhare, G. B., & Moghe, P. G., (1976). Leaf fall disease of Nagpur orange caused by *Phytophthora nicotianae* var Parasitica Waterhouse. *Curr. Sci.*, *45*, 561–562.

Martın, S., Alioto, D., Milne, R. G., Guerri, J., & Moreno, P., (2002). Detection of citrus psoriasis virus in field trees by direct tissue blot immunoassay in comparison with ELISA, symptomatology, biological indexing and cross-protection tests. *Plant Pathol.*, *51*, 134–141.

Masago, H., Yoshikawa, M., Fukada, M., & Nakanishi, N., (1977). Selective inhibitation of *Pythium* spp. from soils and plants. *Phytopathology, 67*, 425–428.

McCain, A. H., Holtsman, O. V., & Trujillo, E. E., (1967). Concentration of *Phytophthora cinnamon* chlamydospores by soil sieving. *Phytopathology, 57*, 1134–35.

Miller, S. A., & Martin, R. R., (1988). Molecular diagnosis of plant disease. *Annu. Rev. Phytopathol.*, *26*, 409–432.

Miller, S. A., Rittenburg, J. H., Petersen, F. P., & Grothaus, G. D., (1990). Development of modem diagnostic tests and benefits to the fanner. In: Schots, H., editor. Monoclonal antibodies in agriculture. center for agricultural publishing and documentation, Wageningen, *The Netherlands*, pp. 15–20.

Mitchell, D. J., Kannwischer-Mitchell, M. E., & Zentmyer, G. A., (1986). Isolating, identifying, and producing inoculums of *Phytophthora* spp. In: Hickey KD, editor. Methods for evaluating pesticides for control of plant pathogens. *American Phytopathological Society*, 63–66.

Morris, P. F., & Gow, N. A. R., (1993). Mechanism of electrotaxis of zoospores of phytopathogenic fungi. *Phytopathology, 83*, 877–882.

Muske, D. N., Anitha Peter, Swetha, Phadnis, S., Jingade, P., & Satish, K. K., (2014). Molecular and serological detection of papaya ringspot virus infecting papaya (*CARICA PAPAYA*). *J. Pl. Dis. Sci. Vol.*, *9*(1), 8–15.

Naqvi, S. A. M. H., (1988). Prevalence of phytophthora species pathogenic to citrus in orange gardens of Vidarbh, Maharashtra, *Indian J. Myco. Pl. Pathol.*, *18*, 274–279.

Naqvi, S. A. M. H., (2000). Distribution of phytophthora spp. And mating types pathogenic to citrus in Vidarbha and Marathwada region of Maharashtra and Northeastern states of India. In: Singh, S., Gosh, S. P., editors. *Proceeding of the "Hi-Tech Citrus Management- Int. Symposium, Citriculture*, Nagpur india", pp. 1073–1080.

National Horticultural Board Database, http://nhb.gov.in. (Accessed Jan 2015).

Paracer, C. S., & Chahal, D. S., (1962). Diseases of root, crown and trunk of Citrus. *Punjab Hort. J., 2*, 92–95.

Sambrook, J., (1989). '*Molecular Cloning: A Laboratory Manual*.' (Eds J Sambrook, EF Fritsch, T Maniatis). (Cold Spring Harbor Laboratory Press: Cold Spring Harbor, USA).

Skaria, M., & Miller, S. A., (1989). A rapid test for detecting *Phytophthora* sp. in citrus. *J. Rio. Grande Valley Hort. Soci., 42*, 63–65.

Sneh, E., (1972). An agar medium for the isolation and macroscopic recognition of *Phytophthora* spp. from soil on dilution plates. *Canadian J. Microbiol., 18*, 1890–1892.

Timmer, L. W., Menge, J. A., Pond, E., Miller, S. A., & Johnson, E. L. V., (1993). Comparison of ELISA techniques and standard isolation methods for *Phytophthora* detection in citrus orchards in Florida and California. *Plant Disease, 77*, 791–796.

Tsao, P. H., & Guy, S. O., (1977). Inhibition of *Mortierella and Pythium* in a *Phytophthora* isolation medium containing hymexazol. *Phytopathol, 67*, 796–801.

Tsao, P. H., & Ocana, G., (1969). Selective isolation of species of *Phytophthora* from natural soils. *Nature Lond, 223*, 636–638.

Uppal, B. N., & Kamath, M. N., (1936). Gummosis on citrus in Bombay. *Indian J. Agric. Sci., 6*, 803–820.

Van Regenmortel, M. H. V., (1982). *Serology and Immunochemistry of Plant Viruses*. Academic Press, New York. 204–205.

Widmer, T. L., Graham, J. H., & Mitchell, D. J., (1988). Histological comparison of fibrous root infection of disease tolerant and susceptible citrus hosts by *Phytophthora nicotianae* and *Phytophthora palmivora*. *Phytopathology, 88*, 389–395.

Zitko, S. E., & Timmer, L. W., (1994). Competitive parasitic activities of *Phytophrhra* parasitica and *P.* palmivora on fibrous roots of citrus. *Phytopathology, 84*, 1000–1004.

CHAPTER 34

DISCRIMINATION OF *PHYTOPHTHORA* AS A DEVASTING FUNGUS OF CITRUS ORCHARD FROM OTHER FUNGI USING MOLECULAR APPROACH

DEEPA N. MUSKE, S. J. GAHUKAR, AMRAPALI A. AKHARE, and R. G. DANI

Dr. Panjabrao Deshmukh Krishi Vidyapeeth, Akola–444104, Maharashtra, India, E-mail: muskedeepa@gmail.com

CONTENTS

ABSTRACT

Citrus is one of the major economic horticultural crops of India. In recent years, the production of citrus is greatly affected because of diseases caused by *Phytophthora*, Citrus Tristeza Virus (CTV), etc. Among them, *Phytophthora* causes 80% to 90% economic losses to grower. It has an extensive host range infecting different crops and weeds. It is very difficult for control management of this pathogen because of its pseudofungus nature. For control management, there is a need of sensitive and accurate diagnosis of the disease. In this study, internal transcribed spacer (ITS) regions are used to discriminate *Phytophthora* from other fungi because ITS region sequences are much more variable, even among very closely related organisms. More than 25 restriction enzymes are used for the digestion of the PCR product. By using restriction digestion of the amplified PCR product of ITS4-ITS6, it was found that there is clear-cut discrimination where it is possible to identify the presence of *Phytophthora* from a bulk of fungi samples. Further, specific amplicon sequence analysis was made for designing specific primer for *Phytophthora*. This primer was used for the detection of *Phytophthora* in the bulk of fungi isolates. By applying this accurate diagnosis method, the grower can follow appropriate control management practices. Further, there will not be mismanagement in disease control. This method is applied to soil as well as nursery plant to detect infection of *Phytophthora*. It is easy for a pathologist to diagnosis disease and develop accurate control management practices.

34.1 INTRODUCTION

Citrus fruits belong to the family *Rutaceae* that are grown all over the world and have numerous therapeutic properties like anticancer, antitumor, and anti-inflammatory. Citrus fruits are cultivated all over the world in around 140 countries including Asia-Pacific countries. India ranks sixth in the production of citrus fruit cultivation in the world. Other major citrus producing countries are Spain, USA, Japan, South Africa, Israel, Brazil, Turkey and Cuba. Citrus occupies third after mango and banana in the cultivation. Citrus fruits are grown in tropical and subtropical regions of southeast Asia, particularly India and china.

In India, major citrus producing states are Andhra Pradesh, Maharashtra, Punjab, Madhya Pradesh, etc., and more than 10 states are involved in citrus production of about 11.14 million MT. In Punjab, citrus occupies 39.19 hectares with annual production of 7.34 million MT. Kinnow occupies 54.9% of the area under citrus. Maharashtra is the second largest producer of citrus (15.79%) in the country and produces 1.76 million MT (out of 11.15 million

MT) of citrus from an area of 0.28 million ha (out of 1.077 million ha) having productivity 6.4 MT/ha (national productivity: 10.3 MT/ha) (NHB, 2015). Maharashtra covers about 26% of the total citrus producing area, among which 60% to 80% area is infected with *Phytophthora* species that cause gummosis (Muske et al., 2015).

Citrus diseases can be broadly categorized as fungal, bacterial, viral, and mycoplasma-like diseases. Among these, fungal diseases are gummosis (causal organism: *Phytophthora* species), pink disease, powdery mildew, citrus scab, anthracnose or wither tip, twig blight or citrus, sooty mold; bacterial diseases like citrus canker; and viral disease. The most damaging ones among these are gummosis and Tristeza virus that cause serious losses in production (Francesco et al., 2015).

The major diseases of concern at the time are *Phytophthora* gummosis, root rot, and CTV. Consequently, trees were grafted onto the highly adaptable *Phytophthora*-resistant sour orange rootstock. Fungi of the genus *Phytophthora* are worldwide known as primary parasites of fine roots and causal agents of root and collar rots and bark cankers on young and mature specimens of hundreds of tree and shrub species. Many *Phytophthora* species belong to the most aggressive and most important plant pathogens of the world. From a global perspective, more than 66% of all fine root diseases and more than 90% of all collar rots of woody plants are caused by *Phytophthora* species. However, *Phytophthora* is very often not detected, leading to wrong diagnoses. The reasons are mainly based on the specific life cycle of *Phytophthora* that requires specific detection methods. Soil-borne *Phytophthora* species are also known as causal agents of root rots and damping-off diseases of seedlings of many deciduous and conifer species, in particular in nursery environment (Fletcher et al., 2001).

The most important species include *Phytophthora parasitica Dastur (P. nicotianae), P. citrophthora,* and *P. palmivora. P. parasitica* has been reported to cause citrus diseases in India (Kamat, 1927; Uppal and Kamat, 1936; Kumbhare and Moghe, 1976, 1982; Naqvi, 1988, 2000b) and Florida (Graham and Menge, 2000; Widmer et al., 1998; Zitko and Timmer, 1994). Other *Phytophthora spp.* viz. *P. boehmeriae, P. cactorum, P. cinnamon, P. citricola, P. citrophthora, P. dreschleri, P. hibernalis, P. megasperma, P. palmivora,* and *P. syngiae* have been reported pathogenic on citrus from different citrus growing areas of the world. There are about 60 species in the genus *Phytophthora*, and all of them are plant pathogens. *Phytophthora*, the "plant destroyer," is one of the most destructive genera of plant pathogens in temperate and tropical regions, causing annual damages of billions of dollars.

Recently, for quick and reliable identification of *Phytophthora* species, internal transcribed spacers (ITS)-based identification have been developed to identify known species of *Phytophthora* and unknown isolates quickly (Lee et al., 1993; Ristaino et al., 1994; Cooke et al., 1996, 2000; Crawford et al., 1996; Cooke and Duncan, 1997; Levesque et al., 2004). An identification website has also been developed to facilitate *Phytophthora* researchers, where user-generated ITS digest profiles of any unidentified taxa may be entered to the search page for comparison with the database. However, some species remain as polyphyletic assemblages awaiting more rigorous analysis (Braiser and Hanson, 1992). There is a need to optimize PCR assays for the detection of different propagule types in soil, if these assays are to be used for precision agriculture application (Ristaino and Gumpertz, 2000). Molecular detection methods of *Phytophthora* and their impacts on plant disease management have been reviewed (Martin et al., 2000).

Phytophthora is a major problem for growing citrus industry not only in Maharashtra but also worldwide; hence, there are needs to develop an appropriate diagnosis method for control management. *Phytophthora* is a psuedofungus that requires different management practices. For control of *Phytophthora*, there is currently a need of correct diagnosis because mismanagement can lead to total destruction of the field by this devastating fungus. Hence, this research was conducted to discriminate *Phytopthora* from other fungi that are majorly present in the citrus orchard by using the following material and methodology.

34.2 MATERIALS AND METHODS

34.2.1 SOURCE OF ISOLATES AND DNA EXTRACTION

Phytophthora spp. isolated from soil samples were collected from various citrus growing orchards in the Vidarbha region of Maharashtra (Nagpur, Varud, Katol, Amravati, Achalpur, and Akola districts).

Soil samples were collected from the rhizosphere of each plant at 5 to 30 cm depth along with feeder roots. The samples were diluted by adding 20 cc soil in 80 mL of sterile distilled water and plated on a selective medium plate containing pimaricin, ampicillin, rifampicin, penta-chloronitrobenzene, and himexazol (PARPH) at a specific concentration of each antibiotic into corn meal agar medium (Kannwischer and Michell, 1978). Typical muddy colonies of *Phytophthora* were observed on the specific medium, and these colonies were then purified on PARPH-CMA selective plate. *Phytophthora* colonies were maintained on corn meal agar plate and in sterile distilled water vials for long-term storage. The majorly found fungus species in citrus orchard isolated in this research program are *Fusarium* spp., *Rhizoctonia* spp., *Trichoderma* spp., *Aspergillus* spp., *Alternaria* spp., and *Sclerotium* spp. used in comparative study with *Phytophthora.*

For DNA extraction, *Phytophthora* were grown in 20 mL culture of a sucrose/asparagine/mineral salts broth containing 30 μg ml^{-1} β-sitosterol (Elliott et al., 1966). After vacuum filtration, the mycelium was freeze-dried for extended storage at −20°C. To extract total DNA, 10–20 mg of dry mycelia were crushed with liquid nitrogen and then suspended in 1:2 of extraction buffer (200 mM Tris-HCl [pH 8], 250 mM NaCl, 25 mM EDTA, 0.2% CTAB), vials were kept in water bath for 60 min at 65°c and centrifuged at 13,000 rpm. The supernatant precipitated with 800 μL of phenol/chloroform/isoamyl alcohol (25:24:1) and centrifuged at 13,000 g for 5 min. The aqueous phase was extracted twice with 800 μL of phenol/chloroform/isoamyl alcohol (25:24:1) and 700 μL of chloroform/isoamyl alcohol (24:1),

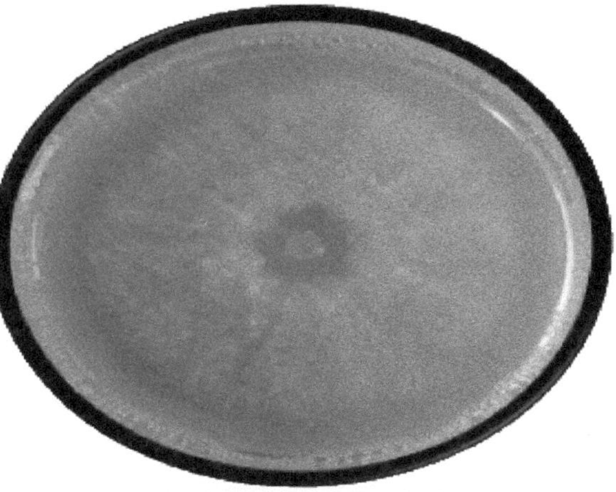

FIGURE 34.1 **(See color insert.)** Pure culture plate of *Phytophthora nicotianae.*

respectively. DNA was precipitated with an equal volume of isopropanol for 1 h at 4°C, and the pellet was then washed with 70% cold ethanol (–20°C), dried, resuspended in sterile distilled water, and stored at −20°C. For routine amplifications, DNA was diluted to 50 ng/μL and maintained at 4°C deep freeze.

34.3 PCR AMPLIFICATION

To ensure the quality of template DNA, all extracts were amplified by PCR with universal primers, i.e., ITS1-ITS4 combination. PCR reactions were performed in a total volume of 25 μL containing 50 ng of genomic DNA, 10 mM Tris–HCl (pH 9), 50 mM KCl, 0.1% Triton X-100, 100 μM dNTPs, 1 mM $MgCl_2$, 1 unit of *Taq* polymerase (Taq DNA polymerase, Promega Corporation, WI, USA) and 2 μM of primers. The PCR reaction was incubated in a programmable thermal cycler starting with 5 min denaturation at 95°C, followed by 35 cycles at 95°C for 30 s, annealing at 56°C for 45s, and extension at 72°C for 1 min. A negative control (no template DNA present in PCR reaction) was included in all experiments. Amplicons were analyzed by electrophoresis in 2% agarose gels in TBE buffer (Sambrook et al., 1989) and visualized by staining with ethidium bromide.

Polymerase chain reaction (PCR) amplification of the ITS region of the template DNA was performed using the abovementioned primer set. Because of the nondiscriminating nature, the PCR products were digested further with restriction enzymes (protocol used according to that given by the manufacturer). The digested PCR products were run on PAGE for better resolution and easy detection. The results are shown in Figure 34.2, and the discriminated band were eluted from the gel and submitted for sequencing to Chromus Biotech (Bangalore) for further genomic research work.

34.4 RESULTS AND DISCUSSION

For disease management purposes, it is frequently desirable to know whether *Phytophthora* spp. are present and if so, at what level. Fruit and leaf baiting methods were developed many years ago for the detection of *Phytophthora spp.* (Klotz et al., 1958; Grimm and Alexander, 1973). These are usually

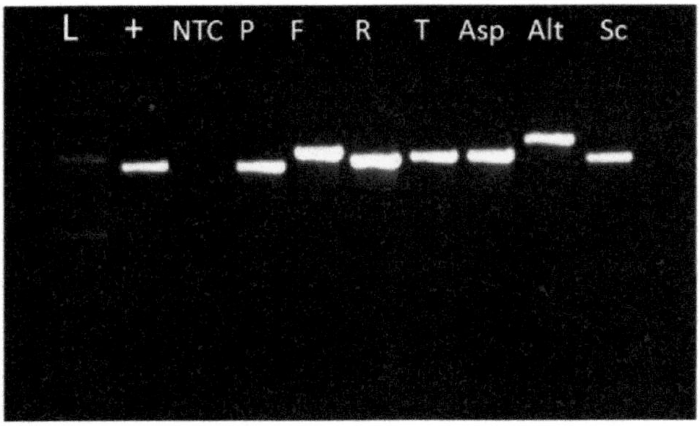

FIGURE 34.2 Amplification of the ITS4-ITS6 region.

considered qualitative tests, but propagule densities can be quantitized by use of more complex techniques. They are relatively simple and require minimal equipment and supplies.

The most commonly used gene targets for the development of species-specific primers are the ribosomal RNA ITS. This is because they can be amplified using universal primers without any prior sequence knowledge and because they contain regions of low intraspecific and high interspecific variability, which are useful for design of species-specific primers (Bruns et al., 1991; Das et al., 2011). The rRNA genes are also present in multiple copies of 100s to 1,000s depending on the species, thereby making the test more sensitive than detection of single copy genes. A comparison of the sensitivities of tests based on different genes shows that those based on ITS regions are more sensitive, although the sensitivities of the different ITS tests span a very wide range (0.1–25,000 fg), perhaps reflecting the influence of primer design (O'Brien, 2008; O'Brien et al., 2009).

PCR would definitely improve the diagnosis of *Phytophthora* species infecting citrus, and in view of the wide host range of *P. nicotianae*, it could be applied to other host pathogen combinations as well. Considering the fact that *Phytophthora* diseases are related to soil inoculums of the pathogens, it could be interesting to develop a protocol that able to estimate the number of propagules of the pathogen using real-time PCR. Here, in this study, the PCR technique followed by restriction enzyme digestion was followed for the identification of *Phytophthora* from the citrus field. The primer used was universal primer that could amplify the ITS region of fungi. The primer was run with *Phytophthora* and other six more prominently present fungi found in citrus orchard as shown in Figure 34.1. To discriminate them, the PCR product was digested with 23 restriction enzymes, of which six enzymes could discriminate *Phytophthora* from other fungi, as shown in Figure 34.2. Basically, the restriction enzymes *Sma*I, *Bsu*I, *Hinf*I, *Msp*I, *Hpa*I, *and Taq*I showed distinct bands that helped in the isolation and correct diagnosis of *Phytopthora* from other fungi (Figure 34.3).

Thus, by using this restriction product, specific primers were designed for *Phytophthora* identification and correct diagnosis to follow accurate control management practices. We need to understand the taxonomy of *Phytophthora* so that we can ensure the specificity of the detection test, e.g., tests for the detection of *Phytophthora* from very closely related fungi. These aspects of the biology can help future research for the accurate diagnosis of *Phytophthora* and thus avoid mismanagement.

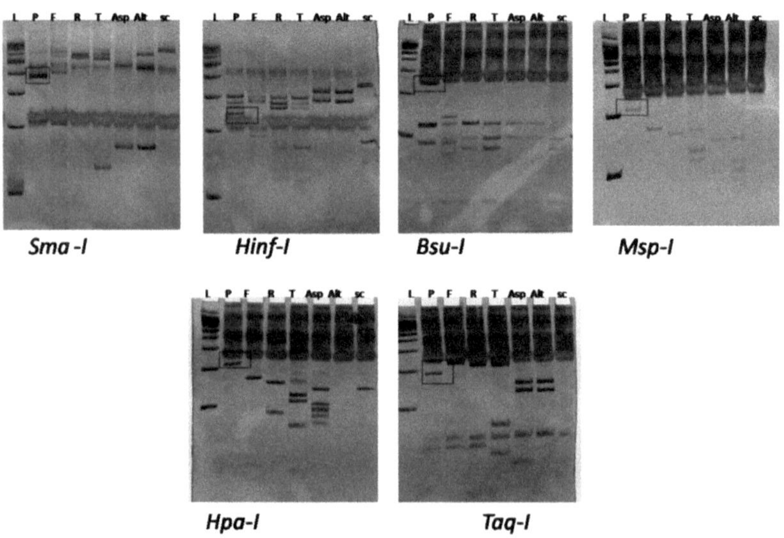

FIGURE 34.3 (See color insert.) PAGE showing differentiation of *Phytophthora* from other fungi (L-100-bp ladder. P-*Phytophthora* spp. F-*Fusarium* spp. R-*Rhizoctonia* spp. T-*Trichoderma* spp. Asp-*Aspergillus* spp. Alt-*Alternaria* spp. Sc-*Sclerotium* spp.)

ACKNOWLEDGMENT

The authors thank RJNF, UGC, Government of India, New Delhi, for providing fellowship in Ph.D. research program.

KEYWORDS

- **citrus diseases**
- **gummosis**
- **ITS (internal transcribed spacer)**
- **PCR**
- ***Phytophthora* spp.**
- **restriction enzyme**

REFERENCES

Brasier, C. M., Kirk, S. A., Delcan, J., Cooke, D. E. L., Jung, T., & Man Intveld, W. A., (2004). Phytophthora alni sp. nov. & its variants: designation of emerging heteroploid hybrid pathogens spreading on Alnus trees. *Mycol. Res.*, *108*, 1172–1184.

Bruns, T. D., White, T. J., & Taylor, J. W., (1991). Fungal molecular systematics *Annual Review of Ecology and Systematics, 22*, 525–564.

Cooke, D. E. L., & Duncan, J. M., (1997). Phylogenetic analysis of *Phytophthora* species based on ITIS1 and ITIS2 sequences of ribosomal RNA gene repeat. *Mycol. Res., 101*, 667–677.

Cooke, D. E. L., Drenth, A., Duncan, J. M., Wagels, G., & Brasier, C. M., (2000). A molecular phylogeny of Phytophthora and related Oomycetes. *Fungal Genet. Biol., 30*, 17–32.

Cooke, D. E. L., Kennedy, D. M., Guy, D. C., Russel, J., Unkles, S. E., & Duncan, J. M., (1996). Relatedness of group I species of *Phytophthora* as assessed by randomly amplified polymorphic DNA (RAPDs) and sequences of ribosomal DNA. *Mycol. Res., 100*, 297–303.

Crawford, A. R., Bassam, B. J., Drenth, A., MacLean, D. J., & Irwin, J. A. G., (1996). Evolutionary relationships among *Phytophthora* species deduced from rDNA sequence analysis. *Mycological Research, 100*, 437–443.

Das, A. K., Kumar, A., Ingle, A., & Nerkar, S., (2011). Molecular identification of Phytophthora spp. causing citrus decline in Vidarbha region of Maharashtra. *Indian Phytopath., 64*(4), 342–345.

Elliott, C. G., Hendrie, M. R., & Knights, B. A., (1966). The sterol requirement of *Phytophthora cactorum. Journal of General Microbiology, 42*, 425–435.

Francesco, A. D., Costa, N., Plata, M. I., & Marıa, L. G., (2015). Improved detection of citrus psoriasis virus and coat protein-derived transgenes in citrus plants: comparison between RT-qPCR and TAS-ELISA. *J. Phytopathol,*. doi: 10.1111/jph.12392, 1–10.

Graham, J. H., & Menge, J. L., (2000). *Phytophthora* induces disease In: Timmer, L. W., Garnesy, S. M., Graham, J. H., editors) "Compendium of Citrus diseases", *American Phytopathological Society, St. Paul, M. N.*, pp. 12–15.

Grimm, G. R., & Alexander, A. F., (1973). Citrus leaf pieces as traps for *Phytophthora parasitica* from soil slurries. *Phytopathology, 63*, 540–541.

Ippolito, A., Schena, L., & Nigro, F., (2002). Detection of Phytophthora nicotianae and P. citrophthora in citrus roots and soils by nested PCR. *Eur. J. Plant Pathol., 108*, 855–868.

Kamat, M. N., (1927). The control of mosumbi gummosis. *Agric. J. India, 22*, 176–179.

Kannwischer, M. E., & Mitchell, D. J., (1978). The influence of a fungicide on the epidemiology of black shank of tobacco. *Phytopathology, 68*, 17–60.

Klotz, L. J., De Wolfe, T. A., & Wong, P. P., (1958). Decay of fibrous root rot. *Phytopathology, 48*, 616–622.

Kumbhare, G. B., & Moghe, P. G., (1976). Leaf fall disease of Nagpur orange caused by *Phytophthora nicotianae* var Parasitica Waterhouse. *Curr. Sci., 45*, 561–562.

Lele, V. C., & Kapoor, J. N., (1982). Phytophthora on citrus in central and peninsular India. *Indian Phytopathol, 35*, 407–410.

Lévesque, C. A., & De Cock, A. W. A. M., (2004). Molecular phylogeny and taxonomy of the genus Pythium. *Mycol. Res., 108*, 1363–1383.

Martin, R. R., James, D., & Lévesque, C. A., (2000). Impacts of molecular diagnostic technologies on plant disease. *Annu. Rev. Phytopathol, 38*, 207–239.

Muske, D. N., Gahukar, S. J., Akhare, A. A., & Deshmuk, S. S., (2015). DNA-based method for detection *phytophthora* species in *citrus spp. PKV Journal Special Issue*.

Naqvi, S. A. M. H., (1988). Prevalence of Phytophthora species pathogenic to citrus in orange gardens of Vidarbh, Maharashtra, *Indian J. Myco. Pl. Pathol., 18*, 274–279.

Naqvi, S. A. M. H., (2000). Distribution of Phytophthora spp. and mating types pathogenic to citrus in Vidarbha and marathwada region of Maharashtra and Northeastern states of India. In: Singh, S., Gosh, S. P., editors. *Proceeding of the "Hi-Tech Citrus management- Int. Symposium, Citriculture*, Nagpur India", pp. 1073–1080.

National Horticultural Board Database, http://nhb.gov.in. (Accessed Jan 2015).

O'Brien, P. A., (2008). PCR primers for specific detection of *Phytophthora cinnamomi Australasian Plant Pathology, 37*, 69–71.

O'Brien, P. A., Williams, N., & Hardy, G. E., (2009). Detecting *Phytophthora. Critical Reviews in Microbiology, 35*(3), 169–181.

Ristaino, J. B., & Gimpertz, M. L., (2000). New frontiers in the study of dispersal and spatial analysis of epidemics caused by species of genus phytophthora. *Ann. Rev. Phytopathology, 38*, 541–576.

Ristaino, J. B., Madritch, M., Trout, C. L., & Parra, G., (1994). PCR amplification of ribosomal DNA for species identification in the plant pathogen genus Phytophthora. *Appl. Env. Microbiol., 64*, 948–954.

Sambrook, J., (1989). '*Molecular Cloning: A Laboratory Manual*.' (Eds. Sambrook, J., Fritsch, E. F., & Maniatis, T.). (Cold Spring Harbor Laboratory Press: Cold Spring Harbor, USA).

Uppal, B. N., & Kamath, M. N., (1936). Gummosis on citrus in Bombay. *Indian J. Agric. Sci., 6*, 803–820.

Widmer, T. L., Graham, J. H., & Mitchell, D. J., (1988). Histological comparison of fibrous root infection of disease tolerant and susceptible citrus hosts by *Phytophthora nicotianae* and *Phytophthora palmivora. Phytopathology,.88*, 389–395.

Zitko, S. E., & Timmer, L. W., (1994). Competitive parasitic activities of *Phytophrhra* parasitica and *P.* palmivora on fibrous roots of citrus. *Phytopathology, 84*, 1000–1004.

CHAPTER 35

GENETIC DIVERSITY AND HERITABILITY IN BITTER GOURD (*MOMORDICA CHARANTIA* L.) GENOTYPES

UMAKANT SINGH, DEVI SINGH, and V. M. PRASAD

Department of Horticulture, Sam Higginbottom Institute of Agriculture, Technology and Sciences, Allahabad–211007, U.P., India

CONTENTS

ABSTRACT

A field experiment was conducted at the Central Research Farm, Department of Horticulture, Sam Higginbottom Institute of Agriculture, Technology and Sciences (Deemed-to-be-University) Allahabad during 2014 and 2015 for "Studies on genetic divergence in bitter gourd (*Momordica charantia* L.) genotypes." The present investigation was undertaken with 40 genotypes of bitter gourd for evaluating their performance for various horticultural characteristics. There was a high significant variation for all the characteristics among the genotypes. High genotypic co-efficient of variation (GCV) was observed for fruit yield (q/ha) in 2014, but in 2015, High genotypic co-efficient of variation (GCV) was observed for fruit yield (kg/plant). In all cases, phenotypic co-efficient variances (PCV) were higher than GCV. High heritability was observed for fruit yield (q/ha) during both years of investigation.

35.1 INTRODUCTION

Bitter gourd (*Momordica charantia* L.) is an important gourd with wide range of uses and is largely cultivated in the tropics and subtropics for its edible fruits. Bitter gourd variability has been studied by many authors, including Heiser (1979), Marimoto and Mvere (2004), Marimoto et al. (2005), etc. Studies in India demonstrated significant regional variability (Sivaraj and Pandravada, 2005). These findings were in line with the results reported by Sakar (2004), who reported fruit size and shape as the most apparently distinguishable morphological traits in the Mediterranean region. Most of the cucurbitaceous crops are monoecious nature and bear more male flowers and less female flowers separately on the same plant. In cucurbitaceous vegetables, only male flowers appear in the beginning and the female flowers appear later, while growth, flowering, and sexual expression was come up by their genetic makeup. Selection may not be effective in population without variability; in terms of variability, it is the genetic fraction of the observed variation that provides measures of transmissibility of the variation under study and responds to selection. The relation of genotypic variability to observed variability represents heritability when the heritability of quantitative characters becomes high.

35.2 MATERIALS AND METHODS

The investigation was carried out at the Central Research Farm, Department of Horticulture, Sam Higginbottom Institute of Agriculture, Technology and Sciences (Deemed-to-be-University) Allahabad during 2014 and 2015 for "Studies on genetic divergence in bitter gourd (*Momordica charantia* L.) genotypes." The experiment was carried out in a randomized block design with three replications. The experiment comprised 40 genotypes, *viz.*, IC-085608(IIVR), IC-085609(IIVR), IC-085610(IIVR), IC-085611(IIVR),

IC-085612(IIVR), IC-085612(IIVR), IC-085613(IIVR), IC-085614(IIVR), IC-085615(IIVR), IC-085615(IIVR), IC-085616(IIVR), IC-085617(IIVR), IC-085618(IIVR), IC-085619(IIVR), IC-085620(IIVR), IC-085621(IIVR), IC-085622(IIVR), IC-085623(IIVR), IC-085624(IIVR), HABG-29(RCER, Ranchi), HABG-30(RCER, Ranchi), IC-085627(IIVR), IC-085628(IIVR), IC-085629(IIVR), IC-085630(IIVR), IC-085631(IIVR), IC-085632(IIVR), IC-085633(IIVR), PBIG-1(Pantnagar), PBIG-3(Pantnagar), PBIG-4(Pantnagar), PBIG-5(Pantnagar), PBIG-6(Pantnagar), PBIG-7(Pantnagar), IC-085634(IIVR), IC-085635(IIVR), IC-085636(IIVR), IC-085638(IIVR), IC-085639(IIVR), and PBIG-2(Check). Observations were recorded for days to 50% seedling emergence, plant height (m), number of branches, days to first appearance of male flower, days to first appearance of female flower, node number at which first male flower appears, node number at which first female flower appears, number of male flowers, number of female flowers, number of fruit per plant, fruit diameter at edible stage (cm), fruit length at edible stage (cm), fruit weight at edible stage (g), days to first fruit harvest, fruit yield (kg/plant), and fruit yield (quintal/ha) along with quality traits such as total soluble solid (Brix°) and vitamin C (mg/100 g). The data were subjected to statistical analysis to obtain information on the mean performance and to assess the association between yield and its components. Phenotypic and genotypic variance of the genotypes was estimated and heritability as described by Mahaveer et al. (2004), Munsi and Acharyya (2005), and Gayen and Hussain (2006), in bitter gourd was estimated. The estimates of heritability in broad sense in the present investigation were classified in three categories, namely high heritability (>75%t), moderate heritability (50–75%), and low heritability (<50%).

35.3 RESULTS AND DISCUSSION

In the present findings phenotypic coefficient of variation (PCV) was observed to the higher than the corresponding genotypic coefficient of variation (GCV) for all the characteristics, except fruit yield. This indicated that genetic factors were predominantly responsible for the expression of these attributes and selection could be made effectively on the basis of phenotypic performance. The finding of Kumar et al. (2007), Kumar et al. (2011), and Yadav et al. (2008) in bitter gourd. High estimates of GCV were recorded for fruit yield (kg/plant) during 2014, but in 2015, GCV was recorded for fruit

TABLE 35.1 Coefficient of Variability and Heritability of Bitter Gourd Genotypes during 2014 and 2015

Characters	Variability				Heritability(%)	
	GCV		PCV			
	2014	2015	2014	2015	2014	2015
Days to 50% seedling emergence	9.26	10.58	11.81	19.81	61.47	68.52
Plant height (m)	10.44	13.96	12.95	16.33	64.99	73.10
Number of branches / plant	8.36	8.74	10.28	10.49	66.07	69.44
Days to first appearance of male flower	7.88	8.81	8.17	9.31	93.16	89.48
Days to first appearance of female flower	8.23	9.17	8.49	9.95	93.82	84.98
Node number at which first male flower appears	13.91	16.14	16.47	20.70	71.31	60.77
Node number at which first female flower appears	8.75	11.14	10.19	13.94	73.63	63.92
Number of male flowers	15.90	15.83	16.86	17.74	88.93	79.66
Number of female flowers	15.53	15.49	15.70	17.06	97.87	82.41
Number of fruit per plant	15.49	15.39	16.83	16.79	84.71	83.94
Fruit diameter at edible stage (cm)	20.74	24.11	22.11	25.23	88.00	91.33
Fruit length at edible stage (cm)	14.72	15.57	16.11	17.32	83.52	80.81
Fruit weight at edible stage (g)	13.81	12.59	15.02	13.31	84.53	89.53
Days to first fruit harvest	8.21	7.70	8.29	8.17	98.18	88.89
Fruit yield (kg/plant)	25.43	30.89	26.48	32.25	92.26	91.71
Fruit yield (q/ha)	26.25	25.22	26.33	25.30	99.34	99.37

yield (q/ha), followed by fruit diameter at edible stage (cm) and number of male flowers, However, low estimates of GCV was noted for characteristics such as days to first appearance of male flower in first year, but in the second year of investigation, low GCV was noted for days to first fruit harvest (Table 35.1).

In the present studies, the heritability estimates (broad sense) were estimates for the characteristics, whereas the genetic advance estimates were measured the unit of character under measurement. The obtained result indicated high heritability for the fruit yield (99.34%) and (99.37%) during 2014 and 2015, respectively. Lower heritability was noted for days to 50% seedling emergence during 2014, but in 2015, lower heritability was measured for node number at which first male flower appears (Table 35.1).

35.4 CONCLUSION

The genetic parameters discussed here are functions of variability; thus, the estimates may differ in other environments. Based on the moderate heritability and high genetic advance shown by the different characters, especially, High genotypic co-efficient of variation (GCV) was observed for fruit yield (q/ha) in 2014 but in 2015, high genotypic co-efficient of variation (GCV) was observed for fruit yield (kg/plant). In all cases, phenotypic co-efficient variances (PCV) were higher than GCV. High heritability was observed for fruit yield (q/ha) during both years of investigation.

KEYWORDS

- **bitter gourd**
- **diversity**
- **fruit yield**
- **genotypes**
- **heritability**
- **phenotypic**

REFERENCES

Heiser, B. C., (1979). *The Gourd Book*. University of Oklahoma Press, Norman and London.

Morimoto, Y., & Mvere, B., (2004). Bitter gourd standley, In: *Plant Resources of Tropical Africa 2. Vegetables*, (Eds. G. J. H. Grubben and O. A. Denton). pp. 353–358.

Kumar, S., Singh, R., & Pal, A. K., (2007). Genetic variability, heritability, genetic advance, correlation coefficient and path analysis in bitter gourd. *Indian J. Hort.*, *64*(2), 163–168.

Kumar, A., Singh, B., Kumar, M., & Naresh, R. K., (2011). Genetic variability, heritability, genetic advance for yield and its components in bitter gourd. *Ann. of Horti.*, *4*(1), 101–103.

Morimoto, Y., Maundu, P., Fujimaki, H., & Morishima, H., (2005). Diversity of landraces of the white-flowered and its wild relatives in Kenya: fruit and seed morphology. *Genet. Resour. Crop Evol.*, *52*, 737–747.

Sarkar, M., (2004). Characterization of the bitter gourd genotypes collected from Mediterranean region. *Master Thesis*, University of Mustafa Kemal, Institute of Natural and Applied Science, Hatay, Turkey.

Sivaraj, N., & Pandravada, S. R., (2005). Morphological diversity for fruit characteristics in bitter gourd germplasm from tribal pockets of Telangana region of Andhra Pradesh, India. *Asian Agri-Hist.*, *9*, 305–310.

Yadav, J. R., Yadav, A., Srivastava, J. P., Mishra, G., Parihar, N. S., & Singh, P. B., (2008). Study on variability heritability and genetic advance in bitter gourd. *Prog. Res. 3*(1), 70–72.

CHAPTER 36

PERFORMANCE STUDY OF TUBEROSE IN THE GANGETIC PLAINS OF WEST BENGAL, INDIA

VANLALRUATI, T. MANDAL, and JENNY ZOREMTLUANGI

Department of Floriculture and Landscaping, Faculty of Horticulture, BCKV, Mohanpur–741252, West Bengal, India,
Mobile: +91-9990881696, E-mail: maruathmar@gmail.com

CONTENTS

ABSTRACT

Performance studies in tuberose were carried out among 11 genotypes for 18 characteristics at the Horticultural Research Farm, Mondouri, Bidhan Chandra Krishi Viswavidyalaya during the year 2009–2010. The research was conducted at the Horticultural Research Farm, Mondouri, Bidhan Chandra Krishi Viswavidyalaya during the year 2009–2010 with 11 tuberose varieties, viz., Calcutta Single, Calcutta Double, Phule Rajani, Swarna Rekha, Prajwal, Vaibhav, Hyderabad Single, Hyderabad Double, Rajat Rekha, Sikkim Selection, and Hybrid CG-T-C-4. The interpretation of the recorded data reveals that double-type cultivars, viz., Vaibhav and Calcutta Double and single-type cultivars, viz., Hyderabad Double and Prajwal, performed comparatively better than other varieties in respect to yield and quality. These varieties could, therefore, be regarded as suitable and promising for cultivation in the Gangetic plains of West Bengal.

36.1 INTRODUCTION

Tuberose (*Polianthes tuberosa* Linn.) is one of the most important bulbous ornamentals of tropical and subtropical areas. It belongs to the Amaryllidaceae family and is a native of Mexico (Bailey, 1919). Its importance among commercially grown flowers is due to its potential for cut flower trade, long-vase life, and essential oil industry (Singh, 1995). Some advantages of growing tuberose are attractive long spikes, pleasant fragrant flowers, high cut-flower yield, and nearly year round yield in tropical and subtropical climates (Benschop, 1993). The white tubular flowers emit the most powerful odor and hence are cultivated on a large scale in some parts of the world for the extraction of its highly valued natural flower oil. The lingering delightful fragrance and its excellent keeping quality are the predominant characteristic of this crop.

Tuberose is cultivated commercially for its valued natural flower oil, floral ornaments, artistic garlands, bouquets, and button holes. The flowers are a good source of high-quality perfumes and fragrant oils (Sadhu and Bose, 1973). The serene beauty of the flower spikes, bright snow-white flowers, sweetness of blooms, and delicacy of fragrance of this ornamental crop transform the entire area into a nectarine and joyous one. It can successfully be grown in pots, beds, and borders. The flowers remain fresh for pretty long time (6.33 to 8.66 days; Nagaraja and Gowda, 1998) and endure long distance transportation. The long flower spikes (88.78 cm; Dalal et al., 1999) are excellent as cut flower for room and table decoration when arranged in bowls and vases.

Tuberose, a native of Mexico, spread to the different parts of the world during the 16th century. The name tuberose is derived from *tuberosa*, which is the tuberous hyacinth as distinguished from the bulbous hyacinth. The name, therefore, is "tuber – ose" not tuberose (Bose and Yadav, 1989). The generic name Polianthes has also been derived from Greek word "*Polis*" means "white" and "*anthos*" means "flower." It was still referred to as *Hyacinthus indicus tuberosus* by the middle of the 18th century. Because of their lingering delightful fragrance and charm, there are adorned with romantic vernacular names in India like Gulchari and Gulshabbo (Hindi), Rajanigandha (Bengali), Sukandaraji (Telugu), Nilasampingi (Tamil), and Sandharaga (Kanarese).

Starting from the long back of its introduction, tuberose is cultivated in tradition fashion with the old age varieties in different pockets of the plains of West Bengal. With the implementation of commercialization of floriculture and improvement for its exportability, it becomes inevitable to follow the international standards and specification. Lengthening of postharvest is another aspect for enhancement of marketability and exportability. Thus, to keep pace with international flower trade, the introduction of improved varieties is of prime importance. But before introduction of any new variety, it is important to have a vivid study of its phenotypical characteristics.

Some of the varieties performed well but not under all geographical and climatic conditions. For commercial cultivation, it is important to know the varietal characteristics, particularly adaptability, productivity, profitability, etc. Conducting experiment using more number of cultivars under different agroclimatic zones and under different conditions will be beneficial to ascertain the best variety that can be grown successfully at any situation.

36.2 MATERIALS AND METHODS

To study the varietal performance of tuberose, research was conducted at the Horticultural Research Farm, Mondouri, Bidhan Chandra Krishi Viswavidyalaya during the year 2009–2010 with 11 tuberose varieties, viz., Calcutta Single, Calcutta Double, Phule Rajani, Swarna Rekha, Prajwal, Vaibhav, Hyderabad Single, Hyderabad Double, Rajat Rekha, Sikkim Selection, and Hybrid CG-T-C-4. Each variety was replicated thrice in a randomized block design comprising 990 number of plants in each replication. Planting was done in May, at a spacing of 20 cm × 30 cm. Representative plants were marked in each replication, and close observations were noted regularly. The observations were recorded for the following vegetative parameters, including height of the plant, number of leaves per clump, leaf area, fresh and dry weight of individual leaf and chlorophyll content of leaf tissue; reproductive parameters including days to spike emergence, days for spike emergence to floret opening, length of flower spike, weight of flower spike, rachis length, girth of flower spike and weight of 10 florets; yield parameters like spike yield per plot (2 m^2) and number of florets per spike; and quality parameters like self-life of spike in the field.

36.3 RESULTS

Significant differences were obtained for vegetative parameters. Prajwal produced the maximum plant height and leaf area, which ranged from 29.14–46.99 cm and 37.60–66.99 cm^2, respectively. This variety also produced the maximum fresh weight (6.01 g) and dry weight (2.42 g) for individual leaf. The maximum number of leaves/clump was found for Hyderabad Single (41.92), followed by Vaibhav (40.19). The highest chlorophyll content of fresh leaf tissue was also found Prajwal (3.71 mg/g), followed by Hyderabad Double (3.52 mg/g), and these were statistically at par. The data represented in Tables 36.1 and 36.2 clearly indicate that reproductive characteristics varied significantly among the genotypes.

Highly significant differences were observed among the genotypes regarding days required for spike emergence. Spike emergence was earliest in Sikkim Selection (167.47), while Swarna Rekha took the maximum number of days (200.67). Maximum days for spike emergence to floret opening were observed in Swarna Rekha (31.33) and minimum days in Sikkim Selection (21.00). The longest flower spike was observed in Prajwal (97.17 cm), followed by Sikkim Selection (95.27 cm) among single-type cultivars, whereas Vaibhav (95.96 cm) recorded the longest flower spike among double-type cultivars followed by Hyderabad Double (90.33 cm). Prajwal (96.33 g) also produced maximum weight of flower spike, which significantly differ from all the varieties. Girth of flower spike was highest in Hyderabad Double (1.79 cm), which differed significantly from all the varieties, and this was followed by Vaibhav (1.78 cm). Prajwal (1.72 g) recorded the maximum girth of flower spike among the single-type cultivars, followed by Calcutta Single (1.69 g). Among the double-type cultivars, Vaibhav (46.39 cm) produced longest rachis, which differed significantly from all the varieties and was followed by Hyderabad Double (42.87 cm), whereas Prajwal (35.70 cm) produced longest rachis among the single-type cultivars.

TABLE 36.1 Performances of Floral Attributes of Different Varieties of Tuberose (*Polianthes tuberosa L.*)

Treatments	Days to spike emergence (DAP)	Days for spike emergence to floret opening	Length of flower spike (cm)	Weight of flower spike (g)	Girth of flower spike (cm)	Length of rachis (cm)
Calcutta Single	174.23	22.44	89.67	48.00	1.69	19.23
Calcutta Double	192.33	27.00	87.67	85.83	1.74	40.67
Phule Rajni	193.00	29.67	71.00	67.00	1.71	28.37
Swarna Rekha	200.67	31.33	72.00	77.36	1.73	37.15
Prajwal	176.11	23.89	97.17	96.33	1.72	35.70
Vaibhav	185.00	25.00	95.96	68.00	1.78	46.39
Hyderabad Single	180.44	24.89	86.61	61.50	1.66	25.07
Hyderabad Double	187.56	26.11	90.33	71.61	1.79	42.87
Rajat Rekha	195.45	30.22	60.00	46.50	1.64	21.33
Sikkim Selection	167.47	21.00	95.27	47.67	1.71	32.07
Hybrid CG-T-C-4	191.67	28.33	69.55	54.67	1.68	22.23
S. Em (±)	1.91	2.21	2.03	2.20	0.02	1.30
C.D. at 5%	5.34	3.12	5.69	6.17	0.05	3.64

TABLE 36.2 Performances of Floral Attributes of Different Varieties of Tuberose (*Polianthes tuberosa* L.)

Treatments	Length of floret (cm)	Diameter of floret (cm)	Fresh weight of 10 florets (g)	Number of florets per spike	Spike yield plot($2m^2$)	Self life of spike in the field (days)
Calcutta Single	5.45	3.59	13.33	36.33	54.95	25.33
Calcutta Double	6.24	4.86	25.38	53.21	75.35	32.05
Phule Rajni	5.69	4.18	17.50	45.67	64.81	17.73
Swarna Rekha	5.94	4.64	24.04	49.33	69.50	30.28
Prajwal	5.81	4.48	21.31	47.17	77.33	28.60
Vaibhav	5.76	4.23	18.30	56.27	84.44	34.43
Hyderabad Single	5.56	3.70	15.39	40.00	62.47	24.92
Hyderabad Double	5.79	4.37	20.53	50.56	78.21	26.61
Rajat Rekha	5.36	3.43	12.67	26.00	52.33	15.41
Sikkim Selection	5.22	3.29	11.42	30.56	66.67	22.03
Hybrid CG-T-C-4	5.62	4.14	16.52	42.00	60.00	20.27
S. Em (±)	0.16	0.11	0.62	1.56	1.22	1.24
C.D. at 5%	0.44	0.32	1.73	4.38	3.42	3.47

With regard to yield parameters, Vaibhav (56.27) produced the maximum number of florets per spike followed by Calcutta Double (53.21). Gurav et al. (2005) also recorded significant variability in number of florets per spike of nine tuberose varieties. Highest spike yield per plot was also recorded in Vaibhav (84.44 nos.) followed by Hyderabad Double (78.21 nos.). Prajwal (47.17) produced maximum florets/spike and maximum spikes/plot (77.33) among the single-type cultivars. Calcutta Double produced the maximum length of floret (6.24 cm), diameter of floret (4.86 cm), and fresh weight of 10 florets (25.38 g), which differ significantly from the other varieties. Self-life of spike in the field was longest in Vaibhav (34.43 days), followed by Calcutta Double (32.05 days), and these were statistically at par. Shortest self-life of spike in the field was found in Rajat Rekha (15.41 days). Generally, self-life is related to carbohydrate (glucose/sucrose) reserve of the stem, which depends upon the photosynthetic activity and photosynthetic efficiency increases as the leaf area or number of leaves per clump increases. Therefore, larger leaf area and leaf numbers might be one of the contributing factors for longer self-life of the variety Vaibhav.

36.4 DISCUSSION

The result and discussion with respect to different vegetative and reproductive parameters based on the statistical analysis are described below.

The marked variations in plant height among nine cultivars of tuberose might be attributed to the inherent genetical characteristics of the individual varieties. Significant variability in plant height among nine varieties of tuberose was also observed by Gurav et al. (2005). The differences in the number of leaves per clump might also be attributed to the inherent genetical characteristics of the individual varieties. Bulb planting method significantly affected the number of leaves, flowering stem height, floret number, fresh and dry weight of flowering stem, quality index, and bulblet number (Nazari et al., 2007). Variations regarding the leaf area of different tuberose might be attributed to the inherent genetical characteristics of the individual varieties. Pal and George (2002) also observed significant variations for individual leaf area among 12 varieties of chrysanthemum. The variations regarding fresh weight and dry weight of individual leaf might be attributed to inherent genetical characteristics of the individual varieties.

The marked variations in days to spike emergence might be due to varietal characteristics of individual varieties that were expressed in their genetical characteristics. Path analysis studies on 25 gladiolus cultivars show that days for first bud to loosen and number of marketable spikes per plant had the highest positive direct effect and days for spike emergence had the highest negative direct effect on spike length (Hedge et al., 1997). Significant differences in days for flowering among four tuberose cultivars were also observed by Banker et al. (1980). Sathyanarayana and Singh (1994) reported that flower spikes from large bulbs emerged and flowered earlier than those from small bulbs and had more florets.

The marked variations in length of flower spike of different tuberose varieties might be due to the inherent characteristics of the individual varieties. Significant variability in length of flower spike among nine varieties of tuberose was also observed by Gurav et al. (2005). Bose et al. (1999) reported that in cv. Calcutta Double, the spike produced during higher temperature is shorter and the flowers are smaller, white in color, and open imperfectly. Significant variations in the girth of flower spike might be attributed to inherent genetical characteristics of the individual varieties. Gonzalez et al. (1992) reported that flower quality, viz., floral spike lengths, number of floret pairs, and flower stem thickness was highest in the spring-planted crop. Significant variations in length of rachis among nine tuberose varieties was also reported by Gurav et al. (2005). The marked variations in fresh weight of 10 florets of different tuberose varieties might be due to the inherent characteristics of the individual varieties. Significant variations in the fresh weight of florets were also reported by Ghosh and Pal (2004).

The variations regarding number of floret per spike of different tuberose varieties might be attributed to the inherent genetical characteristics of the individual varieties. The longest rachis length of Vaibhav (46.39) might be responsible for its maximum florets number/spike. Gurav et al. (2005) also recorded significant variability in the number of florets per spike of nine tuberose varieties. Bulb planting method significantly affected number of leaves, flowering stem height, floret number, fresh and dry weight of fowering stem, quality index and bulblet number (Nazari et al., 2007). The higher spike yield of Vaibhav might be due to the larger leaf area as it is positively correlated. Bose (1988) reported that June–July planting increases duration of flowering and that the number of days required for first flower appearance was minimum in June 5 planting and maximum in December 24 planting. Sathyanarayana and Singh (1994) reported that the duration of flowering was shortest for plants from small bulbs. Significant variations in self-life of spike among four tuberose cultivars were also observed by Banker et al. (1980).

These variations regarding self-life of spike of different tuberose varieties might also be attributed to the inherent genetical characteristics of the individual varieties. Generally, self-life is related to carbohydrate (glucose/sucrose) reserve of the stem, which depends upon the photosynthetic activity that increases as the leaf area or number of leaves per clump increases. Therefore, larger leaf area and leaf numbers might be one of the contributing factors for longer self-life of the variety Vaibhav.

36.5 CONCLUSION

The interpretation of the recorded data reveals that double-type cultivars like Vaibhav, Calcutta Double, and Hyderabad Double and Prajwal among the single-type cultivars performed comparatively better than other varieties with respect to yield and quality. These varieties could, therefore, be regarded as suitable and promising for cultivation in the Gangetic plains of West Bengal.

KEYWORDS

- **cultivars**
- **floral characters**
- **tuberose**

REFERENCES

Bailey, L. H., (1919). *The Standard Cyclopedia of Horticulture*. 3rd ed. MacMillan, London.

Banker, G. J., & Mukhopadhyay, A., (1980). 'Varietal trial on tuberose (*Polianthes tuberosa* L.)'. *South Indian Horticulture, 28*(4), 150–151.

Benschop, M., (1993). Polianthes. In: De Hertogh, A. A., & Le Nard, M., (eds.). *The Physiology of Flower Bulbs*, Elsevier, Amsterdam. pp. 589–601.

Bose, T. K., & Yadav, L. P., (1989). *Commercial Flowers*. Published by Naya Prakash, pp. 519.

Dalal, S. R., Dalal, N. R., Rajurkar, D. W., Golliwar, V. J., & Patil, S. R., (1999). Effect of nitrogen levels and gibberellic acid on quality of flower stalk of tuberose (*Polianthes tuberosa* L.). *J. Soils and Crops, 9*(1), 88–90.

Ghosh, P., & Pal, P., (2004). Influence of temperature and humidity on flowering of *Polianthes tuberosa* L. var. Double. *Horticulture J., 17*(1), 73–78.

Gonzalez, A., Perez, J. G., Fernandez, J., & Banon, S., (1992). Cultivation programming of *Polianthes tuberosa. Acta. Horticulturae, 325*, 357–364.

Gurav, S. B., Singh, B. R., Desai, U. T., Katwate, S. N., Kakade, D. S., Dhane, A. V., & Karade, V. D., (2005). 'Influence of planting time on the yield and quality of tuberose. *Indian Journal of Horticulture, 62*(2), 216–217.

Hegde, M. V., Rajendra Passannavar, & Harish Shenoy, (1997). Path analysis studies in gladiolus. *Advances in Agricultural Research in India, 8*, 37–39.

Nagaraja, G. S., Gowda, J. V. N., & Farooqui, A. A., (1998). Influence of season, bulb size and spacing on bulb production of tuberose cv. Single. *Karnataka Journal of Agricultural Science, 11*(4), 1145–1147.

Nazari, F., Farahmand, M., Khos-Khui, M., & Salehi, H., (2007). Effects of two planting methods on vegetative and reproductive characteristics of tuberose (*Polianthes tuberose* .L). *Adv.* In: *Nat. Appl. Sci., 1*(1), 26–29.

Pal, P., & George, S. V., (2002). Genetic variability and correlation studies in chrysanthemum. *Horticulture J., 15*(2), 75–81.

Sadhu, M. K., & Bose, T. K., (1973). 'Tuberose for most artistic garlands'. *Indian Horticulture, 18*(17), *19*–20.

Sathyanarayana Reddy, B., & Singh, K., (1994). I. Studies on effect of bulb size in tuberose. II. Influence of bulb size on flowering, flower yield and quality of flower spikes. *Advances in Agricultural Research in India, 2*, 123–130.

Singh, K. P., (1995). *Improved Production Technologies for Tuberose* (*Polianthes tuberosa* L.), a review of research done in India. Indian Institute of Horticultural Research, Hessargarhatta, Banglore, India. (*CAB Abst.,* 1996–1998/07).

CHAPTER 37

GENOTYPIC PERFORMANCE ON GROWTH AND YIELD IN BITTER GOURD (*MOMORDICA CHARANTIA* L.) GENOTYPES

UMA KANT SINGH, DEVI SINGH, and V. M. PRASAD

Department of Horticulture, Sam Higginbottom Institute of Agriculture, Technology and Sciences, Allahabad–211007, U.P., India

CONTENTS

ABSTRACT

Evaluation of 40 bitter gourd genotypes collected from various sources was carried out. Observations were recorded for the following traits, viz., plant height, number of branches per plant, days to first appearance of male flower, days to first appearance of female flower, number of male flower, number of female flower, number of fruits/plant, and fruit yield (q/ha), and the data were analyzed statistically. Among the genotypes, IC-085612(IIVR) followed by IC-085616(IIVR) recorded the highest mean value of growth and yield, i.e., plant height, number of branches per plant, days to first appearance of male flower, days to first appearance of female flower, number of male flower, number of female flower, and number of fruits/plant during both years of investigation.

37.1 INTRODUCTION

Bitter gourd (*Momordica charantia* L.) is one of the most important and widely used cucurbitaceous summer vegetables in India. It belongs to the family Cucurbitaceae. Most of the authors agree that India or the Indo-Malayan region is the original home of bitter gourd (Bose and Som, 1986). Bitter gourd is widely distributed in China, Malaysia, India, tropical Africa, and North and South America. In India, Karnataka, Maharashtra, Tamil Nadu, and Kerala are the major bitter gourd growing states. The crop is cultivated over an area of 26,004 ha in India with the production of 1,62,196 tons with a productivity of 6.23 t ha^{-1} (Anon, 2006).

The yield of bitter gourd in India is very low compared to that in the other countries (7.5 to 8 t ha^{-1}). Selection of a high yielding germplasm can therefore significantly increase the production of bitter gourd. Like any other crops, the yield of bitter gourd is a complex component characteristic. In a breeding program for increasing yield, quality and resistance to disease and pests in the exploration of genetic variability for the available germplasms are prerequisites. Therefore, the evaluation of germplasm in terms of local conditions is very important.

37.2 MATERIAL AND METHODS

The investigation was carried out at the Central Research Farm, Department of Horticulture, Sam Higginbottom Institute of Agriculture, Technology and Sciences (Deemed-to-be-University) Allahabad during 2014 and 2015 for "Studies on genetic divergence in bitter gourd (*Momordica charantia* L.) genotypes." The experiment was carried out in a randomized block design with three replications. The experiment comprised 40 genotypes, viz., IC-085608(IIVR), IC-085609(IIVR), IC-085610(IIVR), IC-085611(IIVR), IC-085612(IIVR), IC-085612(IIVR), IC-085613(IIVR), IC-085614(IIVR), IC-085615(IIVR), IC-085615(IIVR), IC-085616(IIVR), IC-085617(IIVR),

IC-085618(IIVR), IC-085619(IIVR), IC-085620(IIVR), IC-085621(IIVR), IC-085622(IIVR), IC-085623(IIVR), IC-085624(IIVR), HABG-29(RCER, Ranchi), HABG-30(RCER, Ranchi), IC-085627(IIVR), IC-085628(IIVR), IC-085629(IIVR), IC-085630(IIVR), IC-085631(IIVR), IC-085632(IIVR), IC-085633(IIVR), PBIG-1(Pantnagar), PBIG-3(Pantnagar), PBIG-4 (Pantnagar), PBIG-5(Pantnagar), PBIG-6(Pantnagar), PBIG-7(Pantnagar), IC-085634(IIVR), IC-085635(IIVR), IC-085636(IIVR), IC-085638(IIVR), IC-085639(IIVR), and PBIG-2(Check). Observations were recorded for days to 50% seedling emergence, plant height (m), number of branches, days to first appearance of male flower, days to first appearance of female flower, node number at which first male flower appears, node number at which first female flower appears, number of male flowers, number of female flowers, number of fruit per plant, fruit diameter at edible stage (cm), fruit length at edible stage (cm), fruit weight at edible stage (g), days to first fruit harvest, fruit yield (kg/plant), and fruit yield (quintal /ha) along with quality traits such as total soluble solid (Brix°) and vitamin C (mg/100 g). The data were subjected to statistical analysis to obtain information on the mean performance and to assess the association between yield and its components.

37.3 RESULTS AND DISCUSSION

Development of high-yielding genotypes of crops requires information about the nature and magnitude of variability present in the available genotypes and depends on judicious assessment of the available data with regard to the phenotypic characteristics linked with yield. Hence, 40 bitter gourd genotypes were evaluated for growth and yield attributes during both the years (Tables 37.1 and 37.2).

Among the bitter gourd accessions, the highest mean value of plant height and number of branches per plant was recorded for the accession IC-085612(IIVR), followed by IC-085616(IIVR) during both years of investigation.

In general, earliness in cucurbits is measured as the days taken for first male and female flower appearance, and number of male flower and number of female flower are considered as desirable traits in any hybrid development program during the investigation period (Singh et al., 2006).

In the present study, maximum first male and female flower appearance, number of male flower, and number of female flower were observed in the

TABLE 37.1 Genotypic Performance on Growth in Bitter Gourd during 2014 and 2015

Genotypes	Plant height (cm)		No. of branches per plant		Days to first appearance of male flower		Days to first appearance of female flower	
	2014	2015	2014	2015	2014	2015	2014	2015
IC-085608(IIVR)	5.09	5.44	18.17	19.47	48.94	50.23	54.16	59.67
IC-085609(IIVR)	5.22	5.43	16.75	17.28	50.38	52.38	55.28	57.50
IC-085610(IIVR)	4.24	4.36	16.90	17.75	51.61	53.67	57.64	59.94
IC-085611(IIVR)	4.14	4.55	17.14	18.02	49.99	51.97	53.60	55.77
IC-085612(IIVR)	6.51	6.86	20.51	21.76	55.03	57.22	59.90	62.27
IC-085612(IIVR)	4.59	5.08	16.10	16.98	48.72	50.66	53.28	52.10
IC-085613(IIVR)	4.54	5.02	18.27	18.96	47.18	49.12	49.10	50.80
IC-085614(IIVR)	4.72	4.57	15.43	16.28	46.47	48.35	51.85	51.27
IC-085615(IIVR)	4.57	5.04	18.55	19.47	47.07	48.98	51.43	51.13
IC-085615(IIVR)	4.54	5.02	17.10	17.97	50.30	52.32	55.86	58.14
IC-085616(IIVR)	6.01	6.29	19.43	20.11	52.49	54.67	58.29	61.55
IC-085617(IIVR)	4.55	5.03	18.00	18.88	43.62	45.35	49.05	54.33
IC-085618(IIVR)	4.96	5.14	16.05	16.57	42.86	44.55	46.06	47.88
IC-085619(IIVR)	4.65	4.80	17.19	18.04	41.13	42.79	45.20	45.31
IC-085620(IIVR)	4.60	4.75	15.88	16.04	43.69	45.40	47.48	50.69
IC-085621(IIVR)	5.08	4.94	16.82	17.69	42.40	44.10	45.97	49.49
IC-085622(IIVR)	5.50	5.72	18.94	19.73	52.39	54.48	57.59	52.32
IC-085623(IIVR)	4.86	5.36	16.36	16.89	43.75	45.52	47.85	50.78
IC-085624(IIVR)	4.71	5.19	16.92	17.43	42.12	43.79	46.23	50.10
HABG-29 (RCER, Ranchi)	4.62	5.11	14.93	15.43	42.41	44.11	48.46	47.11
HABG-30 (RCER, Ranchi)	5.19	5.06	16.64	16.42	41.98	43.64	45.98	49.15
IC-085627(IIVR)	3.94	4.02	17.89	18.44	43.60	45.34	47.36	45.95

TABLE 37.1 (continued)

Genotypes	Plant height (cm)		No. of branches per plant		Days to first appearance of male flower		Days to first appearance of female flower	
	2014	2015	2014	2015	2014	2015	2014	2015
IC-085628(IIVR)	5.33	5.56	16.23	16.68	48.35	50.28	54.16	53.34
IC-085629(IIVR)	4.51	4.96	18.21	18.76	47.96	49.89	52.53	54.65
IC-085630(IIVR)	5.17	5.38	17.54	18.11	43.81	45.54	48.43	53.71
IC-085631(IIVR)	4.13	4.55	18.54	19.39	40.86	42.48	44.29	46.03
IC-085632(IIVR)	4.67	4.82	16.80	16.76	42.33	44.01	48.51	48.83
IC-085633(IIVR)	4.73	5.20	18.01	18.96	44.30	46.06	50.66	49.38
PBIG-1(Pantnagar)	5.24	4.61	14.35	15.67	49.91	47.91	53.38	51.24
PBIG-3 (Pantnagar)	4.50	3.95	15.41	16.23	47.73	45.82	53.97	51.82
PBIG-4 (Pantnagar)	4.23	3.71	13.23	14.79	49.19	47.21	55.46	53.25
PBIG-5 (Pantnagar)	4.25	3.73	13.55	14.02	47.42	45.58	54.31	52.13
PBIG-6 (Pantnagar)	4.03	3.89	16.76	17.44	42.40	40.70	45.08	43.28
PBIG-7(Pantnagar)	4.42	3.88	16.74	16.54	42.49	40.80	46.13	44.28
IC-085634(IIVR)	5.14	4.51	14.32	14.33	45.89	44.05	50.17	48.16
IC-085635(IIVR)	3.81	3.36	17.22	17.25	42.07	40.38	45.50	43.67
IC-085636(IIVR)	4.83	4.24	16.32	16.16	43.22	41.49	47.14	45.25
IC-085638(IIVR)	4.32	3.80	17.17	16.21	49.68	47.70	52.33	50.23
IC-085639(IIVR)	5.05	4.45	18.54	17.33	43.81	42.06	47.87	45.95
PBIG-2(Check)	5.15	4.52	16.39	16.08	46.42	44.57	50.24	48.23
F-test	S	S	S	S	S	S	S	S
SEd ±	0.29773	0.33179	0.82620	0.82462	0.80487	1.15590	0.87428	1.61188
CD (5%)	0.59273	0.66055	1.64484	1.64170	1.60238	2.30121	1.74055	3.20901

TABLE 37.2 Genotypic Performance on Growth and Yield in Bitter Gourd during 2014 and 2015

Genotypes	Number of male flower		Number of female flower		Number of fruits / plant		Fruit yield (q/ha)	
	2014	2015	2014	2015	2014	2015	2014	2015
IC-085608(IIVR)	16.88	18.68	22.00	22.87	21.52	22.80	78.59	76.50
IC-085609(IIVR)	16.95	18.05	26.76	27.80	26.78	29.73	102.08	99.45
IC-085610(IIVR)	15.79	17.18	31.57	33.28	30.24	32.78	117.45	116.30
IC-085611(IIVR)	16.48	17.90	29.12	30.53	28.39	31.27	114.08	111.64
IC-085612(IIVR)	26.88	28.18	39.82	41.01	38.91	42.54	242.52	225.41
IC-085612(IIVR)	20.29	22.20	27.18	28.51	28.28	29.67	134.93	132.83
IC-085613(IIVR)	18.64	20.20	30.82	32.89	30.63	33.38	142.59	144.46
IC-085614(IIVR)	18.49	19.38	32.80	34.54	32.40	34.15	122.48	123.85
IC-085615(IIVR)	20.09	21.06	35.89	39.03	35.12	36.94	151.08	148.29
IC-085615(IIVR)	17.10	18.20	28.84	31.90	29.23	31.00	115.85	114.46
IC-085616(IIVR)	25.67	26.29	39.08	41.91	38.91	39.92	216.74	210.77
IC-085617(IIVR)	14.98	15.61	27.27	28.93	26.14	28.32	101.93	99.65
IC-085618(IIVR)	19.24	20.16	27.95	29.68	26.51	28.85	134.44	124.42
IC-085619(IIVR)	17.75	18.59	22.87	25.20	22.68	23.36	102.67	102.34
IC-085620(IIVR)	17.76	18.62	26.69	29.06	25.71	27.64	123.56	126.49
IC-085621(IIVR)	13.84	15.13	28.85	34.59	28.40	31.29	134.22	133.83
IC-085622(IIVR)	25.28	25.88	37.92	39.43	38.43	39.36	213.33	209.19
IC-085623(IIVR)	16.28	17.00	35.38	37.87	32.13	35.02	132.45	127.13
IC-085624(IIVR)	19.44	20.37	35.86	38.00	37.14	36.49	191.11	192.02
HABG-29(RCER,Ranchi)	14.34	14.99	31.24	32.35	31.30	31.67	171.96	166.88
HABG-30(RCER,Ranchi)	16.25	17.03	32.43	33.53	32.78	32.50	125.26	126.17
IC-085627(IIVR)	17.87	18.70	30.72	32.16	30.15	31.83	109.63	112.57

TABLE 37.2 (continued)

Genotypes	Number of male flower		Number of female flower		Number of fruits / plant		Fruit yield (q/ha)	
	2014	**2015**	**2014**	**2015**	**2014**	**2015**	**2014**	**2015**
IC-085628(IIVR)	17.66	18.50	27.02	29.41	26.34	28.37	134.52	137.39
IC-085629(IIVR)	16.25	16.99	29.04	30.10	29.74	28.76	107.56	103.85
IC-085630(IIVR)	17.61	18.45	30.73	32.45	30.97	31.36	119.70	115.98
IC-085631(IIVR)	17.01	17.83	35.16	34.65	34.49	35.66	148.52	139.53
IC-085632(IIVR)	13.83	14.46	35.81	36.34	35.00	35.82	157.15	153.01
IC-085633(IIVR)	16.68	17.45	23.01	23.32	22.72	24.77	111.07	106.40
PBIG-1(Pantnagar)	15.57	14.67	22.99	22.42	22.31	23.33	105.89	107.80
PBIG-3 (Pantnagar)	17.11	16.13	25.79	25.86	26.77	25.86	110.93	111.55
PBIG-4 (Pantnagar)	16.76	17.17	31.40	30.65	31.13	32.37	147.41	142.40
PBIG-5 (Pantnagar)	15.68	16.16	31.71	30.98	30.69	29.94	148.49	151.41
PBIG-6 (Pantnagar)	17.96	19.36	30.81	30.32	31.16	30.96	142.63	126.37
PBIG-7(Pantnagar)	16.50	15.64	34.22	33.46	34.28	33.13	138.22	145.22
IC-085634(IIVR)	13.49	14.09	23.25	22.72	22.68	21.77	86.73	89.93
IC-085635(IIVR)	17.27	17.99	22.14	22.66	20.89	20.85	77.85	81.00
IC-085636(IIVR)	16.39	16.47	31.34	30.62	30.55	29.60	129.74	134.81
IC-085638(IIVR)	18.32	17.71	31.67	31.64	32.60	31.09	150.37	149.21
IC-085639(IIVR)	17.54	18.23	29.17	28.46	30.36	28.21	141.29	146.41
PBIG-2(Check)	19.18	18.54	35.89	34.86	36.67	34.94	174.85	175.66
F- test	S	S	S	S	S	S	S	S
SEd ±	0.80970	1.20072	0.56677	1.83444	1.61230	1.69979	2.35307	2.18735
CD (5%)	1.61198	2.39046	1.12836	3.65209	3.20984	3.38402	4.68460	4.35468

genotype IC-085612 (IIVR), followed by IC-085616 (IIVR) during the years 2014 and 2015 (Sachan et al., 1989).

In this study, the accession IC-085612 (IIVR) followed by IC-085616 (IIVR) recorded the highest number of fruits per plant and fruit yield (q/ha) (Rahman et al., 1990).

37.4 CONCLUSION

Based on the present study, among the 40 bitter gourd genotypes, the accession IC-085612(IIVR) recorded the highest mean value of fruit yield, followed by IC-085616(IIVR). This study clearly indicated that favorable varieties could be developed with earliness and more number of fruits per plant in bitter gourd fruits to make them suitable for nutraceutical industry.

KEYWORDS

- **bitter gourd**
- **evaluation**
- **female flower**
- **genotypes**
- **vitamin C**
- **yield**

REFERENCES

Anonymous, (2006). Department of Agriculture and Co-operation, Ministry of Agriculture, Government of India. pp. 38.

Bose, T. K., & Som, M. G., (1986). *Vegetable Crops in India*. Naya Prokashan. Bidhansarani, Calcutta, India, *206*, pp. 107–114.

Raghavendra Singh, Sanjay Kumar, & Dixit, S K., (2006). Correlation coefficient and path analysis in bitter gourd (*Momordica charantia* L.). *Progressive Hort., 38*(2), 256–261.

Rahman, M. M., Dey, S. K., & Wazuddin, M., (1990). Yield, yield components and plant characters of several bitter gourds, ribbed gourd, bottle gourd and sweet gourd genotype. *BAU Res. Progress, 4*, 117–127.

Sachan, S. C. P., Mehta, M. L., & Singh, D. P., (1989). Studies on floral biology of bitter gourd. *Prog. Hort., 21*(3–4), 229–234.

INDEX

D

N

Q

R

T

Y

Z